DIRECTIONS IN CHAOS

Volume 2

WORLD SCIENTIFIC SERIES ON DIRECTIONS IN CONDENSED MATTER PHYSICS

World Scientific Series on Directions in Condensed Matter Physics — Vol. 4

DIRECTIONS IN CHAOS

Volume 2

Edited by

Hao Bai-lin

World Scientific
Singapore • New Jersey • Hong Kong

Published by

World Scientific Publishing Co. Pte. Ltd.
P.O. Box 128, Farrer Road, Singapore 9128

U.S.A. office: World Scientific Publishing Co., Inc.
687 Hartwell Street, Teaneck NJ 07666, USA

DIRECTIONS IN CHAOS (Vol. 2)

ISBN   9971-50-361-1
        9971-50-362-X pbk

Printed in Singapore by Kim Hup Lee Printing Co. Pte. Ltd.

FOREWORD

Most review papers in this Volume 2 are related to the authors' contributions to the Beijing Summer School on Chaotic Phenomena in Nonlinear Systems, held on 17–26, August 1987, and sponsored by The Institute of Theoretical Physics, Academia Sinica, The Chinese Natural Science Fundation, The International Center for Theoretical Physics, Trieste, and The Third World Academy of Sciences. Not included in this volume were the following lectures:

1. B. Buti, 3 lectures on stochastic phenomena in plasmas.
2. G. Casati, 3 lectures on quantum mechanics and chaos.
3. M. J. Feigenbaum, 5 lectures on scaling and thermodynamical properties of dynamical objects.
4. Wang Xiao-jing, 3 lectures on coexistence of infinite periodic attractors in homoclinic systems and instabilities in sporadic systems.

Dr. K. Young's review has appeared in Volume 1 of this book. Dr. T. Bohr could not come, but we are happy to have his contributions published in this volume.

I would like to take the opportunity to express my gratitude to all the lecturers and participants of the Summer School as well as the sponsors who helped to make the event a pleasant experience for all of us. Our sincere thanks go to Dr. K. K. Phua and Miss P. H. Tham of the WSPC who have been so kind and patient with the publication of these two volumes of *Directions in Chaos*.

February 10, 1988

Hao Bai-lin
Institute of Theoretical Physics
Academia Sinica
P.O.Box 2735, Beijing 100080
PR China

CONTENTS

CHAOS IN LIQUID CRYSTAL OPTICAL BISTABILITY 46
Hong-jun Zhang, Jian-hua Dai, Peng-ye Wang,
Fu-lai Zhang, Guang Xu & Shi-ping Yang

SIMPLE MATHEMATICAL MODELS FOR COMPLEX
DYNAMICS IN PHYSIOLOGICAL SYSTEMS 90
Leon Glass

THE THERMODYNAMICS OF FRACTALS   194
Tomas Bohr & Tamás Tél

SELF-ORGANIZED CRITICAL PHENOMENA   238
Per Bak, Chao Tang & Kurt Wiesenfeld

AN EXPERIMENTALIST'S INTRODUCTION TO
THE OBSERVATION OF DYNAMICAL SYSTEMS   310
Neil Gershenfeld

Contents

UNIVERSALITY OF DISSIPATIVE SYSTEMS

George Schmidt

Stevens Institute of Technology, Hoboken, New Jersey 07030

INTRODUCTION

The evolution of a dynamical system is described by a set of equations

$$\dot{x}_i = f_i(x_1, x_i, \ldots x_n, K) \tag{1}$$

where f_i is some nonlinear function of the coordinates x_i and K is a parameter. Geometrically one has a flow in $x_1 \ldots x_n$ phase space for a given value of K. A flow line intersects a lower dimensional surface in a sequence of points giving rise to a map on the surface.

Hamiltonian systems in three dimensional space (n = 3), produce two dimensional area preserving maps, which are by now well understood[1-5]. Typically for a given K such a map contains stable and unstable fixed points and periodic orbits, KAM lines, and chaotic regions bounded by KAM lines. The stable orbits are surrounded by island sequences containing complicated structures of stable and unstable orbits, chaotic regions and KAM lines. While islands are generated by stable periodic orbits, chaotic regions arise out of the unstable orbits. Two initially adjacent points in the chaotic region will map into pairs of points, whose distance increases exponentially as one follows the map. This exponential divergence is characterized by the Liapunov exponent, describing the rate of increase in distance between points. This phenomenon of exponential divergence is called "sensitivity to initial conditions" and serves as a definition of chaos. In the regular regions on the islands the Liapunov exponent is nonpositive (negative or zero).

It is customary to set up the parameter K in such a way that for K = 0 the system is integrable, hence the map contains no islands or chaotic regions. These structures emerge for small values of K and undergo changes as K increases. In particular stable periodic orbits become unstable by period doubling bifurcations, KAM lines desintegrate, and chaos spreads as the bounding KAM lines disappear. Since KAM lines are essentially limits of long periodic orbits as the period goes to infinity, the crucial question boils down to understanding the destabilization of periodic orbits. The observed typical destabilization scenario is period doubling, so one is lead to study period doubling in such maps. In order to understand this better, we first turn to the simpler case of period doubling in one dimensional maps.

Consider the quadratic map

$$x' = Kx(1 - x) \tag{2}$$

as an example. When K < 3 iteration of this map from a starting 0 < x < 1 rapidly converges to a fixed point, an attractor. When K is somewhat larger than 3 the fixed point is unstable giving rise to a pair of period doubled attractors, such that the iterated x values oscillate between the two. Further increase in K results in the destabilization of this pair, producing a stable period four, etc. If one designates the critical K value where a period 2^n stable orbit emerges as K_n, one finds that

$$\lim_{n \to \infty}(K_n - K_{n-1})/(K_{n+1} - K_n) = \delta_F = 4.669... \tag{3}$$

so the K_n - s converge geometrically to K_∞, where chaos sets in. The crucial observation made by Feigenbaum was that this behavior is universal[6], i.e. it is the same for a large class of maps not only the one given by Eq. (2). The underlying mathematics, renormalization theory, will be discussed later.

When $K > K_\infty$, two interesting phenomena are observed[7]: band merging and windows. The 2^n period point attractor is replaced by 2^n band chaotic attractor, where the width of the bands increase as K increases. At some critical K_n' pairs of bands merge to yield a 2^{n-1} band chaotic attractor. The K_n' - s again converge geometrically in this inverse bifurcation sequence, with the same factor δ_F as seen in the original bifurcation sequence. There are ranges in the parameter K, where the chaotic bands are replaced by some regular attractor of period m, which goes through its own bifurcation (and inverse) sequence before the chaotic orbit is restored. There are an infinity of such windows in K, and periodic orbits of any period m (not only 2^n) are represented.

It turns out that a typical periodic orbit in a two dimensional Hamiltonian map, undergoes a bifurcation sequence which has universal features of its own. The K_n values converge again geometrically with the universal constant $\delta_H = 8.72...$ toward some K_∞. There is no merging band structure and no windows for $K > K_\infty$. One may again use a renormalization formalism to explain universality[8].

Hamiltonian maps are area conserving with the Jacobian determinant $J = 1$ everywhere. Dissipative two dimensional maps are area contracting with $J < 1$ and can lead to regular or strange attractors. The latter ones are typically chaotic, with a positive Liapunov exponent. Contrary to a chaotic Hamiltonian orbit, which covers a finite area, a strange attractor consists of an infinite set of points situated on a fractal set of zero area. The almost universal example used for illustration is the Henon attractor[9]

$$x' = y + 1 - \alpha x^2$$

$$ (4)$$

$$y' = \beta x$$

with a = 1.4 and β = .3. Note that the Jacobian J = − β. We are
interested in the universal features of strange attractors as the two
important parameters, the strength parameter K (in Eq. (4) K = α) and
dissipation parameter J are both varied. More generally we want to
understand what happens to physical systems as dissipation is
gradually introduced into the originally Hamiltonian dynamics. It
will be shown that universality can be extended over the entire K-J
parameter plane.

DISSIPATIVE STANDARD MAP

Consider the two dimensional map

$$x' = x + y'$$

$$y' = Jy + (K/2\pi)\sin 2\pi x \tag{5}$$

where J is the Jacobian determinant and K the strength parameter[10].
When J = 1 this is the standard map[11], a frequently used example to
study the behavior of area preserving (Hamiltonian) maps. When J = 0
one gets the one dimensional map

$$x' = x + (K/2\pi)\sin 2\pi x \tag{6}$$

whose behavior is similar to Eq. (2). The study of Eq. (5) for 0 < J
< 1 illuminates the relationship between the two limiting cases J = 0
and J = 1.

First, the J = 1 standard map for a given K value shows the
usual complicated structure of area preserving maps. Stable and
unstable periodic orbits, the stable ones surrounded by islands, KAM
curves, and chaotic regions. On the other hand the one dimensional J =
0 limit typically exhibits a simple structure for a given K: point

attractor, 2^n period attractor, 2^n band, or window depending on the choice of K. In order to find out how these limits connect we followed numerically the regions in the J,K parameter plane where different stable periodic orbits exist[10]. Each stable periodic orbit investigated lives in a "channel" in the J,K plane, stretching from J = 1 to J = 0, with the channel narrowing as J decreases (Figure 1).

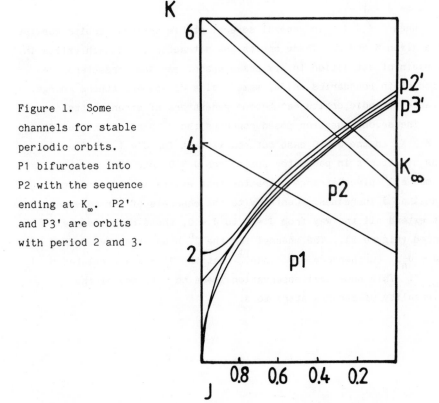

Figure 1. Some channels for stable periodic orbits. P1 bifurcates into P2 with the sequence ending at K_∞. P2' and P3' are orbits with period 2 and 3.

While many channels overlap for J ≈ 1, each channel has separated as J → 0. The stable fixed point for J = 1 becomes the stable attractor for $J = 0$, the bifurcated 2^n orbits of the Hamiltonian map, the 2^n attractor of the 1D map. All other period m stable orbits of the $J = 1$ system become the period m windows of the $J = 0$ map. This observation sheds new light on the origin of windows: these are remnants of the infinity of stable periodic orbits of the Hamiltonian system. One also finds that orbits of long period (m large) become narrow very rapidly as J is reduced. Since KAM lines are limiting cases of long periodic orbits ($m \to \infty$) their channels become infinitely narrow as $J = 1 - \varepsilon$.

When $0 < J < 1$ in general several stable periodic orbits coexist for a given K and J. These orbits are attractors, and each exists in its basin of attraction in x,y phase space. As the parameters are varied basin boundaries shift, some basins disappear, others emerge. Unstable periodic orbits can become generators of strange attractors.

The second question posed concerns the 2^n band structures found for $K > K_\infty$ in the $J = 0$ case but nonexistent for $J = 1$. Following these structures in parameter space from $J = 0$, one observes that each becomes a 2^n piece strange attractor that exists in a narrowing channel as J increases. Contrary to the channels of periodic orbits that extend all the way from $J = 1$ to $J = 0$, these channels are bounded (Figure 2). The channel for the 2^n band stretches from $J = 0$ to $J = J_n$. Furthermore one finds[12] that the J_n - s are related by $J_n = J_{n+1}{}^2$. This numerical observation leads to a theory of the universality of strange attractors.

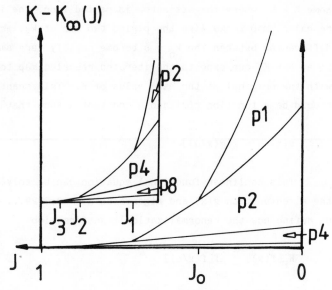

Figure 2.
Phase diagram
in J, K-K$_\infty$
space for
inverse
bifurcations.
Inset shows
details in the
range .6 < J < 1.
(From Ref. 12).

RENORMALIZATION

Consider again one dimensional maps

$$x' = f(x, K) \tag{7}$$

that undergo period doubling as the parameter K is increased (like Eq. (2)). Pick a parameter value, say K = k_1, where iterations converge to a fixed point attractor. As K is increased, one converges to a period two attractor, i.e. two period one attractors of the iterated map f[f(x,K)]. In fact one can find a K = k_2 where the iterated map behaves much the same way near the attractor, as the original map f at k_1, near the period one attractor. The main difference is that in the iterated map x (and x') have to be rescaled x → x/α. Similarly at

some K = k_3 where the attractor is period four, the iterated (and
rescaled) map looks like the period two map at k_2, and so on. The
differences between the k_n - s become rapidly very small for n large.
So at K = K_∞ one expects the iterated rescaled map to be identical
with the original at the same value of K. This means that there
should be a function g(x) and a constant α such that[6]

$$g(x)/\alpha = g[g(x/\alpha)] \qquad\qquad (8)$$

This nonlinear functional equation can be solved numerically and
the eigenfunction g(x) and eigenvalue α =- 2.5029... determined. Let
us define now the renormalization transformation

$$R_F[f(x)] \,\overline{=}\, \alpha f[f(x/\alpha)] \qquad\qquad (9)$$

that transforms the function f(x) into an other function $R_F[f(x)]$ in a
nonlinear way. Geometrically, this is a mapping in function space (of
infinite dimensions), and one may inquire about fixed points of the
map and their properties. But Eq. (8) is just the equation for the
fixed point g(x) in function space.

To determine the stability of g(x), one chooses an infinitesimal
perturbation in function space, and looks for eigenvalues

$$R_F[g(x) + \varepsilon f_n(x)] = g(x) + \varepsilon \lambda_n f_n(x). \qquad\qquad (10)$$

It turns out that with a few exceptions $|\lambda_n| < 1$, so the perturbed
function is attracted toward g(x) upon iteration in these directions.
In particular the space of stable directions corresponds to functions
with K = K_∞. So e.g. the function of Eq. (2) at K_∞ will approach g(x)
upon repeated iterations of the renormalization transformation. This
is the reason for the universality of period doubling; one may chose a
variety of functions at the proper parameter value and they will be

attracted to the same universal function $g(x)$ upon renormalization.

There is one important unstable eigendirection: the one corresponding to $K \neq K_\infty$. A function describing a 2^n period attractor, turns into one describing a 2^{n-1} period attractor when R_F is applied. Repeated renormalizations carry the function further away from $g(x)$, but since the other directions are contracting, the <u>universal line</u> corresponding to the unstable eigendirection of $g(x)$ in function space is approached. The unstable eigenvalue $\lambda = \delta_F = 4.669...,$ is the period doubling rate. When $K > K_\infty$ the same procedure carries the system from the 2^n chaotic band state, to the 2^{n-1} band state.

For area preserving two dimensional maps, the renormalization operator R_H is[8]

$$R_H(T) = B \circ T \circ T \circ B^{-1} \qquad (11)$$

where T is some two dimensional area preserving map like Eq. (5) with $J = 1$, and

$$B = \begin{pmatrix} \alpha & 0 \\ 0 & \beta \end{pmatrix} \qquad (12)$$

is a rescaling matrix. There is again a universal function (map) $T^* = R_H(T^*)$ corresponding to K_∞, the accumulation point of period doubling bifurcations. Most eigendirections around T^* are again attracting, so universality is assured. The unstable eigendirection with $\lambda = \delta_H = 8.72...$ defines a universal line in the space of area preserving maps representing universal maps with 2^n period stable orbits, bifurcated out of the original stable orbit.

<u>Our numerical results for dissipative systems suggested a significant expansion of the renormalization calculation.</u> We enlarge function space around T^* to include maps that are area contracting. The renormalization operator R is like R_H except that T is not restricted to area preserving maps. Solving

$$R(T^* + \varepsilon T_n) = T^* + \varepsilon \lambda_n T_n \qquad (13)$$

leads now to two significant $|\lambda_n| > 1$ values, $\lambda_1 = \delta_H$ and $\lambda_2 = 2$. The second eigenvalue leads to an eigendirection in function space pointing toward area contracting maps including strange attractors. The universal line underline{in} function space underline{is} replaced underline{by} underline{a} universal underline{2D} surface, where each point can be parametrized by the strength parameter $K - K_\infty$, and the Jacobian J. A map corresponding to such a point, defines an infinite set of maps, the result of iterations of R. These maps are universal as well as self similar in the same sense as one dimensional period doubling maps are. Instead of designating the number k_1, k_2, k_3,... k_n, number pairs $(K_n - K_\infty)$ and j_n characterize such a sequence. Sets of such numbers define "fan lines" as shown in Fig. 3.

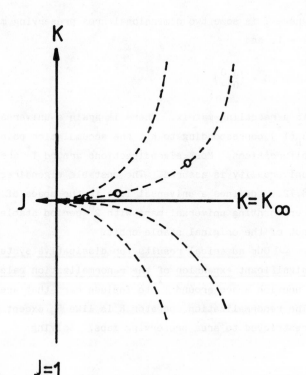

Figure 3.
Renormalization fan lines near the Hamiltonian accumulation point. Circles designate a set of self similiar maps. (From Ref. 12).

The fan lines with $K_n < K_\infty$ define self similar bifurcation
sequences, while those with $K_n > K_\infty$, self similar universal strange
attractors.   In the latter the Liapunov exponents, fractal dimensions,
$f(\alpha)$ curves, etc. follow well defined scaling laws.   Universal scaling
laws govern homoclinic and heteroclinic crises and windows as well.

These results hold not only for maps with constant Jacobians,
but variable Jacobians as well[12]. This is the consequence of the fact
that the formal renormalization calculation yielded only one
dissipative eigendirection, without any constraint imposed as to the
constancy of the Jacobian. Hence these results are very general, and
encompass all dissipative maps, whose area preserving limit belongs to
the T^* class.

SOME EXAMPLES

Consider a compass needle (magnetic dipole) in an oscillating
magnetic field. If the angle between the magnetic axis and the
compass needle is θ, the magnetic field $B = B_o \sin\omega t$, then the torque
acting on the needle is proportional with $B_o \sin\theta \sin\omega t$. In the
presence of damping, and introducing dimensionless variables the
differential equation that describes the motion is

$$\ddot\theta = K\sin2\pi\theta\sin2\pi t - \Upsilon\dot\theta \qquad (14)$$

The variables θ, $\dot\theta$ can be mapped on a time one map ($t = 0, 1, 2,$
etc.). The damping is determined by the parameter Υ, with $J = e^{-\Upsilon}$ and
K is the strength parameter. The same equation describes the motion
of a particle in a standing wave.

We have studied this system experimentally[13] as well as
numerically[14] as the parameters are varied. When K is increased with
fixed Υ a bifurcation sequence leading to K_∞ is obtained. When $K > K_\infty$

a sequence of 2^u piece strange attractors can be observed. This sequence is truncated, depending on the value of J, and the critical J_n - s can be numerically determined.

The driven damped pendulum is described by

$$\dot{x} = y$$

$$\dot{y} = -\gamma y - \sin x - K\sin 2\pi t$$

(15)

The surface of sections map is again the time one map, and was determined analytically for the K,J parameter plane $(J = e^{-\gamma})$.

A ball bouncing vertically on an oscillating table, whose vertical position at time t is $z = K\cos\omega t$, produces a map[15]

$$K\cos\omega t_{n+1} = K\cos\omega t_n + v_n(t_{n+1} - t_n) - g/2 \, (t_{n+1} - t_n)^2$$

$$V_{n+1} = -(1 + b)K\omega\sin\omega t_{n+1} + b[g(t_{n+1} - t_n) - V_n]$$

(16)

where t_n is the time of the n-th bounce, V_n the vertical ball velocity immediately after the bounce, b the elasticity parameter (b = 1 corresponds to elastic bounce), and g the gravitational acceleration. This map, just like the other examples, has a Hamiltonian limit which belongs to the universality class defined by the universal map T^*. As expected, the dissipative examples also produce universal results as determined numerically.

UNIVERSALITY OF CRITICAL JACOBIANS

As a test to check universality, we determined the critical J_n values where the 2^n piece strange attractor is truncated, for the dissipative standard map, driven damped pendulum, particle in a standing wave (or compass in an oscillating magnetic field), and the bouncing ball[14]. The results are exhibited in the table.

It is apparent that not only the $J_n^2 = J_{n-1}$ relation is well satisfied, but the numerical values of the J_n - s are approximately identical. Furthermore, while universal behavior is expected only as $n \to \infty$, the convergence to the universal sequence appears to be extremely rapid.

The law describing universality of bifurcation sequences in one dimensional maps (Eq. (3)) relates ratios of differences in critical K_n values. Since the K values themselves are subject to rescaling, a smooth function $f(K) = K'$ parametrizes the system just as well as K, one does not expect that the K_n values themselves are universal. On the other hand the Jacobian has the well defined meaning of contraction rate of areas upon mapping. In other words J cannot be rescaled and the J_n values themselves are universal.

Table. Approximate values of critical Jacobians J_n, determined numerically for several systems. The correct values are bracketed by the pairs of numbers shown. (From Ref. 14).

Dynamical Systems	J_0	J_1	J_2
Dissipative Standard Map	0.45 -- 0.46	0.66 -- 0.67	0.81 -- 0.82
Driven Damped Pendulum	0.45 -- 0.46	0.66 -- 0.67	0.81 -- 0.83
Particle in Standing Wave Field	0.49 -- 0.50	0.66 -- 0.68	0.81 -- 0.82
Bouncing Ball	0.475-- 0.48	0.665-- 0.67	0.815-- 0.82

It should however be emphasized that the universality of the J_n - s is just an example in a far more general theory. Consider any other well defined point on the K,J plane; e.g. the point of disappearance of the period 5 window. If the corresponding Jacobian is j_5, the points where the period 10 window disappears j_{10}, etc. then the j_5, j_{10}, j_{20}, etc. values will also produce a sequence of universal numbers with $j_{20} \approx j_{10}^2$ and so on. In fact an infinity of universal number sequences can be generated.

The rapid convergence of the J_n - s for different systems as shown on the table is not an accident. Using renormalization theory one can show that if J_n is the value for a given system, and J_n^u the corresponding one on the universal map obtained by expansion around T^*, the ratio

$$(J_{n-1} - J_{n-1}^u)/(J_n - J_n^u) \approx -17 \tag{17}$$

for large n. So the difference between the J_n - s from the universal values reduces very rapidly. The details of the proof are given in Reference 14.

The work presented here is the result of collaboration with C. Chen, G. Györgyi and B. H. Wang. The magnetic dipole experiments were performed in collaboration with H. Meissner. The work was supported by the U.S. Department of Energy DE-FG02-87ER13740 and the Air Force Office of Scientific Research 87-0122.

Last but not least the author wishes to thank the organizers of the Beijing Summer School, in particular Dr. Hao Bai Lin for their hospitality.

References

1. Berry, M. V. in Topics of Nonlinear Dynamics, Ed. S. Jorna, Am. Inst. Phys. Conf. Proc. AIP, New York, 1978 Vol. 46, p. 16.
2. Lichtenberg, A. J. and Lieberman, M. A., Regular and Stochastic

Motion, Springer, New York, 1983.

3. Greene, J. M., J. Math. Phys. 20, 1183 (1979).

4. Schmidt, G., Phys. Rev. A 22, 2849 (1980).

5. Schmidt, G. and Bialek, J., Physica 5D, 397 (1982).

6. Feigenbaum, M. J., J. Stat. Phys. 19, 25 (1978) and 21, 669 (1979) and Physica 7D, 16 (1983).

7. Collet, P. and Eckman, J. P., Iterated Maps of the Interval as Dynamical Systems, Birkhauser, Boston, 1980.

8. Greene, J. M., Mackay, R. S., Vivaldi, F. and Feigenbaum, M. J., Physica 3D, 468 (1981), Bountis, T. C. Physica 3D, 557 (1981), Collet, P., Eckman, J. P., Koch, H. Physica 3D, 457 (1981), Widom, M., Kadanoff, L. P. Physica 5D, 287 (1982).

9. See e.g. in Berge, P., Pomeau, Y. and Vidal, C. Order Within Chaos, John Wiley & Sons, New York, 1984.

10. Schmidt, G. and Wang, B. H. Phys. Rev. A32, 2994 (1985).

11. Chirikov, B. V., Phys. Reports, 52, 265 (1979).

12. Chen, C., Gyorgyi, G. and Schmidt, G., Phys. Rev. A34, 2568 (1986), and A35, 2660 (1987) and G. Schmidt in Chaotic Phenomena in Astrophysics, Annals of the N.Y. Academy of Sciences 497, (1987).

13. Meissner, H. and G. Schmidt, Am. J. Phys. 54, 800 (1987).

14. Chen, C., Gyorgyi, G. and Schmidt, G., Phys. Rev. A to appear.

15. Celaschi, S. and Zimmerman, R. L., Phys. Lett. A120, 447 (1987) and references therein.

Circle Maps, Mode-Locking and Chaos.

Per Bak
Physics Department, Brookhaven National Laboratory
Upton NY 11973 USA
and ITP, UCSB, Santa Barbara CA 93106 USA

Tomas Bohr
NBI, Blegdamsvej 17, Copenhagen, Denmark

Mogens H Jensen
Nordita, Blegdamsvej 17, Copenhagen, Denmark

1. INTRODUCTION.

Many interesting physical phenomena are formulated in terms of angular variables because they fundamentally involve recurring motion. The behaviour of maps of such angular variables - circle maps - is therefore important for the study of many phenomena involving periodic or almost periodic motion as well as their chaotic offspring. In particular, circle maps are of relevance for the description of *resonances* between periodic motions. The present article is an exposition of properties of circle maps which are of importance to the physicist, and indeed many phenomena have been successfully analyzed in terms of circle maps. Many such phenomena have previously been treated by complicated perturbation expansions without much success because the interesting universal properties, which can be understood in terms of circle maps, are all of non-perturbative nature.

Circle maps are the simplest representation of a dynamical system showing the interplay between periodic, quasiperiodic and more complicated "chaotic" behaviour. In many respects circle maps are the dissipative analogs of non-integrable classical Hamiltonian systems where the circle is equivalent to a smooth invariant torus. Thus, like the famous KAM theorem on the existence of invariant tori in Hamiltonian systems, one is, in the theory of circle maps, forced to consider such matters as the distinction between rational and irrational numbers - even how irrational a given number is - questions which any respectable physicist would have considered completely irrelevant a few decades ago.

In dissipative systems a strong degree of universality exists. In contrast to Hamiltonian systems, phenomena involving many degrees of freedom can sometimes be effectively determined by only a few of them and thus be in the same *universality class* as very low dimensional ones. This was strikingly demonstrated by Feigenbaum[1] who represented the period doublings seen in dynamical systems by iteration of a simple map on an interval. The "critical exponents" which emerged from that analysis were later found in diverse experiments in hydrodynamics, optics, electronics etc. For the phenomenology of mode-locking and breakdown of quasiperiodic tori the story is much the same. There is a transition to chaos caused by overlap of resonances which can be characterized by exponents β, η, υ, δ etc. which are completely analogs to the critical exponents occurring near second order phase transitions, and in percolation. The fact that a *dynamical* transition can be described in terms of the same language as a *thermodynamic* phase transition, or the purely *geometric* percolation transition is remarkable and suggests a deeper level of universality encompassing a wide range of phenomena. The analysis of circle maps, which we shall review below, has given results which have later been observed in many experimental situations.

In the following discussion our aim has been to focus on the overall properties of circle maps - to get a feeling for the entire "phase diagram" of such a dynamical system and to point out which parts are known and which parts might be an object of further study. We have tried to avoid technicalities, but give relevant references for the more mathematically inclined reader. The layout is as follows:

In chapter 2 we introduce the circle maps and the appropriate vocabulary for the description of their properties. In chapter 3 we review the critical properties at the transition to chaos, introducing the critical indices. The transition will be studied from two different viewpoints: through quasiperiodicity and through mode-locking. We also compare the predictions with selected experiments. In chapter 4 the structure of the mode locking regions are described, and some scaling behaviour in the supercritical region outlined.

2. CIRCLE MAPS.

A map θ->$f(\theta)$ on the real line can be regarded as a map on a circle $S^1 = [0,1]$ if the image of a point, θ, and the one obtained by translating once around the circle, $\theta+1$, are related in some simple manner. More precisely one defines circle maps of order n as maps satisfying

$$f(\theta+1) = n + f(\theta) . \tag{2.1}$$

Since our purpose is to understand the competition between periodic and quasiperiodic motion we want a class of maps including the pure rotation θ->$\theta+w$ which means that n=1. In the following we shall therefore always assume that our circle maps are of order one. The standard form of such a map is

$$f_\Omega(\theta) = \theta + \Omega + g(\theta) \tag{2.2}$$

where g is *periodic* in θ i. e. $g(\theta+1) = g(\theta)$. Physically, one can think of $f(\theta)$ as the phase of a system at time t+1 given that the phase at time t was θ, with the clock period defined as an internal or external frequency. For instance, one can think of (2.2) as describing the motion of a torsion pendulum subjected to constant torque "Ω" plus a periodic force with period 1 and strength $g(\theta)$.

More generally, any "overdamped" dynamical equation of the form

$$d\theta/dt = \lambda(\theta,t) \tag{2.3}$$

where λ is periodic in both θ and t gives rise to a circle map. Say the periods in θ and t are, respectively, 1 and T, and define $\theta_n = \theta(t=nT)$. Then there exists a function, f, such that

$$\theta_{n+1} = f(\theta_n) \tag{2.4}$$

and $f(\theta)$ is a circle map satisfying (2.1) with n=1.

The fundamental quantity describing the dynamics of such a map is the winding number, w, which is the average rotation per iterate.

The iterates of a given point θ_0 is the sequence $\theta_0, \theta_1, \theta_2, \ldots$ where $\theta_1 = f(\theta_0)$, $\theta_2 = f(f(\theta_0))$ etc. The winding number is defined as the limit

$$w = \lim_{n \to \infty} (\theta_n - \theta_0)/n .\qquad(2.5)$$

As long as f is a *diffeomorphism* (i.e. smooth and invertible) very strong theorems hold[2,3]: The winding number is well defined and does not depend on the choice of initial point, θ_0. If w is rational, $w = P/Q$, there exists a periodic cycle on the circle: θ_0^*, θ_1^*, $\theta_2^* \ldots$

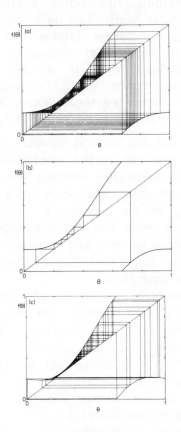

Fig. 1. Iterations for the circle map, eq. (2.8) for $\Omega = 0.2$ and a) K=0.9, b) K=1.0, and c) K=1.1. For K > 1 the map develops local extrema and chaotic behaviour may occur.

..., θ_Q^* such that

$$f^Q(\theta_i^*) = \theta_i^* + P. \qquad (2.6)$$

If w is *irrational* the motion is called quasiperiodic and can be *smoothly conjugated* to a pure rotation, which means that a smooth change of variables on the circle will change f to a pure rotation $\theta \to \theta + w$. Figure 1a shows iterations in a case where the winding number is irrational. These two types of motion exhaust the possibilities for monotonic (invertible) circle maps and we emphasize that no "irregular" motion can be produced.

When the winding number is rational we call the motion mode-locked. Figure 1b shows the iteration for a circle map (the "sine" map to be defined later) in a case where there is mode-locking with winding number w=1/8. For each rational w there exists a whole interval, which we shall denote $\Delta\Omega(w)$, where that winding number is realized. The cycle points $\theta_1^*, \ldots, \theta_Q^*$ change with Ω in this interval, but the winding number (and thus P and Q) are fixed. This means that the relation $w=w(\Omega)$ is highly non-trivial. It is continuous and non-decreasing and has a constant plateau for each rational w. Such an object is often called a "Devil's staircase" and occurs frequently in the physics of competing spatial or temporal periods[4]. One might first find it strange that w is not a smooth function of Ω since the map and its dependence on Ω is, but one has to keep in mind that the winding number is an *asymptotic* quantity defined by the limit (2.5).

In contrast to periodic motion, quasiperiodic motion is only realized for specific $\Omega(w)$. If the given irrational w is approximated to higher and higher order by rationals the corresponding mode-locked intervals $\Delta\Omega$ will shrink to a point. Since mode-locking is realized in a dense set of intervals in Ω and quasiperiodicity only in points one might be lead to the belief that almost all motion must be mode-locked, i. e. that a random choice of Ω will necessarily lead to mode-locking.

That is, however, not true. Consider first the limiting case where the function g in (2.2) vanishes. Then he map is a pure rotation and $w = \Omega$. All the intervals $\Delta\Omega$ have shrunk to zero and almost all choices of Ω lead to quasiperiodic behaviour, since with probability one a real number is irrational. When g is small the mode-locked intervals are also small and decrease rapidly with the denominator of the corresponding rational. In fact, Herman[3] has proved that for any (sufficiently smooth) diffeomorphism the measure of the quasiperiodic Ω's is positive which means that the union of all the mode-locked intervals does not have full measure.

Circle maps can display behaviour which is more complicated than periodic or quasiperiodic. If we relax the demand that f should be monotonic, more interesting behaviour starts to appear and one finds chaos generated by period doublings or intermittency or bistable responses where the winding number depends on the initial condition. To study these phenomena in a systematic way it is important to consider a two parameter family of maps, for instance in the form

$$f_{\Omega,K}(\theta) = \theta + \Omega + K\,h(\theta) \tag{2.7}$$

where, of course, $h(\theta+1) = h(\theta)$. By varying K we can switch between invertible and noninvertible maps. A lot of work[4-12] has been done with the "sine-map" introduced by V. I. Arnold:

$$f_{\Omega,K}(\theta) = \theta + \Omega + (K/2\pi)\sin 2\pi\theta \tag{2.8}$$

which is invertible for K<1. The *phase diagram* for this family is shown in Fig. 2. The "tongues" (so-called Arnold tongues) coming out of the Ω-axis each correspond to a mode-locked region with some rational winding number. They start out with zero width and as K increases they get wider. Above K=1 they start overlapping (the winding number depends on the initial θ). In this region f has two extrema and period doublings take place inside the tongues.

In the next chapter we shall look more closely at this phase diagram and specifically study the behaviour as k->1. We shall see that beautiful scaling appears in this limit and that the "Devils staircase" becomes very interesting.

3. CRITICAL BEHAVIOUR.

3.1. Definition of criticality.

3.1.1. Onset of chaos.

Criticality in the circle map model can be qualitatively described and defined in several ways. The most direct way is to say that the critical parameter values are those where the quasiperiodic orbits become chaotic. The locus of such parameter values will be called the critical surface. For the true circle map (2.2) this is simply related to whether or not the map is monotonic. As discussed in the previous chapter a monotonic circle map can have periodic or quasiperiodic asymptotic motion so that chaos is ruled out. Once the map develops extrema this is no longer true and orbits with positive Liapunov exponents can be found. For the "sine map" (2.8) the locus of critical parameter values is therefore simply the line K=1, where a zero slope appears at θ=0.

The circle map generated by the overdamped equation (2.3) is always *subcritical* : since the differential equation (2.3) can run backwards in time, so can the map (2.4), and therefore it must be monotonic. As we shall see later, higher dimensional systems of differential equations behave much like circle maps. They correspond to higher dimensional *invertible* maps, but their projections into one dimension can behave like circle maps. However, as discussed in section (3.3), at the point where we would expect the circle map to loose monotonicity they "disintegrate", and therefore the criticality condition of having a map with zero slope is difficult to apply in practice.

Another difficulty in locating criticality by looking at the onset of chaos is that it is very important that the motion is quasiperiodic. Within the locked regimes chaos develops later, but the measure of parameter values on the critical surface giving quasiperiodic behaviour is actually zero. The mode-locked regions fill it out so one has to look carefully to find true quasiperiodic orbits.

3.1.2. Overlap of tongues. Hysteresis.

The phase diagram shown in figure 2 indicates that all the locked tongues tend to increase in width as the non-linear parameter K is

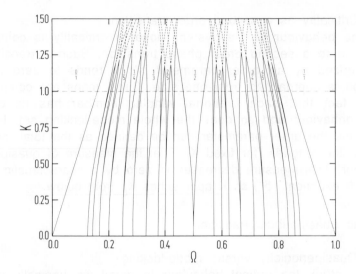

Fig. 2. Phase diagram of the circle map. As K increases the tongues become wider, and eventually they overlap for K > 1.

increased. At some point they therefore start overlapping, and this gives us another way of locating criticality. The pattern of overlaps is quite complicated. Only the very tiniest tongues overlap near the critical surface, whereas the bigger tongues overlap at larger values of K. Also, any given tongue is overlapped for the first time (by the thinnest neighboring tongues) some finite distance above criticality. As soon as two tongues overlap, there is an infinity of tongues in between which are overlapped too since between each pair of rational numbers there are of course an infinity of other rationals. The winding number then becomes non-unique, replaced in general by a *winding interval* [5,6]: different initial conditions can lead to all different winding numbers within an interval.

When an experiment sweeps over the different tongues by variation of a parameter corresponding to Ω, the overlap of tongues will show up as hysteresis effects: the value of the winding number can depend on how the value of Ω is reached. As more and more mode-locked tongues are resolved, the critical surface can be located with increasing precision. This approach has been used in several laboratory and computer experiments.

3.1.3. Criticality as a second order transition.

The behaviour of the system close to criticality is completely analogous to a second order phase transition. Such a transition is characterized by an order parameter which tends to zero as the transition is approached from one side. Circle maps are complicated by the fact that each irrational winding number has its distinct critical behaviour and so does the whole mode-locking set. For the latter case the order parameter can be defined as the total measure between the tongues (for fixed K), i. e. the measure of quasiperiodic behaviour. This measure decreases to zero by a characteristic power law (with exponent "β") as K approaches K_c from below.

3.2. The quasiperiodic transition.

3.2.1. Quasiperiodicity versus mode-locking.

To study the critical behaviour in detail we basically choose between two approaches. Either we study quasiperiodic behaviour or we study mode-locked behaviour. In the first case we must fix the winding number to an irrational number or find a systematic approach to that number. In the latter case we look at rational numbers, i. e. the mode locked tongues and we shall generally study several rational numbers over an interval. It is not obvious which of those two scenarios to choose. Seen from an experimental point of view the study of mode-locked tongues is most appealing since it is quite hard to tune two frequencies to an irrational number. The two approaches are however much related; in trying to study quasiperiodicity one will have to measure the tongues and vice versa. Still the two scenarios are characterized by different critical exponents.

3.2.2. Renormalization group calculation.

A quasiperiodic transition is characterized by a specific irrational number. The most used number is the golden mean $w^* = (\sqrt{5}-1)/2$. This is particularly convenient since it has the simplest continued fraction expansion

$$w^* = 1/(1+1/(1+1/(1+ \dots)) \dots) \qquad (3.1)$$

Successive rational approximants to the golden mean will therefore be on the form F_{n-1}/F_n , where F_n are Fibonacci numbers satisfying $F_{n+1}=F_n+F_{n-1}$, with $F_0=1$ and $F_1=0$. As studied first by Shenker[7] we can calculate the corresponding values of the parameter Ω_n for a tongue with $w = F_{n-1}/F_n$. In that case we have an F_n-cyclic point θ_n^*

$$f^{F_n}(\theta^*) = \theta^* + F_{n-1} . \tag{3.2}$$

In the following we set the initial point as $\theta_0=0$ which is called the superstable point. Shenker observed that the series Ω_n converges geometrically to Ω_∞ with a specific exponent δ (not to be confused with the exponent δ for mode locking to be introduced later!)

$$(\Omega_{n-1}-\Omega_n)/(\Omega_n-\Omega_{n+1}) = -\delta_{GM} \; ; \; \delta_{GM} = 2.833... \; . \tag{3.3}$$

Also, Shenker found that the point nearest to $\theta=0$ on the attractor scales with an exponent α when going from one Fibonacci level to the next

$$f^{F_{n-1}}(\Omega_n,\theta=0)/f^{F_n}(\Omega_{n+1},\theta=0) = \alpha_{GM} \; ; \; \alpha_{GM} = -1.28... \tag{3.4}$$

The most remarkable of Shenker's observations was that these numbers are universal for circle maps with a cubic inflection point and golden mean winding number. This universality was formulated within a renormalization group analysis by Feigenbaum et al[8] and by Rand et al[9]. Basically, one can write down functional equations for a universal function f^* on the form

$$f^*(\theta) = \alpha f^*(\alpha(f^*(\theta/\alpha^2)))$$
$$f''(\theta) = \alpha^2 f^*(\alpha^{-1}f^*(\alpha^{-1}\theta)) \tag{3.5}$$

The solution of these equations demands that $\alpha = \alpha_{GM}$ and their linearization around the fixed point function f^* reveals two relevant eigenvalues, δ_{GM} and α^2_{GM} corresponding to the Ω and K directions, respectively, in full agreement with the numerical observations.

The calculation above was performed at the golden mean winding number. One could now ask the question: what happens at other winding numbers? Another well-known irrational, the "silver mean", $s^* = \sqrt{2} - 1$ has the following expansion

$$s^* = 1/(2+1/(2+1/(2+ \ \dots \ )..)) \ . \tag{3.6}$$

Again one finds geometrical scaling in the parameter space and on the attractor. The scaling numbers are universal, but are slightly different from the numbers at the golden mean[7]. Again one can write down functional equations for the fixed point function. Actually the renormalization group can be formulated for any periodic continued fraction expansion (in the two examples shown here the period was one) and one expects the two scaling numbers only to deviate a few percent from the golden mean scaling numbers[10].

3.2.3.   Ergodic irrationals and strange sets from renormalization.

The irrational numbers with periodic continued fraction expansion (the "noble" numbers) are actually very atypical among the irrationals (have zero measure). The numbers with an ergodic expansion (the typical numbers), such as π, have full measure among the irrationals. Therefore a similar analysis should be performed for such a number. One quickly realizes that there is no geometric scaling in those cases. However, some kind of average scaling can be formulated by the approach suggested by Farmer, Satija and Umberger[11]. Let us consider a general irrational number r,

$$r = 1/(a_1+1/(a_2+1/(a_3+ \ \dots \ )..)) \tag{3.7}$$

where the entries a_n are ergodic. For the successive approximants to r, P_n/Q_n defined by $a_n = \infty$, we can calculate the corresponding parameter values for the super stable cycles, Ω_n. The series

$$\delta_n = (\Omega_{n-1} - \Omega_n)/(\Omega_n - \Omega_{n+1}) \tag{3.8}$$

will not converge in this case. However, Farmer et al found that the geometric average of the δ_n's converges to the value $\overline{\delta} = 16 \pm 1$. Also, the geometric average of the corresponding α_n values converges to the number $\overline{\alpha} = 1.7 \pm 0.2$. In the limit of large n this means that

$$\Delta\Omega_n = |\Omega_{n+1} - \Omega_n| \approx \overline{\delta}^{-n} . \tag{3.9}$$

As discussed above the n'th approximant to r is denoted P_n/Q_n. Using results from number theory[12,13] we know that the typical number of entries for P_n/Q_n is

$$n = 12 \log 2 \, \log Q_n/\pi^2. \tag{3.10}$$

Jensen et al[14] found that for rationals with denominator Q_n the averaged behaviour is[15]

$$\Delta\Omega_n = Q_n^{-\delta} , \quad \delta = 2.29..... \tag{3.11}$$

From the equations above we find $\delta = 12 \log 2 \log \overline{\delta} /\pi^2$ in reasonable agreement with results obtained in ref. 11. Again, for the average α values, ref. 14 found an exponent $\alpha = 0.42$ which similarly can be related to the ergodic number $\overline{\alpha}$ by $\alpha = 12 \log 2 \log \overline{\alpha}/\pi^2$. Again one can try to formulate a renormalization group and write down functional equations. The RG scheme will not lead to a fixed point function. Instead, Farmer and Satija made the interesting discovery that the RG scheme develops a strange set in function space. The periodic orbits in this set correspond to quasiperiodic winding numbers with periodic continued fractions (quadratic irrationals) whereas the gaps are related to the mode locked tongues. The fractal dimension of the set was found to be 1.8 ± 0.1. It should be noted that in the calculation in ref. 11 it was for numerical reasons not possible to use a "real" typical number. Basically it was necessary to use an upper boundary on the entries a_i around 5. Universality properties of such "renormalization group strange sets" is discussed in a recent work by D. Rand[16].

If the RG scheme is formulated such that the first entry in the continued fraction expansion is deleted in each step, the corresponding irrational numbers will jump ergodically in the unit interval[9]. Thus any typical irrational will be visited in this procedure. A natural approach will therefore be to study the global structure of all winding numbers. This is what we shall do in section 3.3.

3.2.4. Multifractal scaling of golden mean attractor.

The attractor obtained from iterating the circle map eq. (2.8) at the critical point (K=1) with golden mean winding number is shown in Fig. 3. At this point the motion is still conjugate to a pure rotation so the points will eventually cover the whole circle (the conjugate function has singularities, however). This implies that the dimension of the attractor is equal to 1. Still, the distribution of points on the unit circle is quite non-uniform as can be seen on the picture.

Recently a formalism has been advanced in order to quantify this variation in density. The main concept of this formalism is the notion of a multifractal[17-19] which is a structure of infinitely many interwoven subfractals. To describe this type of complicated set we partition it into small boxes of size l. Denoting the probability to

Fig. 3. The critical attractor of the circle map with golden mean winding number. Note the varying density of points along the circle.

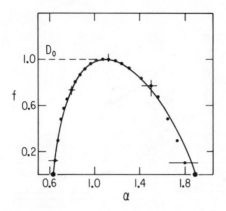

Fig. 4. The f(α) function for the attractor with golden mean winding number shown in figure 3. The dots are experimental values (Ref. 20).

fall in the i'th box as p_i we introduce a scaling index α_i as

$$p_i = l^{\alpha_i} . \tag{3.12}$$

In the limit $l \rightarrow 0$ each point on the attractor is then characterized by a specific value of α, which is called the point-wise dimension. Next, we consider the set of points which are characterized by the same value of α. The dimension of this subset is called $f(\alpha)$[19]. The function $f(\alpha)$ is smooth and is shown in figure 4 for the golden mean attractor in figure 3. The $f(\alpha)$ function is always convex and its maximum value is the dimension of the full set, D_0. The smallest value of α corresponds to the most dense regions of the set whereas the maximal value of α corresponds to the least dense regions. Consequently, the $f(\alpha)$ function is a convenient representation of the entire scaling behaviour of a fractal set, called the multifractality of the set.

The function depicted in figure 4 is universal for critical circle maps with golden mean winding number and has been compared with experiment by Jensen et al[20] and by Gwinn and Westervelt[21]. Jensen et al used a cell filled with mercury which was heated from below.

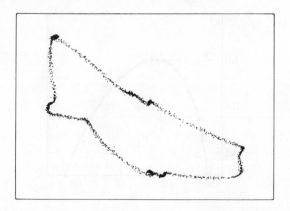

Fig. 5. Experimental attractor with golden mean winding number from a forced
Rayleigh-Benard experiment (Ref. 20). The temperature on top of the
convection box is plotted in intervals of the external period.

At a critical point above the onset of convection there is an
oscillatory instability into a time-dependent state involving an ac
vertical vorticity in the fluid. The period of this oscillation defines
one frequency. The experimentators generated the second oscillator
by applying a small horizontal dc magnetic field parallel to the axis
of the convection cells in the fluid and by applying a vertical sheet
of ac-current through the fluid.[22] The Lorenz force induces an ac
component in the fluid's velocity. This is the second oscillator. By
varying the amplitude and frequency of the current, they were able
to scan a large range of winding numbers and amplitudes, and
obtained a phase diagram with numerous Arnold tongues, similar to
the one shown in figure 2. They identified the critical curve by
tuning the experiment to quasiperiodic states, identifying chaos as
broad band noise in the power spectrum. Figure 5 shows the
experimental attractor obtained at this point, the f(α)-values of
which are shown in figure 4. The agreement with the theoretical
curve from the circle map gives strong support to the notion that
the curve is universal.

3.3. Mode-locking at criticality.

3.3.1 Fractal structure of mode-locking at the critical surface.

To test the conjecture that the mode-locked tongues fill-up the critical line K=1, Jensen et al[14,23] calculated the widths $\Delta\Omega(P/Q)$ for all intervals with Q<100. The intervals were non-zero for all P and all Q. Figure 6 shows the "Devil's staircase" formed by plotting the winding number vs Ω for the sine circle map. When more and more steps are included the Ω-axis becomes more and more filled-up. Of course one can not calculate an infinity of intervals numerically, so Jensen et al. studied the scaling behaviour of $\Delta(P/Q)$ as follows:

Let $S(r)$ be the total width of steps which are larger than a given scale r. We are interested in the space between the steps, $1-S(r)$ and have measured it on the scale r to find the "number of holes" $N(r)=(1-S(r))/r$. $N(r)$ can also be viewed as the density distribution function of locked intervals. Figure 7 shows $\log N(r)$ vs. $\log 1/r$ for several values of r. The points fall excellently on a straight line indicating a power law

$$N(r) \approx r^{-D}, \quad D = 0.8700 \pm 0.0004 \tag{3.13}$$

or, equivalently,

$$M(r) = 1-S(r) \approx r^{1-D}. \tag{3.14}$$

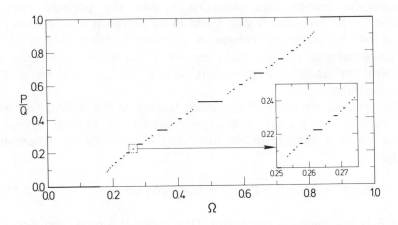

Fig. 6. The staircase formed by mode locked intervals calculated at K=1. Note the self-similarity under magnification.

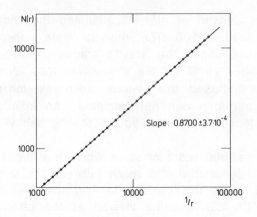

Fig. 7. Plot of logN(r) vs. log(1/r) yielding a straight line with slope $D=0.8700\pm3.7\times10^4$.

Thus the space between the mode-locked steps vanishes as smaller and smaller steps are included. There is no room for quasiperiodic motion at criticality and the staircase is called complete. Indeed, the set defined by removing the mode locked intervals is a Cantor set with dimension $D = 0.8700$. Thus, the mode-locked intervals have measure 1 and the quasiperiodic motion has zero measure at criticality, in contrast to the situation for $K=0$ where the quasiperiodic motion has measure 1 and the periodic motion measure 0. The exponent D was found to be universal, i. e. it does not depend on the particular choice of periodic function $h(\theta)$ in (2.7). This universality is crucial, since otherwise the theory would not be predictive for specific experiments where we do not know the underlying map.

The scaling structure of the mode-locking at the critical line has been verified numerically and experimentally for several systems. The motion of a damped driven pendulum is given by the differential equation

$$Ad^2\theta/dt^2 + Bd\theta/dt + C\sin\theta = D + E\cos\omega t, \qquad (3.15)$$

where B is the damping coefficient, D a constant torque, and $E\cos\omega t$

an oscillating torque. The equation also represents a Josephson junction driven by a constant voltage in a microwave field. Indeed, a mode-locking staircase has been observed in the latter system[24]. This differential equation reduces to a circle map under certain circumstances. Using the frequency ω as a clock, we measure the phase θ_n and its derivative θ'_n with "stroboscopic light" at times $t_n = 2\pi n/\omega$. Both are needed since the differential equation is of second order and their values at time t_{n+1} are (generally unknown) functions of their values at time t_n:

$$\theta_{n+1} = h(\theta_n, \theta'_n) \tag{3.16}$$
$$\theta'_{n+1} = h'(\theta_n, \theta'_n).$$

The map (3.16) must be invertible for the same reason as the map (2.4) and the winding number can be defined again by (2.5),

$$w = <d\theta/dt>. \tag{3.17}$$

Note that for a Josephson junction $d\theta/dt$ is simply proportional to the voltage across it.

If we look at the points of an orbit in the (θ_n, θ'_n)-plane (which has the topology of an annulus) they are not just scattered evenly. The dissipative nature of the motion means that the map (3.16) contracts areas so points (θ_n, θ'_n) must converge on some set of dimension less than 2. If the motion is quasiperiodic, i. e. the power spectrum consists of discrete peaks, the points (θ_n, θ'_n) will tend to a limiting set, an *attractor,* which is a smooth closed curve. This means that after a transient period θ' simply "slaves" θ, i.e.

$$\theta'_n = k(\theta_n) \tag{3.18}$$

where $k(\theta)$ is some smooth function. But then the map (3.16) is effectively one-dimensional

$$\theta_{n+1} = h(\theta_n, k(\theta_n)) \equiv f(\theta_n) . \tag{3.19}$$

Bak et al[25] found that this "dimensional reduction" indeed takes place for certain values of the parameters and by tuning the parameters, critical points can be located where (3.19) breaks down. Fig. 8 shows the return map θ_{n+1} vs. θ_n slightly above criticality as computed numerically from equation (3.10)[25]. In stead of having a single inflection point the whole curve starts to crinkle up and (3.18-19) break down, but until this point there is a smooth circle map. Thus, one has reasons to believe that the mode locking pattern for, say (3.15), should be in the same universality class as circle maps.

Indeed, Yeh et al[26] and Alstrøm, Jensen and Levinsen[27] have studied the mode locking transition by analog numerical simulation. Alstrøm et al determined the critical surface by locating the points where hysteresis first sets in. Figure 9 shows the scaling of mode-locked intervals measured along the critical surface. The resulting fractal dimension was found to be $D \approx 0.87$ in agreement with predictions from the circle map.

Experiments have been performed by Stavans et al[22] on the Rayleigh-Benard system described above, and by Martin and Martienssen[28] on an the ionic conductor BSN. Stavans et al found a

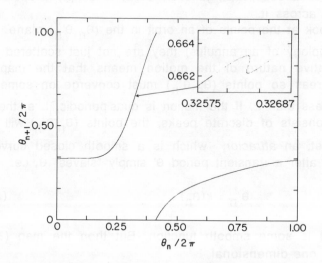

Fig. 8. The return map, θ_{n+1} vs. θ_n, for the Josephson equation (3.15), with A=1, B=1.253, C=1, D=1.2, E=1, and $\omega = 1.76$

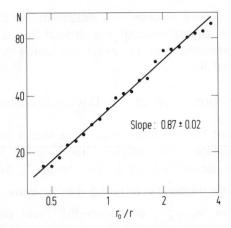

Fig. 9. Plot of log(N(r) vs. log(1/r) for a Josephson analog simulator (Ref. 27). The slope is 0.87±0.02 in agreement with the circle map model. Note that the map is one-dimensional but that small wiggles appear as an indicator of criticality.

complete devil's staircase with fractal dimension 0.86 ± 3%. The ionic conductor BSN is unique in that a constant driving current gives rise to an oscillating voltage. In the experiment by Martin and Martienssen the periodic voltage fluctuations define one frequency. An additional ac current defines another frequency, and one can scan a large range of winding numbers and amplitudes by varying the frequency and strength of the ac current.

So far no accurate experiment has been done where a sizable part of the critical surface is mapped out, which is regrettable since such surfaces might have an interesting structure characteristic of the competition between commensurability and incommensurability. The critical surface has been computed numerically for a simple map of the annulus like (3.16), the "dissipative standard map"[29]:

$$\theta_{n+1} = \theta_n + \Omega + r_{n+1}$$

$$(3.20)$$

$$r_{n+1} = br_n - K/2\pi \sin 2\pi\theta_n.$$

The result is shown in Fig. 10. For each Arnold tongue a "critical point" is found, and when the points are close (i.e. the tongues are thin) the curve becomes smooth. The details of the procedure and the arguments for the smoothness of the critical line are given in refs. 29-30. Further discussion of criticality for maps of the annulus is found in refs. 5 and 9.

3.3.2. Multifractal properties of the Devil's staircase.

The Cantor set for the circle map is in fact a multifractal. Halsey et al[19] have calculated the complete $f(\alpha)$-function for the staircase by locating 1024 locked intervals. The holes l_i between neighbor steps with winding numbers differing by p_i were measured. The largest value of α, α_{max}, describes the most rarefied portion of the staircase which is around the golden mean. Shenker[7] found that in that neighborhood

$$l_i \approx \delta_{GM}^{-n},\qquad(3.21)$$

and the corresponding changes in p_i are

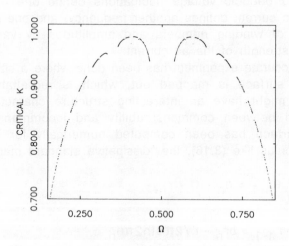

Fig. 10. The critical surface found for the two-dimensional map (3.20) with b=0.25. On small scales the surface becomes smooth, although there is a "dip" for each rational winding number.

$$p \approx (F_n/F_{n+1} - F_{n+1}/F_{n+2}) \approx w^{*-2n}, \tag{3.22}$$

so

$$\alpha_{max} = \log p/\log I = 2 \log w^* / \log \delta_{GM} = 0.924.... \tag{3.23}$$

The most concentrated portion of the staircase is near zero winding number for the 1/Q series where[14] $I' \approx 1/Q^2$ and p' $\approx$ 1/Q so

$$\alpha_{min} = \log p'/\log I' = 1/2. \tag{3.24}$$

Figure 11 shows the function $f(\alpha)$ for the mode locking structure of the circle map. The maximum of the curve is the fractal dimension $D \approx 0.8700$, and the left and right intersections with the α-axis are α_{min} and α_{max}, respectively. The multifractal properties, including the possibility of a phase transition, have recently been studied in more detail by Artuso et al[31]. As K is reduced below K=1 the mode-locked intervals do not have full measure, leaving room for quasiperiodic behaviour. The measure of quasiperiodic motion can thus be used as an order parameter for the transition.

3.3.3. The transition to chaos as a second order phase transition.

Maybe the most beautiful way of viewing the transition is as a second order phase transition. The transition to chaos can be described by a formalism which is completely identical to the one used to describe a percolation transition or a thermodynamic transition. This analogy was discovered by Jensen et al[14,23] and was later elaborated by Alstrøm et al[32].

The analogy goes as follows: at the critical point of a second order phase transition there is a power-law distribution of (magnetic) clusters of all sizes, $N(s) = s^{-D}$ imbedded in a d-dimensional euclidean space. At the transition to chaos there is a distribution of locked intervals of length r = 1/s imbedded in a d=1 dimensional space, $N(s=1/r) = s^{-D}$. Thus, "small intervals" are equivalent to "large clusters" in the analogy. D can be viewed as a *critical index* for the transition.

In a magnetic field h, all the clusters larger than 1/h order, so the magnetization M becomes $M = h^{1-D} \ldots h^{1/\delta}$, and D is related in a

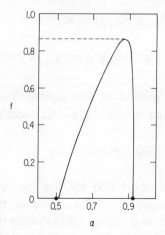

Fig. 11. The $f(\alpha)$ function for the staircase in Fig. 6. The "bare" winding numbers were chosen as the densities p_i , and the "dressed" winding numbers play the role of the length scales l_i (Ref. 19).

simple way to the exponent δ for the magnetic field at criticality. The holes which appear when steps smaller than r^{d-1} (of total length r^{d-D}) are removed correspond to the volume $M(l)$ of magnetic clusters larger than l^d.

Below the critical point, at $T_c-T = t > 0$, clusters which extend beyond the correlation length $\xi = t^{-\upsilon}$ contribute to the order parameter M. For the mode locking case, at $t = 1-K > 0$, intervals smaller than $1/\xi = t^\upsilon$ give contributions to the order parameter. In both cases we have

$$M(t) = \xi^{D-d} = t^{\upsilon(d-D)} \equiv t^\beta. \qquad (3.25)$$

The scaling relation $\beta=\upsilon(d-D)$ can be directly read off this equation. Since the critical index D is related to η and δ through simple scaling relations, this relation can be expressed in more familiar form as, for instance, $d-2-\eta = 2\beta/\upsilon$. Jensen et al found numerically a value of $\upsilon=2.63$. Alstrøm et al found a more accurate value $\upsilon = 2.39$. The scaling relation then yields $\beta=0.31$. This value is very similar to values found for β at second order phase transitions. It would be interesting to measure the exponent independently, by numerical calculation or in an experiment, in order to check the scaling relation.

All other scaling relations known from the theory of second order transitions have their analogs at the mode locking transition. Also, the crossover at K=1 are given by scaling functions such as

$$M(r,t) = t^{\beta}F(r/t^{\upsilon}).\qquad(3.26)$$

The equivalence with critical phenomena encountered in widely different situations suggests a deeper and more global universality which should be explored theoretically.

4. SUPERCRITICAL BEHAVIOUR.

When we cross the critical surface and pass into the supercritical region the strict relation between "real" systems (differential equations like (3.15) or experiments) and 1-d circle maps breaks down. This means that the supercritical behaviour is not only very complicated but also characterized by the appearance of *non-universal* phenomena. This very rich behaviour has been studied in several works[5,6,33-42] but many unresolved questions remain.

Here we shall restrict our attention to the behaviour of the immediate vicinity of the critical surface, where one would expect universality at least regarding qualitative features.

4.1. Anatomy of the Arnold tongues.

To understand the supercritical behaviour we must look more carefully at the structure of a typical Arnold tongue for a circle map, say the tongue corresponding to mode locking with winding number P/Q. As shown in fig. 12, it consists of the entire space between the lines OA_L and OA_R in which tangent bifurcations take place and is defined as the region in which the winding number can have the value P/Q which means that there exists a (perhaps unstable) cyclic point θ^* satisfying (2.6). For concreteness the critical line is taken as K=1 (as for the "sine map" (2.8)); for K<1 the map is monotonic and the winding number is unique. Above the critical line there are at least 3 important lines: First a

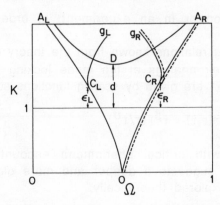

Fig. 12. Structure of typical Arnold tongue. The notation and the various lines are explained in the text (Ref. 36).

hyperbola-like curve with tip at D signifying the first period doubling bifurcation. Period doubling doesn't change the winding number, but the cycle length changes from Q to 2Q and $w = P/Q \to 2P/2Q$. Above this curve there is a complicated set of period doubling cascades.[33,39] Second, there are two lines g_L and g_R from the edges at C_L and C_R. Between these lines the winding number is still unique, but in the regions between OA_L and g_L, and between OA_R and g_R it is possible to find a whole interval of winding numbers, depending on the choice of initial condition θ_0. The lines g_L and g_R are edges of the closest nearby tongues. A nearby tongue with winding number slightly larger than P/Q is shown by the dashed lines in the figure. In the limit $w \to (P/Q)^+$ the left hand edge follows g_R and the right hand edge follows $C_R A_R$. The little tongue doesn't overlap the big one until some finite distance above criticality. The behaviour for winding numbers less than P/Q is completely analogous.

4.2. Scaling of geometry and Liapunov exponents around noble winding numbers.

If we follow a sequence of rationals converging to a quadratic irrational (one with a periodic continued fraction expansion) we expect the tongues to become self-similar. Indeed, that is the

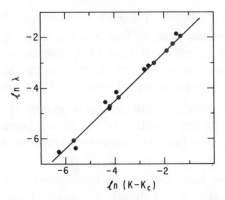

Fig. 13. A plot of $\log \lambda$ vs. $\log(k-k_c)$ for the two-dimensional map (3.20) with b=0.5 close to golden mean winding number. The dots come from numerical calculations and the straight line is the theoretical prediction (Ref. 37).

case[34,36]. At the golden mean the widths scale down with the exponent δ_{GM} (3.3) and the heights d, ε_R and ε_L in fig. 12 scale down by α_{GM}^2 where α_{GM} is given by (3.4). As mentioned below (3.5) they are precisely the eigenvalues of the RG-transformation in those two directions.

This self-similarity manifests itself also in the scaling of the dynamic invariants. If one starts at the golden mean critical point (K_G, Ω_G) and then increases K the Liapunov exponent is found to rise as a power

$$\lambda \approx (K-K_G)^{\gamma} \tag{4.1}$$

where

$$\gamma = \log w^* / \log \alpha_{GM}^2 = 0.948... \tag{4.2}$$

This result was obtained by Jensen and Procaccia[37] for the two-dimensional map (3.20) and conjectured for circle maps by Kaneko[34]. Some care has to be taken since - as for maps of the interval - there are lots of non-chaotic windows where the motion locks into some rational period. A log-log plot of Liapunov exponent vs. K-K$_C$ (shown in figure 13) was obtained by finding "maximally chaotic" states and the straight line has the slope given by (4.2).

It is easy to understand how (4.2) follows from the geometric self-similarity. Going from the n'th to the n+1'th approximant to the golden mean the K-value must be rescaled by a factor α_{GM}^2. But the RG transformation changes the Liapunov exponent in a trivial way - simply according to the number of f's to be composed together in (3.2), namely F$_n$. The ratio of Liapunov exponents thus converges to the golden mean giving the final result (4.2).

One might ask how the system remembers how chaotic it should be when it crosses the network of mode locking tongues. The explanation can be tied to the continued wrinkling of the underlying manifold as seen in fig. 14. Even though the system is locked onto a specific tongue (as in fig. 14 b, d, f) the underlying manifold is still wrinkled. This can be seen by looking at transients for different initial conditions. As soon as we move outside a tongue the system becomes chaotic and the manifold as a whole is traced up (fig. 14 a, c, d). The same kind of reasoning was used by Bohr[30] to argue that the critical surface (fig. 10) becomes smooth around the tongues.

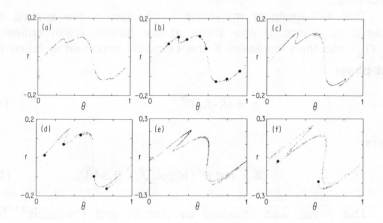

Fig. 14. The continued wrinkling of the manifold as K increases. In b), d), and f) the motion is locked onto a rational tongue (with winding numbers 5/8, 3/5, and 1/2, respectively), whereas in a), c), and e) the motion is chaotic (Ref. 37).

4.3. Scaling around rational winding numbers.

If we inspect a sequence of Arnold tongues with winding numbers converging to a given rational number (for instance the sequence $1/N$, $N = 1, 2, 3...$ converging to 0) we would again expect some form of self-similarity. The width $\Delta\Omega$ of the tongues scales as a power law

$$\Delta\Omega \approx 1/N^3 \qquad (4.3)$$

characteristic of the tangent bifurcation occurring at the edge of each tongue[14]. One would therefore presume that the heights d, ε_R and ε_L in fig. 12 would also approach zero as some power. That is, however, not the case. Bohr and Gunaratne[36] found that those numbers actually converge to finite positive numbers as $N \to \infty$. Thus the tongues become very thin but the heights are not rescaled. The most unexpected feature is that those finite numbers depend on which family of rationals, converging to a given rational, is chosen. Thus, the sequence $2/(2N+1)$, $N = 1, 2 ..$ gives different limiting numbers. Further, these numbers are non-universal - different maps give different results. This means that the "line" characterizing the onset to chaos is very complicated - around the edge of each tongue it is infinitely multiple valued. Friedman and Tresser[38] have studied these properties further and describe them in terms of "comb structures". The hairiness of this boundary should be contrasted with the smoothness of the critical surface.

REFERENCES.

1. Feigenbaum, M. J., J. Stat. Phys. **19**, 25 (1978); **21**, 669 (1979);
2. Arnold, V. I., *Geometric Methods in the Theory of Ordinary Differential equations* (Springer, Berlin 1983).
3. Herman, M. R., Lecture Notes in Mathematics **597**, 217 (Springer Berlin 1977).
4. Bak, P., Phys Today, december1986 , p 39.
5. Aronson, D. G., Chory, M. A., Hall, G. R., and McGehee, R. P., Comm Math. Phys. **83**, 303 (1982).
6. Boyland P., Ph. D. thesis, Boston University (1983).
7. Shenker, S. J., Physica **5**D, 405 (1982).

8. Feigenbaum, M. J., Kadanoff, L. P., and Shenker, S. J., Physica 5D, 370 (1982).

9. Rand, D., Ostlund, S., Sethna, J., and Siggia, E., Phys. Rev. Lett. 49, 132 (1982); Physica 6D, 303 (1984).

10. Shraiman, B. I., Phys. Rev. A 29, 3464 (1984).

11. Farmer, J. D., and Satija, I. I., Phys. Rev. A 31, 3520 (1985); Umberger, D. K., Farmer, J. D., and Satija, I. I., Phys. Lett. 114A, 341 (1986).

12. Khinchin, A., *Continued Fractions* (University of Chicago Press, 1964).

13. Christiansen, P. V., to be published in the proceedings of "chaos in education", Balaton, Hungary (1987).

14. Jensen, M. H., Bak, P., and Bohr, T., Phys. Rev Lett. 50,1637 (1983); Phys. Rev. A 30, 1960 (1984).

15. Lanford, O. E., Physica 14D, 403 (1985).

16. Rand, D., to appear in the Proceedings of Royal Society Meeting on Dynamical Chaos (1987).

17. Mandelbrot, B. B., J. Fluid Mech. 62, 331 (1974).

18. Benzi, R., Paladin, G., Parisi, G., and Vulpiani, A., J. Phys. A 17, 352 (1984).

19. Halsey, T. C., Jensen, M. H., Kadanoff, L. P., Procaccia, I., and Shraiman, B. I., Phys. Rev. A 33, 1141 (1986).

20. Jensen, M. H., Kadanoff, L. P., Libchaber, A., Procaccia, I., and Stavans, J., Phys. Rev. Lett. 55 2798 (1985).

21. Gwinn, E. G., and Westervelt, R. M., Phys. Rev. Lett. 59, 157 (1987).

22. Stavans, J., Heslot, F., and Libchaber, A., Phys. Rev. Lett. 55, 596 (1985).

23. Bak, P., Bohr, T., and Jensen, M. H., Phys. Scripta T9, 50 (1985).

24. Belykh, V. N., Pedersen, N. F., and Sørensen, O. H., Phys. Rev. B 16, 4860 (1978).

25 Bak, P., Bohr, T., Jensen, M. H., and Christiansen, P. V., Sol. State Commun. 51, 231 (1984)

26. Yeh, W. J., He, D.-R., and Kao, Y. H., Phys. Rev. Lett. 52, 480 (1984).

27. Alstrøm, P., Jensen, M. H., and Levinsen, M. T., Phys. Lett.103A, 171 (1984); Alstrøm P. and Levinsen, M. T., Phys. Rev. B 31, 2753 (1985).

28. Martin, S. and Martienssen, W., Phys. Rev. Lett. 56, 1522 (1986).

29. Bohr, T., Bak, P., and Jensen, M. H., Phys. Rev. A **30**, 1970 (1984).
30. Bohr, T., Phys. Rev. Lett. **54**, 1737, (1985); Phys. Scripta T**3**, 124 (1986).
31. Artuso, A., Cvitanovic, P., and Kenny, B. K., Niels Bohr Institute preprint (1987).
32. Alstrøm, P., Hansen, L. K., and Rasmussen, D. R., Phys. Rev. A **36**, 828 (1987).
33. Glass, L., and Perez, R., Phys. Rev. Lett. **48**, 441 (1982).; Belair, J. and Glass, L., Physica **16**D, 143 (1985).
34. Kaneko, K., Prog. Th. Phys. **72**, 1089 (1984).
35. Schell, M., Fraser, S., and Kapral, R., Phys. Rev. A **28**, 373 (1983).
36. Bohr, T. and Gunaratne, G. H., Phys. Lett. **113**A, 55 (1985).
37. Jensen, M. H. and Procaccia, I., Phys. Rev. A **32**, 1225 (1985).
38. Friedman, B. and Tresser, C., Phys. Lett. **117**A, 15 (1986).
39. MacKay, R. S. and Tresser, C., Physica D **19**, 206 (1986).
40. Chen, C., Georgyi, G., and Schmidt, G., Phys. Lett. **122**A, 89 (1987).
41. Glaxier, J. A., Jensen, M. H., Libchaber, A., and Stavans, J., Phys. Rev. A **34**, 1621 (1986).
42. Gunaratne, G. H., Jensen, M. H., and Procaccia, I., to appear in Nonlinearity **1** (1988).

CHAOS IN LIQUID CRYSTAL OPTICAL BISTABILITY

Hong-jun Zhang, Jian-hua Dai,Peng-ye Wang
Fu-lai Zhang, Guang Xu, and Shi-ping Yang

Institute of Physics, Academia Sinica
P.O.Box 603, Beijing, China

ABSTRACT

Chaos behavior in liquid crystal hybrid optical bistability (LCOB) with one and two delay times in the feedback loop is reviewed. Varied manifestations of chaos in the system with one delay time are discussed: period doubling, split bifurcation, transient oscillation and the critical behavior. The results of numerical simulation with the iteration equation and the relaxation equation are presented in detail. Some analytical results are also obtained. In the two-delay system, the frustrated instability is observed. The frequency-locking structure is given. The route to chaos in this system is analysed by means of the Lyapunov exponents and correlation dimension. The influence of the feedback strength is discussed.

CONTENTS

1. Introduction

Optical bistability is attracting ever more attention in recent years. Especial attention has been paid to the instability and chaos behavior in optical bistability lately because chaos is related to some fundamental questions in physics such as turbulence. Chaos may be the bridge connecting the deterministic physics and the statistical physics. Chaotic behavior exists in many nonlinear systems such as fluids, plasmas, electric circuits, chemical reactions, celestial movements, acoustical systems, ecosystems and so on. Optical bistability is a typical example of a nonequilibrium physical system. The hybrid optical bistable device is a powerful experimental device for the study of optical turbulence.

The occurrence of chaos in a ring cavity was proposed by Ikeda[1] in 1979. He pointed out that bifurcation and chaos would appear in an optical bistable device with a time delayed feedback. The first experimental verification was provided by Gibbs et al.[2] and Hopf et al.[3] with a hybrid optically bistable device containing an electro-optic crystal. In this paper, the study on chaos in hybrid liquid crystal optical bistability (LCOB) is reviewed. In the system with one delay time in the feedback loop, the period doubling bifurcation to chaos is obtained. The split bifurcation is analysed. The transient oscillation and the critical behavior are discussed. The results of numerical simulation with the iteration equation and the relaxation equation are presented in detail. Some analytical results are also obtained. In the two-time-delayed system, the experimental results of frequency locking, hysteresis and frequency beating are given. The competition between these feedback mechanisms causes the system to fall into a state of frustrated instability. The route to chaos is analysed by means of the Lyapunov exponents and correlation dimension. The structure of frequency locking is obtained. The influence of the feedback strength is observed.

2. Chaos in LCOB with One Delay Time

2.1 General description of the system[4]

The liquid crystal hybrid optical bistable device consists of a twisted nematic liquid crystal (LC) cell (as a nonlinear medium) and a feedback loop with a variable delay time t_R. The delay is realized by means of a microcomputer. The experimental setup is shown in Fig.2.1.

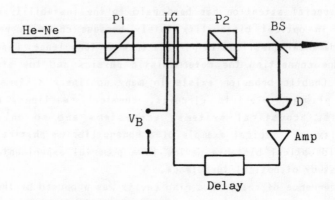

Fig.2.1. Experimental setup.
   He-Ne: laser; P_1, P_2: polarizers; LC: liquid
   crystal cell; BS: beam splitter; D: photo-detector;
   Amp: amplifier; V_B: bias voltage; and Delay: delay
   system.

The electrical signal of the photo-detector is proportional to the output intensity I_{out}. This voltage signal, suitably amplified and delayed, is fed back to the LC cell, where it is added to a constant, but adjustable, bias voltage V_B. The LC cell is placed between a polarizer P_1 and analyzer P_2 oriented at $+45°$ and $-45°$, respectively, relative to the surface molecular orientation. The transmittance of the system[5] can be approximated by the following functional relation

$$T(V)=H\sin^2(\pi V/2V_H),\tag{2.1}$$

where V is the total voltage applied to the LC cell and $|V|\leqslant2V_H$, V_H is the half-wave voltage, and H is the maximum transmittance. Therefore, if we omit the relaxation time of the medium, the output of the system is given by

$$I_{out}=I_{in}H\sin^2(\pi V/2V_H),\tag{2.2}$$

where I_{in} is the input intensity. The total voltage V is composed of the feedback voltage V_f and the bias voltage V_B, namely

$$V = V_f + V_B. \tag{2.3}$$

The feedback loop makes the voltage V_f a linear function of the output intensity at time $t-t_R$, i.e.

$$V_f = KI_{out}(t-t_R), \tag{2.4}$$

where K is an instrumental constant.

Equations (2.2), (2.3) and (2.4) can be combined to yield the following equation:

$$X(t) = A\sin^2[X(t-t_R) - X_B], \tag{2.5}$$

where $X(t) = \pi KI_{out}(t)/2V_H$, $A = \pi KHI_{in}/2V_H$, $X_B = -\pi V_B/2V_H$. Let $t = (n+1)t_R$ (n is an integer) and $X_n = X(nt_R)$, Eq.(2.5) can be cast into the form:

$$X_{n+1} = A\sin^2(X_n - X_B) = F(\mu, X_n), \tag{2.6}$$

where μ stands for A or X_B. This is a one-dimensional iteration equation containing two control parameters A and X_B. The dynamical behavior of the system can be described by Eq.(2.6) very well in the limit where the delay time t_R is much longer than the relaxation time of the medium. Let X^* be the stationary solution of Eq.(2.6). The stability of the stationary solution X^* is determined by the slope of the curve $F(\mu, X) = A\sin^2(X-X_B)$. The condition to be stable is $\left|\frac{d}{dX}A\sin^2(X-X_B)\right|_{X=X^*} < 1$. Therefore, if $|A\sin(2X^*-2X_B)| > 1$, the system is unstable which results in bifurcation and chaos. If the condition $t_R \ll 1$ (τ_0 is the relaxation time of the nonlinear medium) is not satisfied we can not use Eq.(2.6) to describe the system any more. The dynamical behavior of the hybrid optical bistable device with time delayed feedback can be described by a relaxation equation of the following type

$$\tau_0 \frac{dX(t)}{dt} = -X(t) + A\sin^2[X(t-t_R) - X_B]. \tag{2.7}$$

The equation can also cause the bifurcation and chaos behavior.

2.2 Bifurcation diagrams

Equation (2.6) obtained from the system is a bimodal mapping with two independent control parameters in the range of $|X-X_B|<\pi$. It can show some interesting bifurcation and chaos structures.These structures can be seen from the bifurcation diagrams.[6] Now we consider two kinds of bifurcation diagram corresponding to the two control parameters.

A typical X-A bifurcation diagram is shown in Fig.2.2. The solid line is the steady state solution of Eq.(2.6). It is a typical "S" curve of optical bistability. Many characteristics of the bifurcation diagram are similar to those of unimodal logistic maps. The period-doubling bifurcation to chaos , self-similarity structure, tangent bifurcation windows and attractor crisis can be seen clearly.

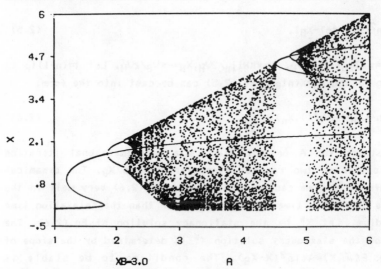

Fig.2.2. X-A bifurcation diagram.

Another kind of bifurcation diagram corresponding to the parameter X_B is shown in Fig.2.3. Although it is quite different from the X-A bifurcation diagram, we can also find the universal properties. The period-doubling bifurcations always exist.

The phenomena of period-doubling in LCOB is easily verified in the experiment. The typical experimental results for the waveforms and the power spectrum are shown in Fig.2.4. It can be seen that $T=2t_R$ in Fig.2.4(a), $T=4t_R$ in Fig.2.4(b), and chaos appears in Fig.2.4(c).

The above stated properties are similar to the universal properties

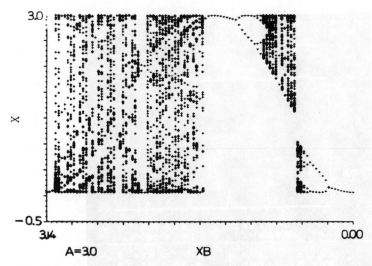

Fig.2.3. X-X_B bifurcation diagram.

of the logistic map. We should note that Eq.(2.6) is a bimodal mapping
and has bistability and multistability. The structure of bifurcation
and chaos will therefore be affected, which will give rise to some
significant differences from the logistic map. This will be discussed
in the following.

〈1〉 A bifurcation window with period 1 appears at the transition
point (A=4.52) from the optically bistable state to the multistable
state as shown in Fig.2.2. The transition from the bistable state to
the multistable in the stationary regime takes place in the form
similar to a discontinuous jump in phase transition . In the chaotic
case, however, the transition appears as a tangent bifurcation. When
A>3.0, some new windows in addition to the one at A=4.52 can be
observed such as a period 2 window at A=3.6735. This window and the
window at A=5.6127 are both caused by the bimodal map.

〈2〉 The bifurcations not only take place in the upper branch of the
bistability but also in the lower branch, as shown in Fig.2.5. The
discontinuous change (crisis) and hysteresis of chaotic attractors due
to the bistability can be observed. This phenomena can also be
observed in the X-X_B bifurcation diagram as shown in Fig.2.6. As an
example, the interpretation of this phenomena may be found in Fig.2.7.
At X_B=0.60 the 45° bisector d separates the chaotic region into a peak
and a valley. At the peak, $(X_n)_{max}$=A=3, $(X_n)_{min}$=Asin^2(A-X_B)=1.369. The

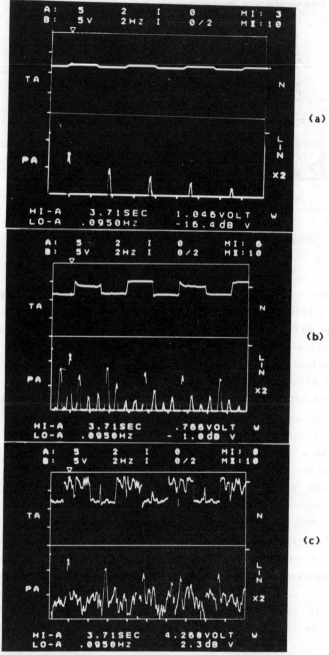

Fig.2.4. Experimental results of period-doubling bifurcation
to chaos.

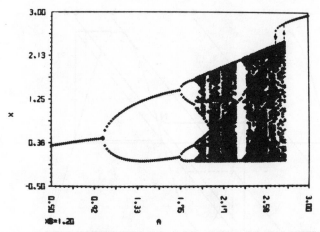

Fig.2.5. Bifurcation diagram in the lower state of bistability.

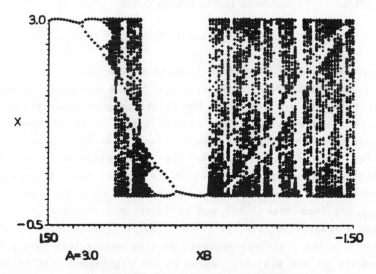

Fig.2.6. Hysteresis and discontinuous jump in $X-X_B$
bifurcation diagram.

value of $\{X_n\}$ is therefore confined within the 1.369 to 3.0 region, as
shown by the solid line at the peak in Fig.2.7. The chaotic region at
the valley has $\{X_n\}_{min}=0$ and $\{X_n\}_{max}=A\sin^2(0-X_B)=0.956$, and the value
of $\{X_n\}$ is confined between 0 and 0.956, as shown by the dotted line
at the valley in Fig.2.7. Hence, the region of X_n between the two

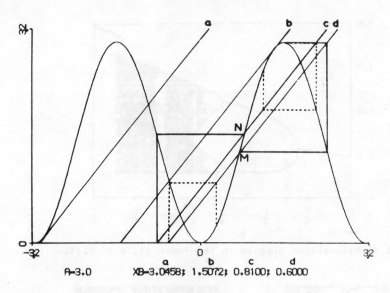

$$A=3.0 \qquad X_B=3.0458; \ 1.5072; \ 0.8100; \ 0.6000$$

Fig.2.7. F(X) at A=3.0. The abscissa is X-X_B.

chaotic regions (1.369-0.956=0.413) is a forbidden zone in the phase space. This forbidden zone is inaccessible to the iteration values. When X_B is slightly less than 0.6, the chaotic region jumps suddenly from the peak to the valley. This sudden change is established via transient states in the forbidden zone. In fact, when X_B=0.6, the peak $(X_n)_{min}$=1.369 exactly coincides with the intersection M of the 45° bisector and the curve. The properties of the M point keep the (X_n) of the peak within the 1.369-3.00 range. When X_B is slightly less than 0.6, $(X_n)_{min}$ is less than 1.369 and is lower than the M point. The system makes a transition into the valley chaotic region (shown by the dotted line) via the transient process. We also notice that X_B=0.6 is located exactly in the bistable region of the right-hand peak of the transmission curve. Hysteresis is therefore obtained. When iterations are performed with increasing X_B and decreasing X_B, as shown in Fig.2.6, the hysteresis can be seen in the 0.60-0.81 range of X_B. The arrows indicate the direction of the discontinuous jump. Iteration in the reversed direction shows a sudden change of the chaotic region at X_B=0.81. This is attributed to exactly the same reason for the effect at X_B=0.6 in the forward iteration. In Fig.2.7 the straight line C is a 45°bisector for X_B=0.81 and its intersection N with the curve is the

upper limit of the valley chaotic
region. It should be pointed out
that the fundamental reason for
the hysteresis is the coexistence
of two attractors. In order to
observe the whole structure in
the parameter plane, we plotted
the $A-X_B$ phase diagram as
shown in Fig.2.8. 2P, 4P and 8P
are the periodic regions and C is
the chaotic region. No window
positions are shown in the
chaotic region because they are
too small. The dotted lines
indicate the bistable region of
the steady states. The $A-X_B$
phase diagram serves as a guide
to the selection of experimental
parameters.

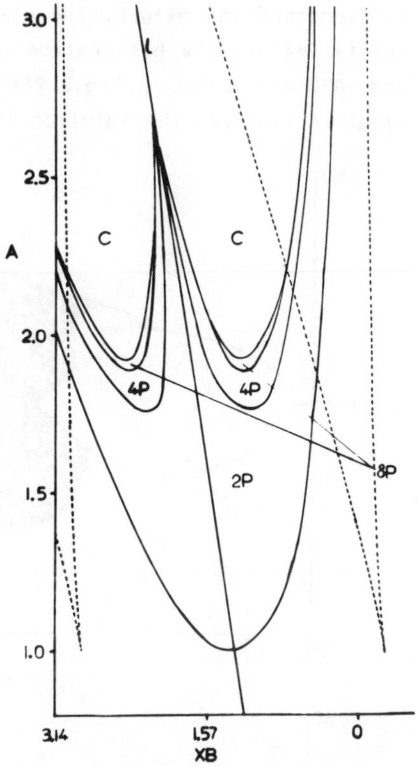

    (3) Split bifurcation[7]
can be obtained when the iteration
is performed along the straight
line L in Fig.2.8. Making the X
coordinate transformation
$Y_n=X_n-A/2$, and selecting

Fig.2.8. $A-X_B$ phase diagram.
Dotted line indicate
the bistable region.

$$X_B=(1/2)(A+\pi/2) \tag{2.8}$$

we obtain the iteration equation

$$Y_{n+1}=-(A/2)\sin(2Y_n). \tag{2.9}$$

This symmetry property is similar to that of the cubic map.[8]
Equation (2.9) displays only one control parameter (A), but we note
that X_B varies linearly as A varies in Eq.(2.8). In the phase diagram
shown in Fig.2.8, Eq.(2.8) is a straight line, L, that passes between
the two chaotic bands.

    We obtained the bifurcation diagrams by iterating Eq.(2.9) and

noticed that the bifurcation diagrams have a strong dependence on the
initial value. The bifurcation diagrams with the initial value $X_0=A/2$
and $-A/2$ are shown in Fig.2.9(a) and Fig.2.9(b), respectively. In the
range of $1<A<2.25$ the solution is confined within the period 2 domain

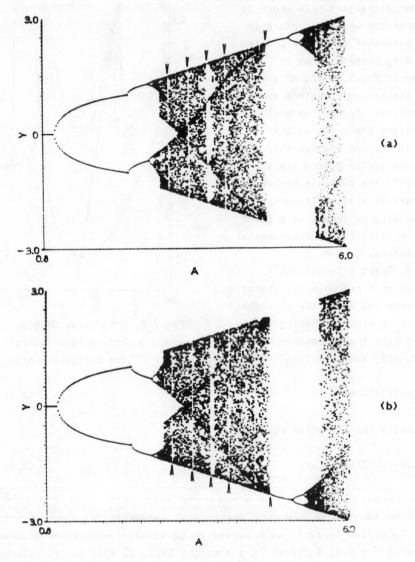

Fig.2.9. Split bifurcation diagram for (a) $X_0=A/2$
and (b) $X_0=-A/2$.

of the pitchfork bifurcation. The split bifurcation begins to develop at A=2.25. The first split bifurcation at A=2.25 corresponds to the tangent of curve $F^{(2)}=F(F(X;\mu);\mu)$ with the 45° bisector at two inflection points i.e. $F^{(2)\prime}=1$ at these points. When A increases, these two points become unstable and two intersection points appear beside each inflection point. These four intersection points are stable as shown in Fig.2.10. The only difference between the pitchfork bifurcation and the split bifurcation is that, for pitchfork bifurcation, all the intersections belong to the same attractor, while in the case of split bifurcation, the points of intersection do not belong to the same attractor. These points correspond to two period 2 attractors with the two intersection points located to the right and to the left of the other symmetric inflection point. Similar explanations can be given to the split bifurcations in other regions.

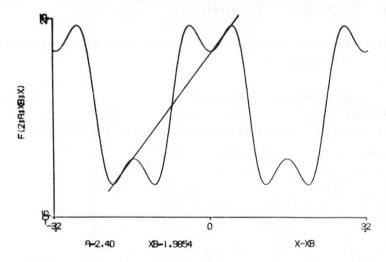

Fig.2.10. $F^{(2)}(X,\mu)$ versus $X-X_B$.

(4) In the bifurcation diagrams, the parameter value for a given superstable period can be determined from the corresponding word in the symbolic dynamics made of four letters. The boundaries and darklines in the chaotic region can be obtained analytically[9]. We first write down the inverse function of Eq.(2.6)

$$F^{-1}(y)=X_B+(1/2)\cos^{-1}(1-2y/A).\qquad\qquad (2.10)$$

Equation (2.10) represents only one branch of the inverse function, as labelled N in Fig.2.11. Thus we can define a function

$$N(y)=F_N^{-1}=X_B+(1/2)\cos^{-1}(1-2y/A).$$
(2.11)

Analogously, we have for the other branches:

$$L(y)=F_L^{-1}=N(y)-\pi =X_B-\pi +(1/2)\cos^{-1}(1-2y/A),$$
(2.12)

$$M(y)=F_M^{-1}=2X_B-N(y)=X_B-(1/2)\cos^{-1}(1-2y/A),$$
(2.13)

$$R(y)=F_R^{-1}=M(y)+\pi =X_B+\pi -(1/2)\cos^{-1}(1-2y/A).$$
(2.14)

The iteration equation (2.6) has four critical points:

$X_1=X_B-\pi/2,$ $X_3=X_B+\pi/2,$ (peaks)

$X_2=X_B,$ $X_4=X_B+\pi.$ (valleys)

A superstable orbit starts at some critical point X_i and returns to the same X_i, i=1,2,3,4, i.e.

$$\underbrace{F \circ F \circ \ldots \ldots \circ F(X_i)=X_i,}_{N\ times} \qquad (i=1,2,3,4)$$
(2.15)

where the symbol "∘" represents the composition of the function.

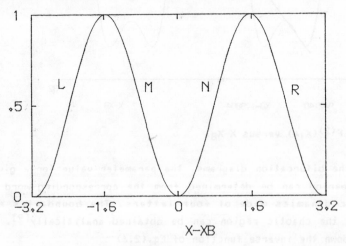

Fig.2.11. The bimodal mapping F(x). For the labels L,M,N, and R see text.

Let us transfer (N-1) functions F from the left hand side of Eq.(2.15) to the right hand side. To retain uniqueness in taking the inverse one must indicate which branch it belongs to by attaching a subscript to F^{-1}, i.e.

$$F(X_i)=F_a^{-1} \circ F_b^{-1} \circ \ldots \ldots \circ F_m^{-1}(X_i), \quad (i=1,2,3,4) \qquad (2.16)$$

where a, b, ..., m stand for one of the letters R, L, M, N. Putting together the subscripts, we obtain a word W composed of these letters,

$$W=ab...m. \qquad (2.17)$$

The superstable orbit, Eq.(2.15), is charaterized by the word W in the symbolic dynamics of four letters. We can lift the word W to a function W(X) by using the definitions Eq.2.11-14, to define a composite function

$$W(X)=a(b(...m(X)...)), \qquad (2.18)$$

where a(X), b(X), ..., m(X) stand for one of the functions in Eq.2.11-14. Therefore, Eq.(2.16) now reads

$$F(X_i)=W(X_i), \quad (i=1,2,3,4) \qquad (2.19)$$

which is an equation for the parameters A and X_B. The parameters A and X_B for every superstable orbit can be obtained by solving Eq.(2.19). The A-X_B phase diagram obtained by plotting Eq.(2.19) for periods 1, 2, and 3 are shown in Fig.2.12. It is in agreement with Fig.2.8 which is obtained by numerical simulation.

Now we turn to another characteristic feature of the bifurcation

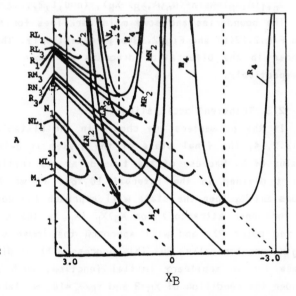

Fig.2.12. Analytically obtained A-X_B phase diagram. Only periods 1, 2, and 3 are shown.

diagrams, namely, the darklines going through the chaotic region and
the boundaries of the chaotic bands.

Suppose we use homogeneously distributed X_n as input for the map.
After just one iteration there will be a singularity in the
distribution of X_{n+1} located at the iteration image of each X_c. This
also gives the extreme points of the map, i.e. the boundaries of
chaotic output at the given parameter value. The singularity will show
up in subsequent iterations as local maxima in the distribution of X,
i.e. the loci of darklines embedded in the chaotic region. From what
has just been said the definition of all darklines follows:

$P_0(\mu)=X_c$,

$P_{n+1}(\mu)=F(\mu,P_n(\mu))$.   (2.20)

In our case, we have two sets of composite functions using the map
Eq.(2.6)

$P_0(A,X_B)=X_B \pm \pi/2$,

$P_{n+1}(A,X_B)=A\sin^2(P_n(A,X_B)-X_B)$, $(n=0,1,2,...)$ (2.21)

and

$Q_0(A,X_B)=X_B$ or $X_B+\pi$,

$Q_{n+1}(A,X_B)=A\sin^2(Q_n(A,X_B)-X_B)$, $(n=0,1,2,...)$ (2.22)

The boundaries and some main darklines for $X_B=3$ and $A=3$ are shown
in Fig.2.13(a) and Fig.2.13(b), respectively. They are consistent with
those in the bifurcation diagrams as shown in Fig.2.2 and Fig.2.3,
respectively.

2.3 Transient oscillation[10,11]

If the parameters are chosen in the bistable region, as shown in
Fig.2.14, the final state of the system is determined by the initial
value of X. Because of the time delay the initial value is a function
of t defined in the interval $[0,t_R]$. If we let the function be a
constant X_0 then the state will approach the upper stable fixed point
H (or upper attractor) for $X_0>X_p$ (X_p is the X value at the unstable
fixed point P), and will approach the lower stable fixed point for
$X_0<X_p$ monotonically. But this process will be different from the above
case for an arbitrary initial function, such as a linear function.
Under the condition of $X_B=3$ and $t_R/\tau_0=10$, we take the initial function
as a linear function whose value is in the range of $[0,0.9]$. The final
state is determined by the parameter A. By numerical simulation with
Eq.(2.7) we find that there is a critical value $A_c=1.3350625$. When

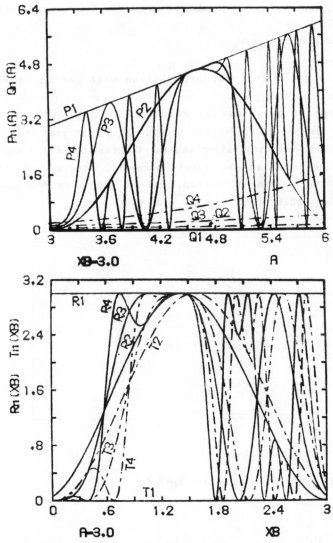

Fig.2.13. (a) $P_n(A)=P_n(A,3.0)$ and $Q_n(A)=Q_n(A,3.0)$ versus A,
(b) $R_n(X_B)=P_n(3.0,X_B)$ and $T_n(X_B)=Q_n(3.0,X_B)$ versus X_B.

$A>A_c$ the final state is at the upper branch while for $A<A_c$ the final
state is at the lower branch. The transient process is a decaying
oscillation with period $T=t_R$, as shown in Fig.2.15. In the limit of

$t_R/\tau_0 \to \infty$, the system can be described by the difference equation

$$X(t)=A\sin^2[X(t-t_R)-X_B].\qquad\qquad (2.23)$$

We can see below that the oscillation with period t_R will last indefinitely in this case.

The typical experimental results for $t_R/\tau_0 \gg 1$ are shown in Fig.2.16. The period is $T\dot{=}t_R=10$ second. It can be seen that the width of the square wave oscillation and the fluctuation of the amplitude on the peak increase as the input intensity increases. Finally, the period-t_R oscillation is blurred, and one has fully developed chaos. The $I_{out}-\dot{I}_{out}$ phase portraits are also given.

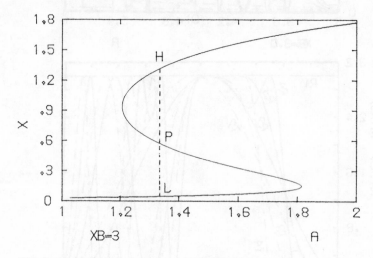

Fig.2.14. Bistable region for $X_B=3.0$.

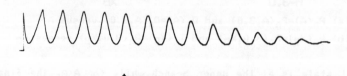

Fig.2.15. Transient oscillation with period $T\dot{=}t_R$.

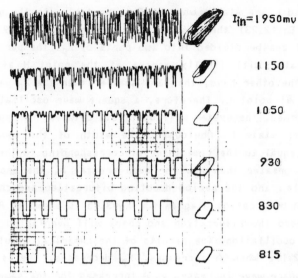

Fig.2.16. Experimental results of period-t_R oscillation.

In order to give a further interpretation of this phenomenon, we consider Eq.(2.23). Figure 2.17 shows the nonlinear function $y=A\sin^2(X-X_B)$ for the parameters $A=1.4$ and $X_B=3.05$. The fixed points H, P and L are given by the intersections of y and the 45° straight line. The stability of the fixed points is determined by the slope of y at these intersections. If $\langle dy/dX\rangle|_\alpha \langle 1$ (α =H, P or L), the intersection is stable; otherwise it is unstable.

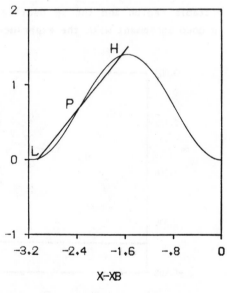

Fig.2.17. F(X) versus $X-X_B$ for $A=1.4$ and $X_B=3.05$.

Therefore, H and L are stable whereas P is unstable. In the numerical
simulation of Eq.(2.23) the linear initial function [0,0.9] in the
interval [0,t_R] can be divided into two parts by X_p. The part greater
than the X_p value will finally arrive at the point H after some
iteration. On the other hand, the part smaller than the X_p value will
finally arrive at point L. Therefore, a square wave oscillation with
period t_R is formed. The system will oscillate between the upper state
H and the lower state L. Obviously, the width of the square wave
oscillation is equal to the time interval corresponding to the initial
value interval greater than X_p. For even greater A, the upper state
will be unstable and the period doubling bifurcation and chaos will
appear. The X–A bifurcation diagram for X_B=3.05 is shown in Fig.2.18.
As A is increased the bifurcation and chaos will overlap on the peak
part of the t_R oscillation. The results of the numerical simulation is
shown in Fig.2.19. When A increases, X_p decreases. Therefore, the
width of the square wave increases as A increases for the same initial
function. We noted that for A=2.80 the chaotic domain is outside the
bistable region and the t_R oscillation disappears. These results are
in good agreement with the experimental ones.

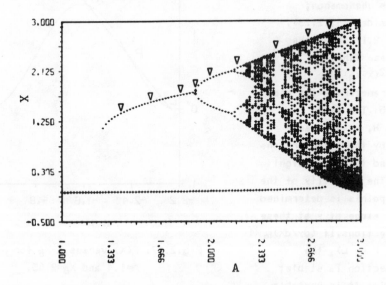

Fig.2.18. X–A bifurcation diagram for X_B=3.05. The arrows
indicate the values of A in Fig.2.19.

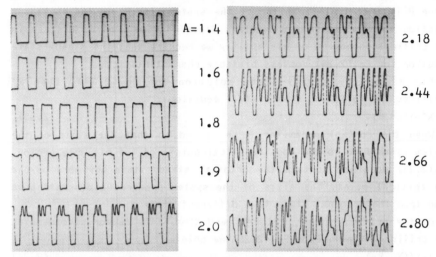

Fig.2.19. Waveforms of the numerical simulation for $X_B=3.05$ and various A.

2.4 Critical behavior[12]

In this section with a general time-delayed feedback relaxation equation as starting point, the consistency of the critical slowing down at the edges of the bistable region with the divergence of the time duration of intermittency is discussed. The critical exponent is 1/2. We also find that, as distinct from the critical points at the edges of the bistable region, there is consistency with the period-doubling bifurcation points or split bifurcation points at the cusp point in the catastrophe model of bistability. The critical exponent is 1. The results possess universal properties. Under limiting conditions, i.e. when the delay time $t_R \to 0$, the present results are in agreement with the results reported in Refs.[13-17]; when $t_R \to \infty$, the present results are consistent with those reported in Refs.[18-20].

⟨1⟩ The critical slowing down at the edge of the bistable region

In order to give a general discussion, the relaxation equation of a time-delayed feedback nonlinear system is written in a generalized form

$$\tau_0 \frac{dX(t)}{dt} = -X(t) + f[X(t-t_R), P],$$ (2.24)

where P is a control parameter of the system. $f(X,P)$ is the nonlinear function for the particular system. Since what we want to discuss is the critical phenomena of bistability we require that the steady-state solution for Eq.(2.24) possess bistable characteristics.

Let X^* be the steady-state solution for Eq.(2.24), then X^* satisfies the following steady state equation:

$$X^*=f(X^*,P).\tag{2.25}$$

When the control parameter P is varied so that the state of the system is shifted to the edge (critical point) of the bistable region, Eq.(2.24) does not contain the original steady state any more beyond the critical point. The state of the system produces a discontinuous jump from the original state to a different state.

At the critical point, let $P=P_c$ and $X^*=X_c$ (the subscript c denotes the critical point). From Eq.(2.25) we obtain

$$X_c=f(X_c,P_c).\tag{2.26}$$

Obviously, since the curve $f(X,P)$ is tangential to the 45° straight line at the critical point, the following is true:

$$\frac{\partial f(X_c,P_c)}{\partial X}=1,\tag{2.27}$$

and

$$\left.\frac{dP}{dX}\right|_c=0.\tag{2.28}$$

In the vicinity of the critical point $dX/dt \longrightarrow 0$ any finite temporal change can be considered as an infinitesimal quantity. The difference term in Eq.(2.24) can be approximated as a differential term. Thus we obtain

$$(1+ \tau_0 /t_R)(dX/dt)=-X+f(X,P).\tag{2.29}$$

Under the conditions specified in Eqs.(2.26), (2.27) and (2.28), the right-hand side of Eq.(2.29) can be expanded in the vicinity of the critical point up to the second-order terms in X (since the first-order term is zero) and first-order terms of P. After the expansion, the equation is then integrated near the critical point to obtain the time required for a discontinuous jump of the state:

$$t_s = \frac{\tau_0'}{\sqrt{\frac{1}{2}\left(\frac{\partial f}{\partial P}\right)_c \left(\frac{\partial^2 f}{\partial X^2}\right)_c}(P-P_c)} \, \text{arctg} \frac{X-X_c}{\sqrt{2\left[\left(\frac{\partial f}{\partial P}\right)_c \left(\frac{\partial^2 f}{\partial X^2}\right)_c\right]}(P-P_c)}\Bigg|_{X_1}^{X_2}, \qquad (2.30)$$

where $\tau_0' = 1 + \tau_0/t_R$, $X_1 < X_c < X_2$.

When $P \rightarrow P_c$, the arc tangent term in the above equation approaches π; and then

$$t_s \propto \tau_0' |P-P_c|^{-1/2}. \qquad (2.31)$$

These are all results under the condition of time delay with a critical exponent 1/2. It can be seen from Eq.(2.24) that the ratio τ_0/t_R can be used as a measure to determine whether the system equation is closer to the case of zero time-delay ordinary differential equations or to the case of iterative equations. When $t_R \rightarrow 0$, then $\tau_0' \rightarrow \tau_0/t_R$. This then is the condition for bistability without time delay. Matters are even more interesting when $\tau_0/t_R \rightarrow 0$, i.e. $t_R \rightarrow \infty$, then $\tau_0' \rightarrow 1$ in Eq.(2.31). Under this limiting condition, we can combine the critical phenomenon along the edge of the bistable region and the intermittence in chaos, since they are both the case that the 45° straight line is tangent to the curve $f(X,P)$ at the critical point. Eq.(2.31) then represents the critical phenomena not only for the bistability but also for the intermittency. They have the same critical exponent 1/2. This result is in agreement with that reported in Refs.[13-17] and in Refs.[18-19] in the limiting conditions $t_R \rightarrow 0$ and $t_R \rightarrow \infty$, respectively.

(2) The critical slowing down at the cusp point of the bistability (the catastrophe model[21])

At the cusp critical point, the two edges of the bistable region merge into one; the width of the bistable region is zero. The tangent point between the 45° straight line and the curve $f(X,P)$ overlaps with the inflection point of the curve $f(X,P)$. We note that the cusp represents a different critical point which differs entirely from the edge of the bistable region. Let $y(t)=X(t)-X^*=C\exp(-t/t_s)$ (where C is a constant and $t_s > 0$). Expanding $f(X,P)$ near $y=0$ and expanding the terms including P near P_c to the first order, we obtain

$$t_s = \frac{\tau_0'}{\frac{\partial^2 f(X_c,P_c)}{\partial X \partial P}(P-P_c)}. \qquad (2.32)$$

As mentioned above, we obtain a value of 1 for the critical exponent.

For $\tau_0/t_R \to 0$, we discovered that Eq.(2.32) is consistent with the results about the critical slowing down at the period-doubling bifurcation point given in Ref.20. The reason is that at the N period doubling point there are N inflection points of f(X,P) tangent to the 45° straight line.

⟨3⟩ The numerical simulation

For our LCOB system, the function f(X,P) is

$$f(X,P)=A\sin^2(X-X_B).$$ (2.33)

Near the upward jump edge of the bistable region ($A_c=1.81502547$), the numerically calculated results for different values of the time delay t_R are shown in Fig.2.20. The parameter X_B is fixed, $X_B=3$. The log-log plots are used in Fig.2.20. The solid lines are the results of linear least-squares fitting. The slopes of the fitted straight lines are very close to $-1/2$ (error<0.004). Near the cusp point, the results are shown in Fig.2.21. The slopes of the fitted straight lines are very close to -1 (error<0.007). The numerical simulation is in good agreement with the analytical results.

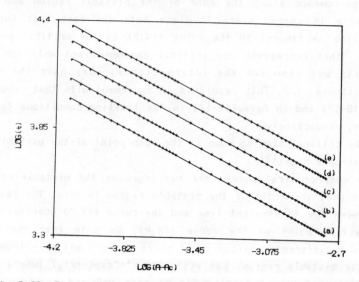

Fig.2.20. The critical slowing down at the edge of the bistable
region. a, $t_R/\tau_0=0$; b, $t_R/\tau_0=1$; c, $t_R/\tau_0=3$; d, $t_R/\tau_0=10$;
e, $t_R/\tau_0=\infty$.

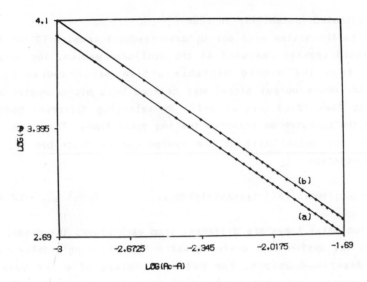

Fig.2.21. The critical slowing down at the cusp point for $t_R/\tau_0=0$ (curve a) and $t_R/\tau_0=\infty$ (curve b).

3. Instability in Two-Delayed LCOB

There is much current interest in the dynamical behavior of physical systems with competing interactions. Recently, Ikeda and Mizuno[22] demonstrated that an optical bistable device with two delayed feedback loops can be the source of competing effects which force the system to fall into a state of frustrated instability. The oscillations are highly sensitive to the parameter

$$\omega = \left| \frac{t_1 - t_2}{t_1 + t_2} \right| \tag{3.1}$$

where t_1 and t_2 are the delay times. In fact, even a small change in ω causes significant variations in the oscillation patterns.

In the liquid crystal bistable system designed with two delayed-feedback loops we observed the frustrated instability experimentally. The route to chaos, frequency locking, and the influence of the feedback strength have been observed and analysed.

3.1 Frustrated instability in LCOB[23]

Similar to the system with one delayed-feedback loop, a 90° twisted nematic liquid crystal was used as the nonlinear medium. The output intensity from the hybrid bistable system was measured by a photodetector whose output signal was delayed by a microcomputer and fed back to the liquid crystal cell. By selecting different buffer lengths of the computer we adjusted the two delay times. The dynamical behavior of the output X(t) of the system can be described by the relaxation equation

$$\tau_0 \frac{dX(t)}{dt} = -X(t) + A\sin^2[X(t-t_1) + X(t-t_2) - X_B].\tag{3.2}$$

If the two delay times are different from each other, the dynamical behavior of the system will differ greatly from the single-delay case which we described before. For rational values of ω the output pulsations display only one fundamental frequency f which depends on the values of delay times t_1 and t_2, the bias voltage applied to the liquid crystal cell, and the input intensity. The two competing frequencies $f_1=1/2t_1$ and $f_2=1/2t_2$, in this case, disappear and are replaced by a single frequency of oscillation. The appearance of frequency locking is indicative that the phase space trajectories which are evolving on the surface of a two-torus may collapse onto a limit cycle[24]. A typical waveform and the corresponding power spectrum are shown in Fig.3.1. We can see the single-oscillation frequency and its harmonics from the power spectrum.

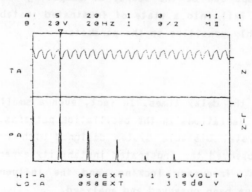

Fig.3.1. Typical waveform and its power spectrum in the case of single-frequency oscillation. $\omega=0.9$, $I_{in}=560mv$, $V_B=0.4v$.

Keeping the control parameters A and X_B fixed, we measured the power spectrum for many values of ω in order to study the oscillation frequency versus ω. The results are shown in Fig.3.2 where τ_0 =0.14 sec. and t_1=2.5 sec. It can be seen that the single oscillation frequency is locked over certain intervals $\Delta\omega$. The locking intervals are of the order of 0.1, in agreement with the characteristic width $\Delta\omega=2/\tau$, where $\tau =(t_1+t_2)/\tau_0$, predicted in Ref.22. For small values of ω the two delay times t_1 and t_2 are almost equal. Therefore, the dynamical behavior of the system for small values of ω is similar to that of the system with single time delay, as it should be.

For large values of ω single-frequency oscillation dominates. For intermediate values, quasi-periodic

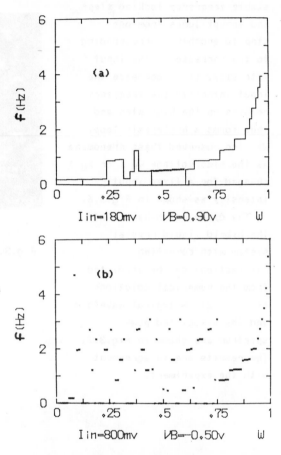

Fig.3.2.Oscillation frequency versus ω for t_1=2.5sec. (a) I_{in}=180mv, V_B=0.9v; (b) I_{in}=800mv, V_B= -0.5v. The dots indicate quasi-periodic motion.

motion can be observed. Figure 3.3 shows the waveform and the power spectrum of the quasi-periodic motion (f_1=0.525Hz, f_2=0.1375Hz). We can only obtain the rational value due to the limited resolution of the spectrum. Figure 3.4 shows that for values of ω contained between the two frequency steps. The strong competition between the two closest modes generates a beat pattern. In Fig.3.5 we show extremely

stable frequency locking steps
and abrupt jumps from one
step to another, corresponding
to the increase in the input
intensity. If we decrease the
input intensity the frequency
remains on the high step and
thus forms a hysteresis loop.
We also observed these phenomena
as the bias voltage V_B (or X_B)
changed for a fixed input
intensity as shown in Fig.3.6.

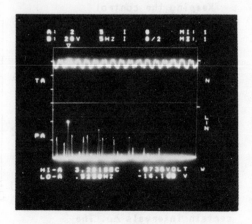

The dynamical behavior of
the hybrid liquid crystal
system with competing
interactions can be predicted
from the numerical solutions
of Eq.(3.2). A typical waveform
and the associated power
spectrum are shown in Fig.3.7.
The results are in agreement
with the experiment.

Fig.3.3. Waveform and its power
         spectrum for quasi-
         periodic motion.
         I_{in}=180mv, V_B=0.9v,
         ω=0.5

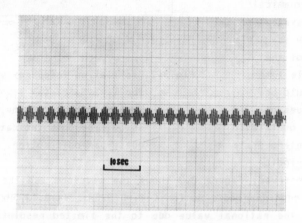

Fig.3.4. Beat pattern for ω=0.9, I_{in}=180mv, and
         V_B=1.9v.

3.2 Routes to chaos[25]

Under proper conditions, the chaotic output can be obtained in this system with two delayed-feedback loops. The route to chaos is similar to that in the circle map. In the A-t_2 phase diagram obtained by calculating Eq. (3.2) numerically, we can find that the frequency locking regions form a series of Arnold tongues, as shown in Fig.3.8. Only the tongues corresponding to f_1/f_2 =1/5, 2/9, and 1/4 are given. The dotted line is the boundary of the single frequency oscillation. The tongues are situated in the region of [0/1,1/2]. We find that the route to chaos is as follows: Single frequency oscillation (limit cycle) ⟶ quasi-periodic motion (two dimensional torus) ⟶ frequency locking (limit cycle) ⟶ chaos (strange attractor). In order to observe this route clearly, we

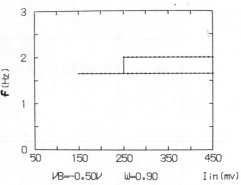

Fig.3.5. Frequency locking versus I_{in} for fixed V_B.

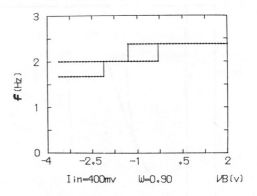

Fig.3.6. Frequency locking versus V_B for fixed I_{in}.

calculated four Lyaponov exponents in our system along the dot-and-dash line in Fig.3.8. The results are shown in Fig.3.9, where X_B=-1, t_1/τ_0 =10, and t_2/τ_0 =1.8. Let LE_i (i=1,2,3,4) be these Lyaponov exponents and $LE_i>LE_j$ for i<j. Then $LE_1=0$ and $LE_2<0$ for a limit cycle, $LE_1=LE_2=0$ and $LE_3<0$ for a two-dimensional torus, and $LE_1>0$ for chaos. Therefore, it can be seen that in Fig.3.9 the quasi-periodic state

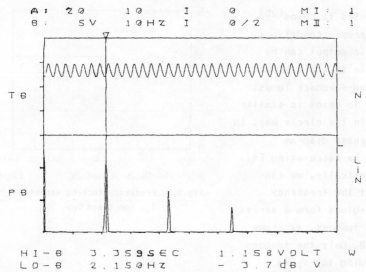

Fig.3.7. Typical numerical results for $\omega=0.9$, A=2.5, and $X_B=-1$.

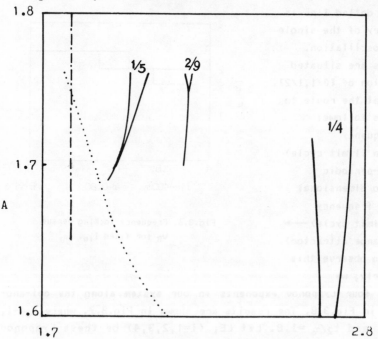

Fig.3.8. Frequency locking tongues in the $A-t_2$ diagram.

appears at A=1.76. When A=1.825, frequency locking $(f_1/f_2=1/6)$
appears; A=1.85, a quasi-periodic state appears; A=1.855, frequency
locking $(f_1/f_2=5/31)$ appears; and A=1.86, chaos is obtained. We also
calculated the correlation dimension (D_2) along the dot-and-dash line
in Fig.3.8. The results are shown in Fig.3.10.

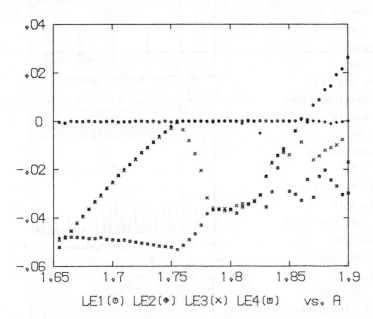

Fig.3.9. Four Lyapunov exponents versus A.

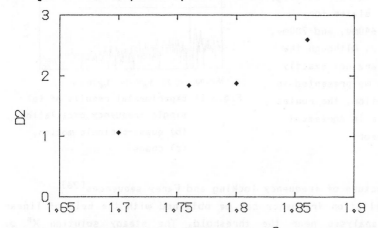

Fig.3.10. Correlation dimension versus A.

We can find that $D_2 \cong 1$
for single frequency
oscillation, $D_2 \cong 2$ for
quasi-periodic motion,
and $D_2 > 2$ for chaos. The
route to chaos in a
particular frequency
locking tongue is
period-doubling
bifurcation; and the
route from quasi-periodic
state to chaos is
accompanied by
frequency lockings as
shown above. The
experimental results are
shown in Fig.3.11. We
can clearly see the route
to chaos with the
increase of the input
intensity I_{in}.
Fig.3.11(a),(b),(c)
correspond to the quasi-
periodic, frequency locking,
and chaotic states for
I_{in}=540mv, 640mv, and 700mv,
respectively. Although the
parameters are not exactly
the same as we presented in
the calculation, the routes
to chaos are in agreement
with the theory.

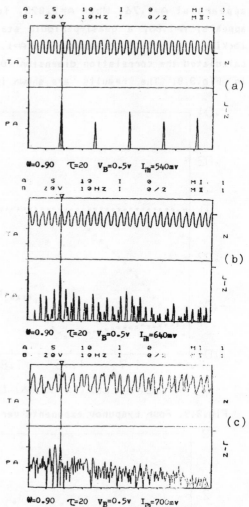

Fig.3.11.Experimental results of (a)
single-frequency oscillation,
(b) quasi-periodic motion,
(c) chaos

3.3 Structure of frequency locking and Farey sequences[26]

The oscillation frequency can be obtained with the help of linear
stability analysis near the threshold. The steady solution X^* of
Eq.(3.2) satisfies the following equation

$$X^* = A\sin^2(2X^* - X_B) \tag{3.3}$$

Expanding Eq.(3.2) near the steady solution X^*, and letting $|X-X^*| \propto \exp(\lambda t)$, we obtain the characteristic equation

$$\lambda = -1 + A\sin(4X^* - 2X_B)[\exp(-\lambda t_1) + \exp(-\lambda t_2)] \tag{3.4}$$

where λ is the eigenvalue.

Letting $\lambda = \alpha + i\beta$, then the stability of the system is determined by α. If $\alpha > 0$, the system is unstable, the imaginary part β of the eigenvalue measures the oscillation frequency. Under the condition of $\tau = (t_1 + t_2)/\tau_0 \gg 1$, we find that at the threshold of instability the oscillation frequency has the discrete values, i.e. $\beta = \beta_q = 2\pi q(1 - 2/\tau)/\tau$ ($q = 1, 2, \ldots\ldots$). From Eq.(3.4) we obtain

$$[2A\sin(4X^* - 2X_B)]^{-1} = (-1)^q \cos(2\pi q/\tau)\cos[\omega\pi q(1 - 2/\tau)] \tag{3.5}$$

The right hand side of Eq.(3.5) is a function of q. Under the condition of $\tau \gg 1$, we define $\psi(q) = (-1)^q \cos(2\pi q/\tau)\cos(\omega\pi q)$. Detailed analysis shows that $\psi(q) > 0$ corresponds to the negative slope on the bistable curve. Therefore, $\psi(q)$ is confined within the region of $-1 < \psi(q) < 0$.

As the parameter A increases, the first unstable mode $q = q^*$ is the one which minimizes the function $\psi(q)$. We can prove that the maximum value of q^* is $q^*_{max} = [\tau/4]$. Under the above conditions, we calculate the curves of the mode q^* versus .

When $\omega(1 - 2/\tau)$ approaches 1, the steady state X^* is unstable only when $A^* \to \infty$. Therefore, the stable range of a bistable system with one feedback can be increased by adding another feedback with very short time delay. Fig.3.12. shows A^*, q^* and $(q^*)^{-1}$ versus ω for $\tau = 1000$. A series of frequency locking steps can be found. The location of these steps can be characterized by a rational number $\omega_s = P_s/Q_s$ for the oscillation mode $q^* = Q_s$. The steps corresponding to $\omega_s = 1/2$, $2/3$, $3/4, \ldots$ and $\omega_s = 1/2$, $1/4$, $1/6, \ldots$ are indicated in Fig.3.12(c) by the arrows with (I) and (II), respectively. The width of each step is given by $\Delta\omega_s = C/q^*$, where q^* corresponds to the mode and C is a constant. We also find that the characteristic fraction P_s/Q_s of each step takes the form odd/even, and then the sum of P_s and Q_s is an odd number. In Fig.3.12(b), we can find the holes in which the oscillation mode takes minimum. The locations of these holes can also be characterized by a rational value P_h/Q_h. The characteristic fraction

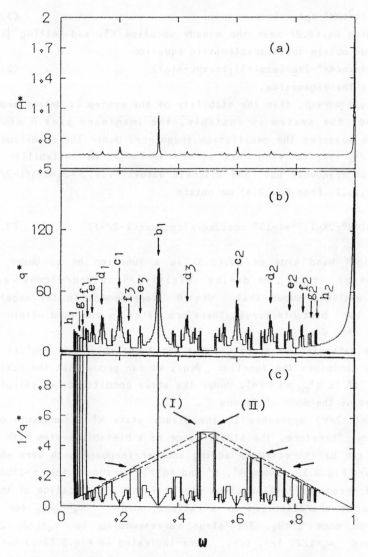

Fig.3.12. Theoretical results of (a) the threshold A^*, (b)
the first oscillation mode q^*, and (c) $1/q^*$ versus ω .

P_h/Q_h is odd/odd, and then the sum of P_h and Q_h is an even number. The
locations for each hole indicated with the arrows in Fig.3.12(b) are
listed in Table (I).   The holes forms two sequences corresponding to

the position indicated by b_1, c_1, ... h_1 and b_1, c_2, ... h_2 in Fig.3.12(b), respectively. Any two consecutive fractions P/Q and R/S of these two sequences do not satisfy the unimodularity condition $|PS-QR|=1$. Usually, the quantity $|PS-QR|$ is called the modularity of P/Q and R/S. Therefore, the modularities of our two sequences are 2 and 4, respectively. The positions of these two sequences on the Farey tree are shown in Fig. 3.13. The numbers in the solid circles and in the dotted circles indicate the sequences of b_1, c_1, ... h_1 and b_1, c_2, ... h_2, respectively.

Table (I). The values of ω and their rational expressions for b_1, c_1, ..., h_1 and b_1, c_2, ..., h_2 in Fig.3.12(b)

	ω values	rational expressions		ω values	rational expressions
h_1	0.067	1/15	h_2	0.868	13/15
g_1	0.077	1/13	g_2	0.848	11/13
f_1	0.091	1/11	f_2	0.820	9/11
e_1	0.111	1/9	e_2	0.779	7/9
d_1	0.143	1/7	d_2	0.716	5/7
c_1	0.200	1/5	c_2	0.601	3/5
b_1	0.334	1/3	b_1	0.334	1/3

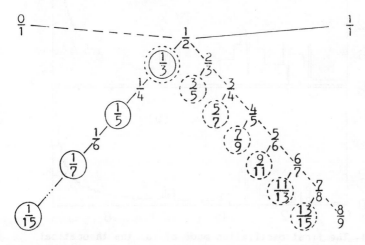

Fig.3.13. The Farey tree. See the text.

The jumping interval of the mode Δq^* is closely related to the holes. There are one series of jump-up steps and one series of jump-down steps outside each hole. The jumping interval of these modes is $\Delta q^* = Q_h$. The ω value of the jumping point from one step to the next is the Farey sum of these two steps. We also find that the highest step beside each hole is equal to $[\tau/4Q_h]$. There are also a series of jump-up and jump-down steps inside each hole, but they are much narrower than those outside the hole. So it is difficult to observe these steps. After ω falling into the hole, the oscillation mode is very sensitive to the change of ω. In addition, we discover from Fig. 3.12(a) that the thresholds of the oscillation modes in the holes are relatively higher.

There is no significant change on the frequency locking structure when τ is increased except that the width of the steps will get smaller and more steps and holes will emerge.

Our experiment was performed with a LiNbO$_3$ crystal as the nonlinear medium instead of the liquid crystal in order to get a shorter relaxation time to fit the theoretical condition $\tau \gg 1$. The transmittance property is not very different from LC.

The experimental result of q^* versus ω is shown in Fig.3.14(b) for

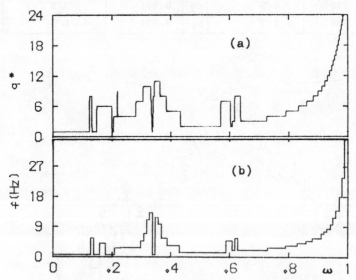

Fig.3.14. The first oscillation mode of (a) the theoretical result and (b) the experimental result.

τ=130 and X_B=-1. The corresponding theoretical result is shown in Fig. 3.14(a). For even larger τ (=1000), the experimental results of the hole structure near ω=1/3 are shown in Fig. 3.15. The experimental results are in agreement with the theory.

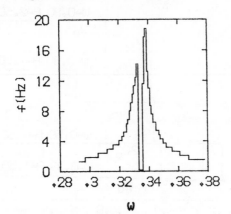

Fig.3.15. Experimental result of the hole structure near
ω=1/3 for τ=1000.

3.4 Influence of the feedback strength[27]

The feedback strength of the system can be varied with the microcomputer in the feedback loop. The frequency locking, quasi-periodic motion and chaos can be observed by varying the feedback strength.

The relaxation equation including the varied feedback strength is written in the following form:

$$\tau_0 \frac{dX(t)}{dt} = -X(t) + A\sin^2[FX(t-t_1) + GX(t-t_2) - X_B] \tag{3.6}$$

where F and G are the strengths of the feedback loops with the delay time t_1 and t_2, respectively.

The numerical results of Eq.(3.6) for ω=0.9 and X_B=-1 are shown in Fig. 3.16. Fig. 3.16(a) shows the relation between the oscillation frequency Ω and the feedback strength G for various ω and fixed F=1. We can see that when the feedback strenth G is weak, the oscillation is the same as that in one delay case with a fundamental frequency Ω_0= 1/2t_1. As G increases, the frequency jumps from one step to another with an interval of the odd harmonics of the fundamental frequency. In

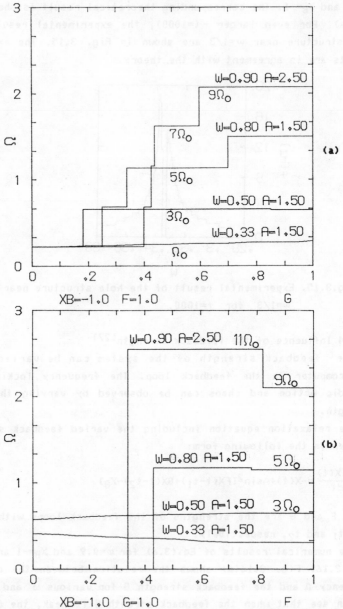

Fig.3.16. Oscillation frequency (a) versus the feedback strength G
for fixed F=1.0, and (b) versus the feedback strength F
for fixed G=1.0.

the same way, the oscillation frequency versus the feedback streegth F
for fixed G is shown in Fig.3.16(b).

The F-G phase diagram is shown in Fig.3.17, where $\omega = 0.90$, A=2.5
and X_B=-1. It can be seen that for small G, the oscillation is similar
to that in the one delay case. The inset of Fig.3.17 shows that there
is no triple point in the phase diagram. This implies that the
coexistence of three oscillation modes is not possible in this system.
The F-G phase diagram for $\omega = 0.5$ is shown in Fig.3.18. For a fixed
value of F(=0.80), we can see two frequency locking regions Ω_0 and $3\Omega_0$,
a quasi-periodic region and a chaotic region as G increases. This
result implies that the feedback strength can be viewed as a control
parameter which can drive the system to fall into chaos through the
Ruelle-Takens-Newhouse route[28]. The route from period-doubling to
chaos can be observed in the F-G phase diagram for $\omega = 0.1$ as shown in
Fig.3.19.

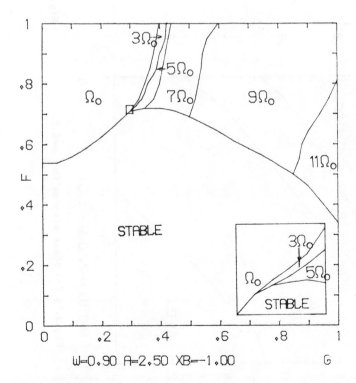

Fig.3.17. F-G phase diagram for $\omega = 0.9$.

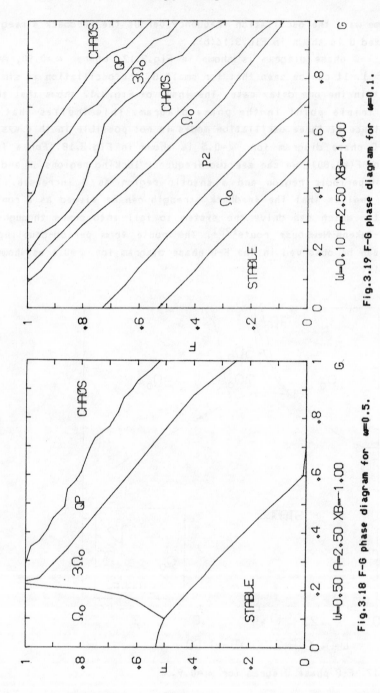

Fig.3.19 F-G phase diagram for ω=0.1.

ω=0.10 A=2.50 XB=-1.00

Fig.3.18 F-G phase diagram for ω=0.5.

ω=0.50 A=2.50 XB=-1.00

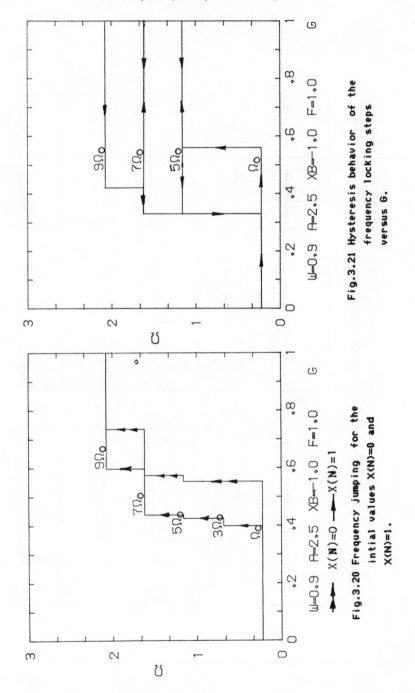

Fig.3.20 Frequency jumping for the intial values X(N)=0 and X(N)=1.

Fig.3.21 Hysteresis behavior of the frequency locking steps versus G.

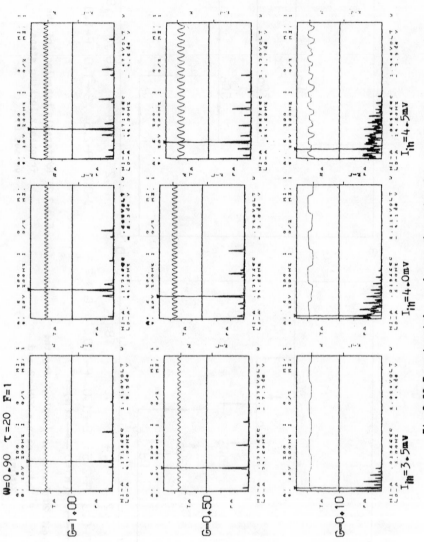

Fig.3.22 Experimental waveform and spectra for various G and Iin.

We also studied the dependence of the oscillation frequency on the initial value. The numerical result is shown in Fig.3.20. We can see that for different initial values the frequency jumps at different locations. This phenomenon indicates that at least two attractors coexist. The hysteresis loops are obtained when tracing techniqes are used, i.e. the final calculated value is selected as the initial value for the next step of the changing parameter. The results are shown in Fig. 3.21.

The experimental results are shown in Fig. 3.22. Again we used LiNbO$_3$ crystal instead of LC. It can be seen from the spectrum that the locked frequencies are Ω_0, $7\Omega_0$ and $9\Omega_0$ for G equal to 0.1, 0.5 and 1.0, respectively, when w=0.90, X_B=-1, τ=20, and I_{in}=3.5mv. As the input intensity increases, the quasi-periodic motion and chaos will appear, as shown in Fig.3.22. As an example, for I_{in}=4.5mv, we can find quasi-periodic motion, frequency locking, and chaos for G=1.0, 0.5, and 0.1, respectively. The experimental results are in agreement with the numerical simulation.

4. Conclusions

We have presented a summary of the instability and chaos behavior in hybrid liquid crystal optical bistablity. The hybrid liquid crystal optical bistable device is a convenient system for studying chaos, because the parameters are easily changeable and controllable. But in essence the dynamical behavior of the system is far from totally understood. Some questions are still open. A question, for example, about the system with one delayed feedback loop is how the relaxation process can be adiabatically approximated, so that the relaxation equation of the system can be transformed to a iteration equation (one dimensional mapping). In the two-delayed system, further physical interpretation is needed for the observed phenomena both in the experiment and in the numerical simulation. More universalities are expected in our system.

References

1. K.Ikeda, Opt. Commun. 30, 257(1979).
   K.Ikeda, H.Daido, and O.Akimoto, Phys. Rev. Lett. 45,709(1980).
2. H.M.Gibbs, F.A.Hopf, D.L.Kaplan, and R.L.Shoemaker, Phys Rev. Lett. 46, 474(1981).
3. F.A.Hopf, D.L.Kaplan, H.M.Gibbs, and R.L.Shoemaker, Phys. Rev. A25, 2172(1982).
4. Hong-jun Zhang, Jian-hua Dai, Peng-ye Wang, and Chao-ding Jin, JOSA, B3, 231(1986); Acta Phys. Sin. 33, 1024 (1984).
5. Hong-jun Zhang, Jian-hua Dai, Jun-hui Yang, and Cun-xiu Gao, Opt. Commun. 38, 21(1981).
6. Jian-hua Dai, Hong-jun Zhang, Peng-ye Wang, and Chao-ding Jin, Acta Phys. Sin. 34, 992(1985); Chinese Phys. 6, 14(1986).
7. Jian-hua Dai, Hong-jun Zhang, Peng-ye Wang, and Chao-ding Jin, Opt. Commun. 57, 207(1986).
8. R.M.May, Ann.N.Y. Acad. Sci. 316, 517(1979).
   J.Testa and G.A.Held, Phys. Rev. A28, 3085(1983).
9. Hong-jun Zhang, Jian-hua Dai, Peng-ye Wang, Chao-ding Jin, and Bai-lin Hao, Chinese Phys. Lett. 2,5(1985).
10. Peng-ye Wang, Jian-hua Dai, and Hong-jun Zhang, Acta Phys. Sin. 34, 581(1985).
11. Hong-jun Zhang, Jian-hua Dai, and Peng-ye Wang, Chinses Phys. Lett. 2, 129(1985).
12. Peng-ye Wang, Hong-jun Zhang, and Jian-hua Dai, Acta Phys. Sin. 34, 1233(1985). Chinese Phys. 6, 336(1986).
13. E.Garmire, J.H.Marburger, S.D.Allen, and H.G.Winful, Appl. Phys. Lett. 34, 374(1979).
14. Hong-jun Zhang, Jian-hua Dai, Jun-hui Yang, and Cun-xiu Gao, Acta Phys. Sin. 30, 810(1981).
15. Yong-gui Li, Hong-jun Zhang, Jun-hui Yang, and Cun-xiu Gao, Acta Phys. Sin. 31, 446(1982).
16. G.Grynberg, and S.Cribier, J. Physque Lett. 44, L449(1983).
17. S.Cribier, E.Giacobino, and G.Grynberg, Opt. Commun. 47, 170(1983).
18. P.Manneville and Y.Pomeau, Phys Lett. A75, 1(1979).
19. J.E.Hirsh, B.A.Huberman, and D.J.Scalapino, Phys. Rev. A25, 519(1982).
20. B.L.Hao, Phys. Lett. A86,267(1981).

21.T.Poston, and I.Stewart, "Catastrophe Theory and its Applications", (Pitman Press, London, 1978).

22.K.Ikeda, and M.Mizuno, Phys. Rev. Lett. 53, 1340(1984).

23.Hong-jun Zhang, and Jian-hua Dai, Opt. Lett. 11, 245(1986).

24.H.Haken, "Advanced Synergetics", (Springer-Verlag, Berlin, 1983), pp.42 and 239.

25.Guang Xu, Jian-hua Dai, and Hong-jun Zhang, (to be published).

26.Fu-lai Zhang, Jian-hua Dai, and Hong-jun Zhang, (to be published).

27.Jian-hua Dai, and Hong-jun Zhang, in Technical Digest, IQEC XV (1987) p.132. (invited paper).

28.S.Newhouse, d.Ruelle, and F.Takens, Commun. Math. Phys. 64, 35(1978).

SIMPLE MATHEMATICAL MODELS FOR COMPLEX DYNAMICS IN PHYSIOLOGICAL SYSTEMS

Leon Glass, McGill University, Montreal

1. INTRODUCTION

Key physiological systems, such as the cardiovascular, respiratory, neuromuscular, and hormonal systems, display intrinsic oscillatory behavior. These systems interact with one another and the outside environment. Moreover, there are innumerable feedback loops acting to maintain key physiological variables within normal limits.

One approach to study this complex situation is to formulate mathematical models which capture essential aspects of the hypothesized physiological mechanisms. Such models are not intended to be realistic (in the sense that detailed quantitative comparisons with experiment are possible), but caricatures which may be useful for fixing ideas. In some cases, such caricatures are sufficiently interesting that a detailed mathematical analysis seems warranted. The point of this review is to provide a brief summary of simple mathematical models for physiological dynamics that have been considered by myself and co-workers over the past several years. As all of the material here has been previously published, I provide a brief summary of the motivations and conclusions but do not discuss the technical details. I also indicate open questions which require further analysis. I will give key historical references and some references to work by others, but for a much more complete bibliography the original papers should be consulted. Further discussion of much of this material, along with extensive biological examples can be found in the recent book "From Clocks to Chaos: The Rhythms of Life" by myself and M.C. Mackey[1].

2. PHASE-LOCKING OF NONLINEAR OSCILLATIONS

In response to periodic stimulation, a nonlinear oscillator may become entrained or phase locked to a periodic stimulus. Calling τ the period of the periodic stimulation, then a phase locking rhythm will be called an N:M rhythm if it repeats with a periodicity of Nτ and there are M cycles of the forced oscillator during this time. We now consider phase locking of simple mathematical models of biological oscillators.

2.1 Phase Locking of Limit Cycle Oscillators

A stable limit cycle is a periodic solution of a differential equation which is reached in the limit t→∞ for all points in its neighborhood. The concept of a limit cycle was introduced into mathematics by Poincaré who considered a simple example in which the vector field in a polar coordinate system is radially symmetric. We consider an equation frequently used as a model in theoretical biology[2-5] that is close (but not identical) to Poincaré's equation. Differential equations with radially symmetric vector fields and stable limit cycles have been given different names, viz. λ-w systems[6] and radial isochron clocks[4,5], but a less obscure and perhaps more appropriate name is the Poincaré oscillator.

We consider the equations,

$$\frac{dr}{dt} = ar(1-r),$$

$$\frac{d\phi}{dt} = 2\pi,$$

(1)

where a is a positive parameter. Since $\frac{dr}{dt} > 0$ for $r < 1$ and $\frac{dr}{dt} < 0$ for $r > 1$, in the limit t→∞ all points with the exception of the origin approach the limit cycle at r=1. A perturbation to this system is assumed to be a horizontal translation of magnitude b. Then if a perturbation is delivered at a phase ϕ, immediately after the perturbation the oscillator is at a new phase $g(\phi)$ which can be found from the expression

$$\cos 2\pi g(\phi,b) = \frac{b + \cos 2\pi\phi}{(1 + 2b \cos 2\pi\phi + b^2)^{\frac{1}{2}}} \qquad (2)$$

The function g (ϕ,b) is sometimes called the phase transition curve (PTC). It is a map of the circle into itself $g:S^1 \to S^1$. The PTC is of topological degree 1 for $|b| < 1$ and of topological degree 0 for $|b| > 0$. This has been called type 1 and type 0 phase resetting, respectively by Winfree[2].

Now consider periodic stimuli delivered to the Poincaré oscillator with a time interval τ between stimuli (Fig. 1). Call ϕ_i the phase just before the ith stimulus. In the limit a$\to\infty$

$$\phi_{i+1} = f(\phi_i,b) = g(\phi_i,b) + \tau \qquad (\text{mod } 1) \qquad (3)$$

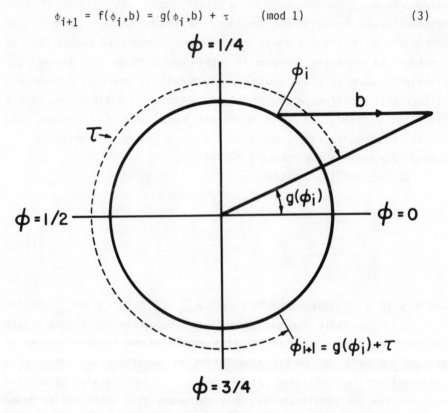

Fig. 1: Phase-locking of the Poincaré oscillator (Reprinted from Ref. 1).

where g is the PTC defined above. For $|b| < 1$, $f(\phi_i, b)$ is a 1:1 invertible map of the circle. For this case the dynamics of (2) are well understood and were described by Arnold[7]. For any irreducible fraction M/N with b fixed, $|b| < 1$, there is in general an interval of values of τ such that all initial conditions asymptotically approach a stable N:M phase-locked rhythm. The organization is schematically shown in Fig. 2. For $|b| > 1$, f is a map of topological degree 0 (it

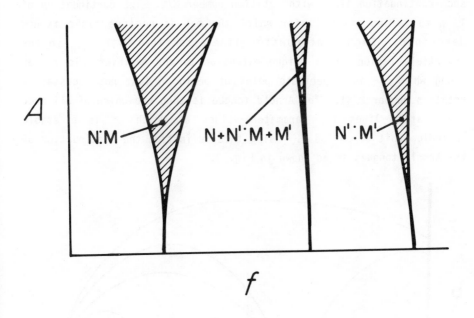

A

N:M   N+N':M+M'   N':M'

f

Fig. 2:   Schematic organization of the Arnold tongues for invertible circle maps. A corresponds to an amplitude and f a frequency of periodic forcing (reprinted from Ref. 1).

is non-invertible), and the bifurcations are much more complex. For this case it was shown that period-doubling bifurcations and chaos can be found[3],[4]. A detailed analysis of the bifurcations which uses techniques suggested by kneading theory is in Ref. 5. This paper enables us to show certain results concerning the continuity of the Arnold tongues for $|b| > 1$. The following definition was proposed in Glass and Bélair[8]. For parameter values for which the map, (3), is invertible any given point ϕ^* will be a periodic point of period N and rotation number M/N along some line $\tau = \tau(b)$. For values of (b,τ) for which the map is invertible the Arnold tongue $T_{M/N}$ is the union of all such continuation lines with rotation number M/N. The continuation of $T_{M/N}$ to values of (b,τ) for which the map is noninvertible is now described. Each continuation line associated with $T_{M/N}$ in the invertible region has a unique extension to the noninvertible region along which ϕ^* is a periodic point of period N (but not necessarily rotation number M/N). The Arnold tongue is then the union of all such continuation lines for parameter values at which ϕ^* is a stable periodic orbit of period N. Using this definition the continuation of the Arnold tongues is as shown in Fig. 3.

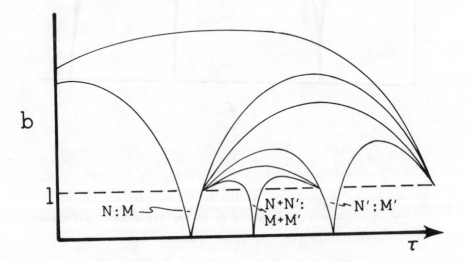

Fig. 3: Schematic organization of the Arnold tongues for the periodi-cally forced Poincaré oscillator (reprinted from Ref. 8).

It seems to me that the Poincaré oscillator is an equation of unexpected depth and beauty. A complete description of the symbolic dynamics of this system (with a→∞) under periodic stimulation has not been given. There are also no studies of the bifurcations found during periodic stimulation when a in (1) is finite. Piro and Gonzalez studied another mathematical model for periodic forcing of nonlinear oscillations that has properties similar to the Poincaré oscillator[9].

2.2   Phase Locking of Integrate and Fire Models

In many situations it is useful to think of a biological oscillator as an "activity" which rises to a "threshold" and then "fires". The activity then relaxes to a second lower threshold and the cycle starts again. We consider the integrate and five model shown in Fig. 4, in which there is a linear rise and fall to a sinusoidal

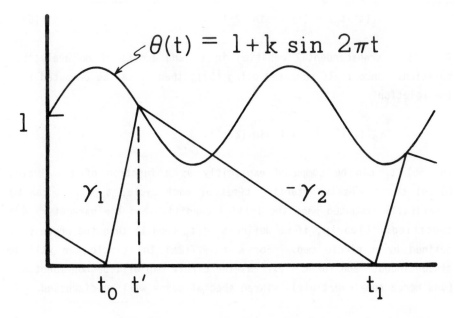

Fig. 4:   The integrate and fire model (reprinted from Ref. 8).

function. Our presentation here is adopted from Glass and Bélair[8]. We call γ_1 the rising slope and $-\gamma_2$ the falling slope and for convenience assume the sinusoidal threshold is given by

$$\theta(t) = 1 + k \sin 2\pi t. \qquad (4)$$

There are three parameters, the slopes of the rising and falling phases of the activity (denoted γ_1 and $-\gamma_2$, respectively) and the modulation of the sine wave, k. In the limit of no modulation k=0, the intrinsic frequency of the oscillator is given

$$f_0 = \gamma_1\gamma_2/(\gamma_1 + \gamma_2). \qquad (5)$$

Assume that t_0 is known (Fig. 4). Then t', the time the activity first reaches threshold can be found by solving the equation

$$\gamma_1(t'-t_0) = 1 + k \sin (2\pi t'). \qquad (6a)$$

This is a transcendental equation in t' and admits of no analytical solution. Once t' is computed using (5a), then t_1 can be computed from the relation

$$\gamma_2(t_1-t') = 1 + k \sin (2\pi t'). \qquad (6b)$$

In (6b) t_1 can be computed explicitly as a function of t'. Using (6a,b) the successive starting times of each cycle, $t_1, t_2, \ldots$ can be numerically computed once the initial condition and the parameters are specified. Clearly, if we define $\phi_1 = t_i$ (mod 1) then the process is defined by a circle map. For $k > \gamma_1/(2\pi)$ the circle map will be discontinuous, and for $k > \gamma_2/(2\pi)$ the circle map will be non-monotonic (and hence non-invertible). Three special cases will be discussed.

Case 1. Consider the limit in which $\gamma_2 \to \infty$ (Fig. 5). In this case, there is a slow rise to the threshold and a vertical fall. Early

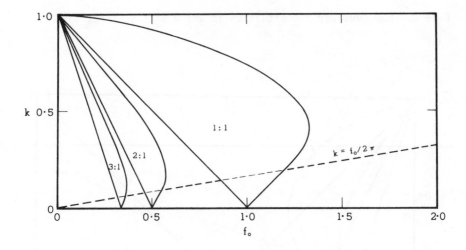

Fig. 5:   Phase locking zones for the integrate and five model with
$\gamma_2 = \infty$, $f_0 = \gamma_1$ (reprinted from Ref. 8).

studies analyzed this model in the context of periodically forced
relaxation oscillations[10]. The model was subsequently proposed in the
context of entrainment of biological oscillators[11,12].   For $k < f_0/2\pi$
the dynamics are described by a continuous invertible map of the unit
circle, and the discussion above concerning Arnold tongues is
applicable.   For $k > f_0/2\pi$ the dynamics are given by a discontinuous,
non-invertible monotonic function.   Mathematical analysis of this
problem has been undertaken by Keener and colleagues[13-15].   There are
still N:M phase locking for all integers N and M relatively prime.
There are also parameter values which give rise to aperiodic dynamics.
There are however two differences between the properties of the
aperiodic dynamics in the case in which the dynamics are described by
the discontinuous piecewise monotonic maps, and the invertible maps.
First, the successive iterates no longer form a dense orbit on the unit
circle, but rather a Cantor set.   Second, the probability that one will
choose a set of parameter values associated with the aperiodic dynamics
is now zero (this means the Lebesgue measure associated with the
aperiodic dynamics is zero).

Case 2. Now consider the situation in which $\gamma_1 = \gamma_2$ (Fig. 6)[8].

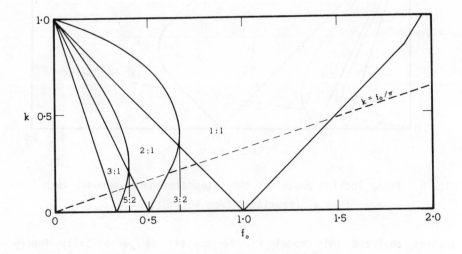

Fig. 6:   Phase locking zones for the integrate and five model with
$\gamma_1 = \gamma_2$, $f_0 = \gamma_1/2$ (reprinted from Ref. 8).

For $k < f_0/\pi$ the map is invertible. The phase locking ratios N:2 where
N is an odd integer, are present only for $f_0 = 2/N$ (i.e., the Arnold
tongue in this case does not extend over an interval of f_0 with k
fixed). In this degenerate situation, all points on the unit circle
are periodic and neutrally stable. In the non-invertible case,
$k > f_0/\pi$, the map is both discontinuous and non-monotonic. The
boundaries of the N:1 zones overlap giving rise to bistability. The
N:2 zones (N odd) vanish at a finite value of k which can be explicitly
computed. This vanishing is a result of the periodic points being lost
into the discontinuity of the circle map as k increases. There is a
connection between periodically forced van der Pol oscillators and
integrate-and-fire models[16].

Case 3. Now assume $\gamma_1 \to \infty$ (Fig. 7). As originally shown by Perez and Glass[17] we obtain

$$\phi_{i+1} = \phi_i + \tau + b \sin 2\pi \, \phi_i \qquad (7)$$

where $b = k/\gamma_2$ and $\tau = 1/\gamma_2$. For $k > f_0/2\pi$ the map is still continuous

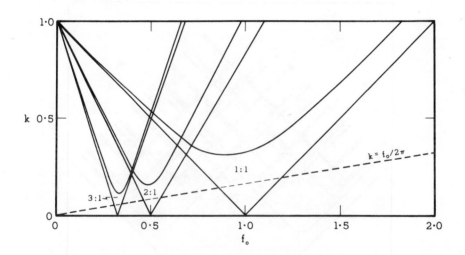

Fig. 7: Phase locking zones for the integrate and five model with $\gamma_1 = \infty$, $f_0 = \gamma_2$.

but it is no longer monotonic and hence it is non-invertible. Recently, there has been interest in the transition from invertibility to non-invertibility, at the value $b = 1/(2\pi)$[18]. It is also of interest to examine the dynamics in which the map is non-invertible, and several papers deal with this question[17, 19-22].

Each of the Arnold tongues present for $k < f_0/2\pi$ extends to the region $k > f_0/2\pi$. Each tongue, however, splits into two branches. Once again, the Arnold tongues can cross leading to a situation in which two different rotation numbers can be found for the same values of the parameters. The other main feature is that there are complex sequences of bifurcations which are present in the Y-shaped regions formed by each Arnold tongue[21]. The structure of the bifurcations in each of the Arnold tongues is shown in Figure 8.

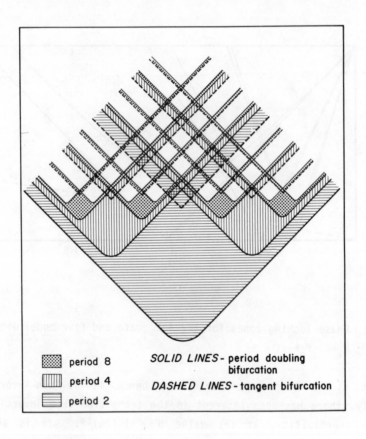

Fig. 8: Schematic picture of the period doubling structures found
         inside each Arnold tongue in Fig. 7 (reprinted from Ref. 21).

2.3   Significance of Simple Models

In recent years there have been intensive studies of the "transitions to chaos" via "quasiperiodicity" ("overlapping of resonances")[18]. A circumstance that has been considered analytically is the one in which a 1:1 circle map of topological degree 1 develops a cubic nonlinearity and becomes noninvertible (Case 3 of the integrate and fire models above). It does not seem well appreciated that although 1:1 invertible maps seem to be appropriate to describe periodically forced nonlinear oscillators for low stimulation strengths, at higher strengths the invertibility can be destroyed in a number of different ways. In particular, in the examples considered here the map undergoes a transition to topological degree 0 (Fig. 3), becomes discontinuous (Fig. 5), and is topological degree 1 but noninvertible (Fig. 7). There is not any reason to expect that scaling properties along the line in parameter space at which invertibility is lost will be the same in all these different cases. Finally, the examples show that the global structures of the bifurcations in these different examples are different. This has important implications for experimental studies of phase locking. In general, as the amplitude of the periodic perturbation is increased, the global organization of bifurcations may differ widely in different systems. Further detailed experimental and theoretical studies are needed to unravel the global organization for each particular case.

3.   DYNAMICS IN DELAY-DIFFERENTIAL EQUATIONS

In physiological systems there are frequently delays between the sensing of a disturbance and the mounting of some appropriate physiological response. In addition, delays have been introduced into experimental systems in a number of different ways. The delays in physiological systems are modelled by differential delay equations.

First, consider the equation

$$\frac{dx}{dt} = h(x_\tau) - \gamma x \tag{8}$$

where $x_\tau = x(t-\tau)$ and γ is a positive constant. When h is a monotonically decreasing function, then (8) can display either a single stable steady state or a stable limit cycle oscillation. If h is a non-monotonic function then the dynamics are much more complex. Mackey and Glass[23,24] considered the situation in which

$$h(x_\tau) = \frac{\lambda x_\tau}{\theta^n + x_\tau^n} \tag{9}$$

where λ, θ and n are positive constants as a mathematical model for the control of blood cell production. Other workers have considered other functional forms for the non-monotonic function h[25-28].

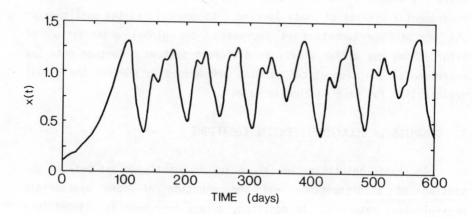

Fig. 9:   Chaos in a time delay differential equation (8,9) (reprinted from Ref. 23).

One of the interesting features of (9) with a non-monotonic feedback is that it displays complex bifurcations and chaotic dynamics (Fig. 9). For example, Mackey and Glass[23, 24] described period doubling bifurcations and chaos in this system. Detailed numerical and theoretical studies of the bifurcations in this system are just getting started. Scaling in the period-doubling regime appears to follow Feigenbaum's scenario[29]. A proof of chaos has been made for the situation in which h is piecewise constant for a continuous range of values (but not at the origin or at infinity) but zero elsewhere[30].

Farmer found that increasing the time delay in (8,9) leads to high dimensional strange attractors[31], and this observation has resulted in recent years in numerous studies in which this equation was used to test algorithms for the computation of the dimension of strange attractors[32-34].

In (8) there is only a single feedback function with a single time delay. A more usual circumstance in physiology is that there are multiple systems controlling key variables such as ventilation[35] and blood pressure[36]. Surprisingly, there is very little work done in trying to analyze dynamics in multi-looped nonlinear feedback systems. A first attempt to study the dynamics in such a system was undertaken by Glass, Beuter and Larocque[37]. They considered a system in which there are multiple feedback paths controlling a variable P via the equations

$$\frac{dP}{dt} = \frac{1}{\varepsilon} \left[F(x_1, x_2, \ldots, x_N) - P \right] \tag{10}$$

$$\frac{dx_i}{dt} = f_i(P_{\tau_i}) - \gamma x_i, \qquad i = 1, 2, \ldots, N$$

where the variables x_i are nonlinear functions of P at a time τ_i in the past, and the value of P tends to $F(x_1, x_2, \ldots x_N)$ with a time constant ε. With one variable in the limit $\varepsilon \to 0$, this reduces to equation (8).

We now consider a special case with multiple delays. Assume that:

1) $\varepsilon = 0$.

2)   F is the average value of the x_i

3)   Each of the control functions $f_i(P_{\tau_i})$, is a simple step

function. It is one if $P(t-\tau_i) < \theta_i$ but is zero otherwise. θ_i is
called the threshold for variable i.

4)   All the decay constants are equal.

With these assumptions, the equations of interest are

$$P = \frac{1}{N} \Sigma_i \; x_i$$

$$\frac{dx_i}{dt} = 1-x_i \quad \text{if } P(t-\tau_i) < \theta_i \tag{11}$$

$$\frac{dx_i}{dt} = -x_i \quad \text{if } P(t-\tau_i) > \theta_i$$

If at time t, $P(t-\tau_i) > \theta_i$ then we say that x_i is off at time t;
otherwise it is called on. These equations can be readily integrated.
The dynamics will either approach a steady state, in which case a
subset of the x_i will be on and the others off, or the dynamics will
fluctuate. In the case that the dynamics are fluctuating, P will
continue to cross one or more of the thresholds. If a threshold i is
crossed at time t, then at time $t + \tau_i$, x_i will be switched off if P
was increasing when the threshold was crossed. If P was decreasing
when the threshold was crossed then x_i is switched on at a time τ_i
afterwards. Therefore, each time a threshold is crossed this will lead
to a discontinuity in P after a delay.

One of the interesting properties of (11) is that it can show
highly complex dynamic behaviors and bifurcations as parameters are
varied. For example, even after 3500 threshold crossings, it is still
not possible to find periodic orbits for many values of the parameters.
This is illustrated in Fig. 10. The bifurcations which are observed
here do not appear to follow previously described bifurcation sequences

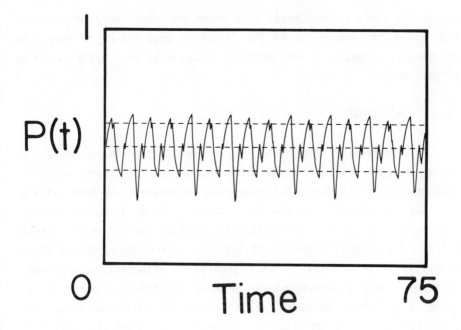

Fig. 10: Aperiodic behavior for an equation with multiple time delays
(11). θ_1=0.4, θ_2=0.5, θ_3=0.6, τ_1=0.56, τ_2=2.2, τ_3=0.87.
Dynamics after a transient of 450 time units are shown
(reprinted from Ref. 37).

and the global structure of these bifurcations is not now understood.
Although we do not have any proof that the dynamics in this example are
"chaotic", it is clear that there are very long cycles which are found
over very narrow ranges of parameters. Further, for some parameter

values, if there is not chaos, then there are either very long periods (we looked for cycles up to 500 threshold crossings), or very long transients, or both. A more detailed analysis of this system would be of interest.

4.   CONCLUSIONS

Physiological systems, both in experimental studies and in various normal and pathological circumstances display complex dynamic behavior. Mathematical models for such systems must by necessity be formulated as nonlinear equations. It has been striking to me that the simplest imaginable models for physiological systems nevertheless display sufficiently complex dynamics, that a full understanding of the simple models is often a research level problem in mathematics. Moreover, there is often a correspondence between the bifurcations which are found in the simple mathematical models and the living systems. This indicates that the qualitative properties of complex, realistic, nonlinear mathematical models of the physiological systems should show a correspondence with the qualitative properties of the simple caricatures that have been considered in this article. Elucidation of the dynamics in the simple and complex models should prove to be of mathematical interest, and should also set the stage for the understanding of the qualitative dynamics of health and disease.

ACKNOWLEDGEMENTS

The models in this paper have been developed in collaboration with a number of colleagues. In particular I have benefited greatly from my collaborations with M.C. Mackey, M.R. Guevara, R. Perez, J. Keener, J. Bélair, A. Beuter, D. Larocque. My research has been supported by the Natural Sciences Engineering and Research Council (Canada) and the Canadian Heart Foundation.

REFERENCES

1. Glass, L. and Mackey, M.C., From Clocks to Chaos: The Rhythms of Life. Princeton: Princeton University Press (1988).
2. Winfree, A.T., The Geometry of Biological Time. New York: Springer-Verlag (1980).
3. Guevara, M.R. and Glass, L., J. Math. Biol. 14, 1 (1982).
4. Hoppensteadt, F.C. and Keener, J.P., J. Math. Biol. 15, 339 (1982).
5. Keener, J.P. and Glass, L., J. Math. Biol. 21, 175 (1984).
6. Kopell, N. and Howard, L.N., Studies in Appl. Math. 52, 291 (1973).
7. Arnold, V.I., Translations A.M.S., Series 2, 46, 213 (1965).
8. Glass, L. and Bélair, J., In: Lecture Notes in Biomathematics No. 66, p. 222. Berlin: Springer-Verlag (1986).
9. Gonzalez, D.L. and Piro, O., Phys. Rev. Lett. 50, 344 (1983).
10. Builder, G. and Roberts, N.F., A.W.A. Technical Review 4, 165 (1939).
11. Glass, L. and Mackey, M.C., J. Math. Biol. 7, 339 (1979).
12. Glass, L., Graves, C., Petrillo, G.A., and Mackey, M.C., J. Theor. Biol. 86, 455 (1980).
13. Keener, J.P., Trans. A.M.S. 26, 589 (1980).
14. Keener, J.P., J. Math. Biol. 12, 215 (1981).
15. Keener, J.P., Hoppensteadt, F.C. and Rinzel, J., SIAM J. Appl. Math. 41, 503 (1981).
16. Petrillo, G.A. and Glass, L., Am. J. Physiol. 246, (Regulatory Integrative Comp. Physiol. 15) R 311 (1984).
17. Perez, R. and Glass, L., Phys. Lett. 90A, 441 (1982).
18. Jensen, M.H., Bak, P., and Bohr, T., Phys. Rev. 30A, 1960 (1984).
19. Glass, L. and Perez, R., Phys. Rev. Lett. 48, 1772 (1982).
20. Schell, M., Fraser, S. and Kapral, R., Phys. Rev. 28A, 373 (1983).
21. Bélair, J. and Glass, L., Physica 16D, 143 (1985).

22. Mackay, R.S. and Tresser, C., Physica 19D, 206 (1986).

23. Mackey, M.C. and Glass, L. Science 197, 287 (1977).

24. Glass, L. and Mackey, M.C., Ann. N.Y. Acad. Sci. 316, 214 (1979).

25. Lasota, A., Asterisque 50, 239 (1977).

26. Perez, J.F., Malta, C.P. and Coutinho, F.A.B. J. Theor. Biol. 71, 505 (1977).

27. Gurney, W.S.C., Blythe, S.P. and Nisbet, R.M., Nature 287, 17 (1980).

28. May, R.M., Ann. N.Y. Acad. Sci. 357, 267 (1981).

29. de Oleivara, C.R. and Malta, C.P., Phys. Rev., in press, 1987.

30. an der Heiden, V. and Mackey, M.C., J. Math. Biol. 16, 75 (1982).

31. Farmer, J.D., Physica 4D, 366 (1982).

32. Grassberger, P. and Procaccia, I., Physica 9D, 189 (1983).

33. Kostelich, E.J. and Swinney, H.L., In: Chaos and Related Nonlinear Phenomena, I. Procaccia and M. Shapiro, Eds. New York: Plenum (1987).

34. Le Berre, M., Ressayre, E., Tallet, A., Gibbs, H.M., Kaplan, D.L. and Rose, M.H., Phys. Rev. 35A, 4020 (1987).

35. Goodman, L., IEEE Trans. Biomed. Eng. BME-11, 82 (1964).

36. Akselrod, S., Gordon, D., Ubel, F.A., Shannon, D.C., Barger, A.C. and Cohen, R.J. Science 213, 220 (1981).

37. Glass, L., Beuter, A. and Larocque, D. Math. Biosci., submitted (1987).

MOST STABLE MANIFOLDS AND TRANSITION TO CHAOS
IN DISSIPATIVE SYSTEMS WITH COMPETING FREQUENCIES

Yan Gu

Department of Physics, Lanzhou University

Lanzhou, Gansu, China

ABSTRACT

The phase space of a dissipative dynamical system can be
foliated into families of submanifolds called "most stable
manifolds". For such a system, the loss of smoothness of an
invariant torus is a consequence of tangency of the torus
with one of the most stable manifolds. With the help of the
numerically calculated foliations of most stable manifolds,
the transition to chaos of an invariant circle for a two-
dimensional map (dissipative standard map) is discussed.

1. INTRODUCTION

It was remarked by Poincare[1] in 1899 that the generation of chaos
in near-integrable Hamiltonian systems is closely related to transversal
intersections (homoclinic and heteroclinic crossings) of stable and
unstable manifolds in a surface of section of phase space. The recogni-
tion that chaotic motion can occur in dissipative systems and can also
be related to these transversal intersections is quite recent, dating
from the pioneering work of Lorenz[2] and Smale[3]. However, within the
last few years research has elucidated the patterns of transition to
chaos from an invariant circle in two-dimensional dissipative maps[4-9],

which shows that the loss of smoothness of invariant circles in two-dimensional dissipative maps is, in general, not induced by homoclinic or heteroclinic crossings. The objectives of this lecture are to elucidate the mechanism involved in the collapse of tori in dissipative dynamical systems and to report on a computer assisted study of transition to chaos in a dissipative phase-locked dynamical system with two competing frequencies. In particular, we will present a practical numerical method for obtaining the precise parameter value at which an invariant circle loses its smoothness.

We start by demonstrating that the phase space of a dissipative dynamical system can be foliated into families of submanifolds called "most stable manifolds" (MSM's)[10]. The foliation of these MSM's provides a global characterization of the contraction of the phase space onto attractors of lower dimensionality. Then, in Sec. 3, we discuss how to calculate the foliation of MSM's. In Sec. 4, we show that the loss of smoothness of an invariant torus is, in general, a consequence of tangency of the torus with one of these MSM's. In Sec. 5, a simple numerical method is given to pin down the critical parameter value at which an invariant circle loses its smoothness. Finally, in Sec. 6, the evolution of strange attractor through the phase-locked regimes of two-dimensional dissipative standard maps is studied by the aid of computer.

2.   WHAT IS A MSM?

A MSM is a submanifold in the basin of an attractor such that all the points on this submanifold will converge to a single (generally non-fixed) point with the highest possible exponential rate (i.e. the most negative Lyapunov exponent of the attractor). The existence of MSM can be most easily seen in a two-dimensional continuous dynamical system with a generic stable node (i.e. a fixed point with two distinct negative eigenvalues $\lambda_2 < \lambda_1 < 0$). All the nearby orbits will converge to the sink, but only two of them approach it with the highest exponential rate $|\lambda_2|$. These two orbits plus the sink form an invariant MSM. Besides this invariant MSM, there exists a continuum of non-invariant MSM's which, in the neighborhood of the sink, are represented by curves approximately parallel to the invariant MSM. These curves have the property that they move toward the invariant MSM with exponential rate $|\lambda_1|$ while the length of each moving curve contracts with the higher rate $|\lambda_2|$.

Now, consider an arbitrary attractor (denoted by A) of an m-dimensional dissipative dynamical system

$$\dot{x} = F(x), \qquad x \in R^m,$$

with its m Lyapunov exponents denoted by

$$LE_1 \geq LE_2 \geq \cdots \geq LE_m .$$

For any two points x and y within the basin B of the attractor A, one may define an equivalence relation:

$$x \sim y \qquad \text{iff} \qquad \lim_{t \to \infty} \frac{1}{t} \ln |f^t(x) - f^t(y)| \leq LE_m ,$$

where f^t represents the flow of the system, i.e.

$$x(t) = f^t(x(0)) .$$

Then, using this equivalence relation, the basin B can be foliated into equivalence classes:

$$B = \bigcup_x M_x ,$$

where x is a representative point of the equivalence class M_x.

Since for dissipative dynamical systems LE_m is always negative, the equivalence class M_x will consist of all the phase points converging to $f^t(x)$ as $t \to \infty$, with an exponential rate higher than or equal to $|LE_m|$. Accordingly, the point set M_x is called the MSM of the attractor A through x.

For discrete time dynamical systems (i.e. iterated maps) the above definition of MSM holds again, with t replaced by integer n. It should be noted that although a single M_x need not be invariant, the foliation of MSM's is time invariant, i.e. the members of this family of MSM's are taken into each other by the action of the flow:

$$f^t(M_x) = M_{f^t(x)} .$$

3.   A METHOD FOR COMPUTING MSM'S

When the MSM's of an attractor are smooth curves, they can be computed by first determining the direction of maximal contraction at each point within the basin of attraction, and then integrating this direction field. The calculation of the direction of maximal contraction is, however, closely related to the calculation of Lyapunov exponents.

Consider a vector w_0 in the tangent space of an arbitrary phase point x. The evolution of the tangent vector w_0 is governed by

$$\dot{w} = \hat{J}(f^t(x)) w , \qquad (1)$$

where $\hat{J}$ is the Jacobi matrix with elements

$$J_{ij}(x) = \partial F_i(x)/\partial x_j , \quad i,j = 1,\cdots,m. \qquad (2)$$

Integrating eq. (1), we obtain

$$w(t) = \hat{U}_x(t) w_0 , \qquad (3)$$

where $\hat{U}_x(t)$ is a m x m matrix with $\hat{U}_x(0)$ being the identity matrix. The matrix $\hat{U}_x(t)$ can be written as the product, $\hat{O}_x(t) \hat{S}_x(t)$, of an orthogonal ($\hat{O}$) and a non-negative ($\hat{S}$) matrix, whose actions are to induce rotations and dilatations separately in the tangent space. Then, for each x $\in$ B, the following limiting matrix exists[11]:

$$\hat{L}_x = \lim_{t \to \infty} \frac{1}{t} \ln \hat{S}_x(t) . \qquad (4)$$

$\hat{L}_x$ is a symmetric matrix whose eigenvalues are just m Lyapunov exponents ($LE_1 \geq LE_2 \geq \cdots \geq LE_m$) of the attractor[12] and are independent of x within the basin of attraction. On the other hand, the eigendirections of $\hat{L}_x$ may vary with x. Assume that $LE_{m-1} > LE_m$, then, for each point x $\in$ B, there is a unique eigendirection for the most negative eigenvalue of the matrix $\hat{L}_x$. This eigendirection is the direction of maximal contraction at point x.

Figs. 1 and 2 show some numerically calculated MSM's of the Henon map[13] in a stable regime (a=.3, b=.3) and a chaotic regime (a=1.4, b=.3). All our numerical results seem to point out that MSM's are smooth curves even if the attractor is chotic.

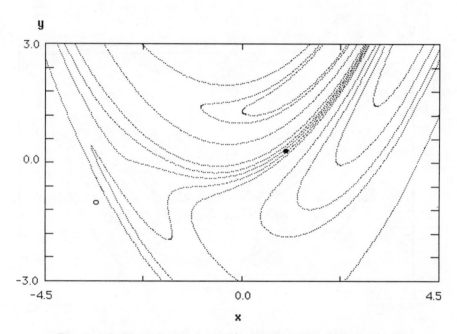

Fig. 1. Stable node and its MSM's for Henon map (a=0.3, b=0.3).
Filled circle represents the sink.

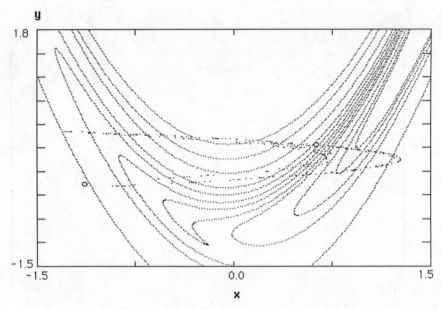

Fig. 2. Strange attractor and its MSM's for Henon map (a=1.4, b=0.3).
Open circles represent saddles.

4. DESTRUCTION OF TORI IN DISSIPATIVE DYNAMICAL SYSTEMS

In what follows we will focus our attention to the role played by the MSM in the destruction of invariant tori in dissipative dynamical systems. Consider a two-dimensional dissipative map which has an invariant circle denoted by S. We shall not consider the case in which S is the basin boundary and assume that at least a segment of S intersects transversally some MSM's of an attractor A. Since a neighborhood of this segment will be contracted toward S, the attractor A must be a subset of S. Then, when S is smooth, the attractor within S is either periodic or quasiperiocic, i.e. either S consists of the outsets (unstable manifolds) of the unstable periodic points, or S coincides with A. Moreover, a smooth S is transverse to each MSM it intersects[7]. In fact, any point of S tangent to a MSM will map to a point of S which is again tangent to some MSM; thus any tangency between S and the foliation of MSM's invariably leads to an infinite set of tangent points between them. These tangent points will either accumulate at periodic sinks if the attractor is periodic, or be densely distributed on S if it is quasiperiodic. Then, assuming that MSM's are smooth curves, the invariant circle must either lose smoothness at all its periodic sinks or collapse completely immediately after its tangency with MSM's. Since similar conclusions can also be reached for higher-dimensional dissipative dynamical systems, the destruction of torus in dissipative dynamical systems is, in general, a consequence of its tangency with MSM's.

In Figs. 3 and 4 we show computer simulations of the unstable manifold from a distant saddle point approaching a sink for the two-dimensional dissipative map (5). In plotting these two figures, we have displaced the sink to right lower corner and transformed its two eigendirections to rectangular axes by a linear transformation so that MSM's are represented to a very good approximation by vertical lines (not shown in Figs. 3 and 4). As seen in Fig. 3, the invariant circle (i.e. unstable manifold of the saddle) has infinitely many recurrent cubic tangencies with vertical lines (MSM's) in approaching the sink, which cause the loss of smoothness of the invariant circle at the sink.

Yan Gu

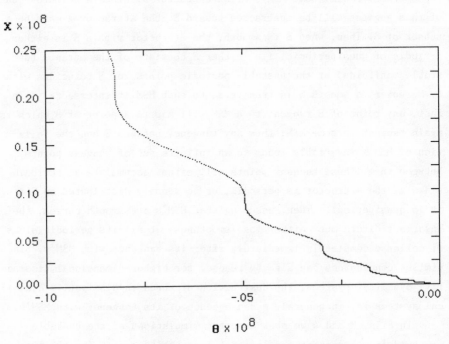

Fig. 3. Recurrent cubic tangencies between invariant circle and MSM's

for the dissipative standard map (b=0.25, D=0.87, A=0.3618298).

$\mathbf{X} \times 10^8$

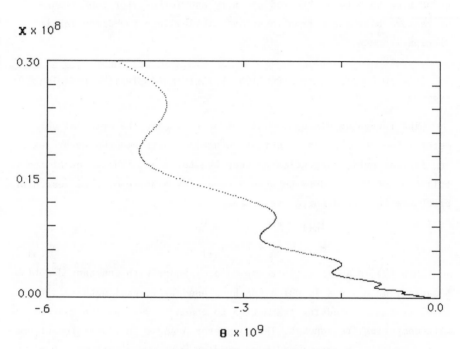

$\mathbf{\theta} \times 10^9$

Fig. 4. Folding structures of the invariant circle in the neighborhood of the sink for the dissipative standard map (b=0.25, D=0.87, A=0.36184).

Fig. 4 shows that immediately after the critical tangency the invariant circle develops an infinite set of folds, each successive fold becomes flatter than the previous one. These folding structures are caused by the quadratic tangencies between invariant circle and MSM's as well as the different contracting rates in vertical and horizontal directions under each iteration. It is interesting to note that similar folding structures have been observed by Curry and Yorke[4] for the strange attractors bifurcated from invariant circles in a two-dimensional dissipative map.

5. HOW TO LOCATE THE DESTRUCTION OF TORI IN DISSIPATIVE DYNAMICAL SYSTEMS

The foregoing discussion allows us to locate the critical parameter value at which an invariant torus loses its smoothness by searching for the cubic tangencies between the torus and the leaves of the foliation of MSM's. Here we show a numerical example of this searching procedure for the dissipative standard map:

$$x_{n+1} = b\, x_n - A \sin \theta_n$$

$$\theta_{n+1} = \theta_n + E + x_{n+1} \; . \tag{5}$$

Map (5) is an invertible map on a cylinder with constant Jacobian b satisfying $o < b < 1$, which has been used by several authors[6-8] as a model system in studying transition to chaos in dissipative systems with competing frequencies. The parameter A and E in (5) represent nonlinear coupling constant and bare winding number respectively. For weakly nonlinear coupling, map (5) has a unique attracting invariant circle the motion on which is either phase-locked or quasiperiodic. As the parameter A increases, chaotic motion may occur after the destruction of the invariant circle. Since in phase-locked regimes the invariant circle consists of the outsets of the unstable periodic points, we will locate the destruction of the invariant circle by calculating both the tangent of the outsets and the tangent of MSM's at each point on the outsets. This is done as follows:

(i) Calculation of the tangent of the outset of a saddle. For concreteness, we discuss the phase-locked regimes of map (5) with the winding number $0/1$. The invariant circle $x = g(\theta)$ satisfies the functional equation

$$b\, g(\theta) - A \sin\theta = g(\theta + E + b\, g(\theta) - A \sin\theta). \tag{6}$$

On the other hand, in the neighborhood of the saddle point (θ_s, x_s), the outset can be described by

$$x = g(\theta) = x_s + \sum_{j=1}^{\infty} a_j\, (\theta - \theta_s)^j, \tag{7}$$

where $a_1 = g'(\theta_s)$ is chosen to specify the unstable direction of the saddle and all the other coefficients (i.e. a_2, a_3, $\cdots$) are calculated in recurrence by substituting (7) in (6). Having computed a segment of the outset and its tangents in the neighborhood of the saddle from eq. (7), the entire curve $x = g(\theta)$ and its tangents $u/v = g'(\theta)$ can be obtained by iterating the known segment with map (5) and its differential map

$$\begin{pmatrix} u_{n+1} \\ v_{n+1} \end{pmatrix} = \hat{U}(\theta_n) \begin{pmatrix} u_n \\ v_n \end{pmatrix} = \begin{pmatrix} b & -A \cos\theta_n \\ b & 1 - A \cos\theta_n \end{pmatrix} \begin{pmatrix} u_n \\ v_n \end{pmatrix} \tag{8}$$

respectively.

(ii) Calculation of the tangent of MSM. Let $(\theta_1, g(\theta_1))$ be any point on the outset at which the tangent of MSM is to be calculated. Consider a vector w_1 in the tangent space of this point. After n iterations, we have

$$w_{n+1} = \hat{U}_n\, w_1 \, ,$$

where

$$\hat{U}_n = \hat{U}(\theta_n) \cdots \hat{U}(\theta_1),$$

with matrix $\hat{U}(\theta)$ given by eq. (8). According to discussions carried out in Sec. 3, the tangent of MSM at $(\theta_1, g(\theta_1))$ can be obtained by calculating the eigenvector for the smallest eigenvalue of the limiting matrix

$$\hat{L} = \lim_{n \to \infty} \frac{1}{2n} \ln [\, \hat{U}_n^T \hat{U}_n \,],$$

where $\hat{U}^T$ denotes the transposed matrix of $\hat{U}$. In practical calculations, it is more convenient to evaluate matrix

$$\hat{M} = \lim_{n \to \infty} \hat{U}_n^T \hat{U}_n / Tr[\hat{U}_n^T \hat{U}_n], \tag{9}$$

which obviously shares the same set of eigenvectors with $\hat{L}$. (A nor-
malizing factor such as $Tr[\hat{U}_n^T \hat{U}_n]$ is indispensable to the existence of
a non-trivial $\hat{M}$.) Moreover, the convergence of the r.h.s. of (9) is,
in general, so good that we can expect to have an accuracy of 10
decimals for n not greater than 100.

Fig. 5 shows part of the phase-locked horn in the E-A parameter
plane (b fixed at 0.25) where map (5) has one sink and one saddle. The
solid curve through O in Fig. 5 is a hyperbola described by

$$E^2 (1-b)^2 = A^2 - D (1-\sqrt{b})^4 \tag{10}$$

with D = 1, on which the Jacobi matrix at the sink has two equal eigenvalues ($\lambda_1 = \lambda_2 = \sqrt{b}$). The straight line through B is the saddle-
node line ($\lambda_1 = 1$, $\lambda_2 = b$) described by eq. (10) with D = 0. Since
the outset from the saddle will spiral into the sink if the eigenvalues
λ_1 and λ_2 become complex, the hyperbola with equal eigenvalues gives
the maximum value of A for which the outset is smooth. However, below
this hyperbola the outset of the saddle is not necessarily smooth[5-9].
We have searched along several hyperbolae (10) with different D values
chosen from the interval (0, 1] and found that cubic tangencies between
the outset and MSM's occur on two arcs (dashed arcs CB and CF shown in
Fig. 5). In fact, below the hyperbola with equal eigenvalues, two
regions with different types of smooth outsets have been found: (i) the
region below arc CB (vertically hatched region) with two smooth
branches of the outset approaching the sink from both sides; (ii) the
region surrounded by hyperbola CF and dashed arc CF (horizontally
hatched region) with two smooth branches of the outset forming a cusp
at the sink (see Fig. 6).

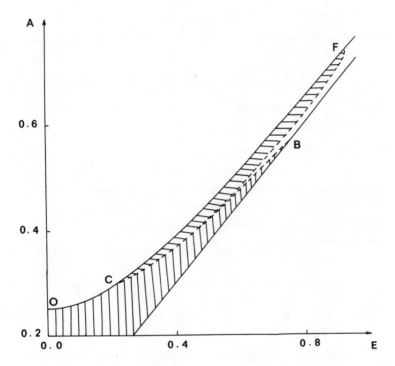

Fig. 5. Phase-locked region with winding number 0/1 for the map (5) with b fixed at 0.25. Cubic tangencies between the invariant circle and MSM's occur on the dashed curves CB and CF. The hatched region indicates where the invariant circle is smooth.

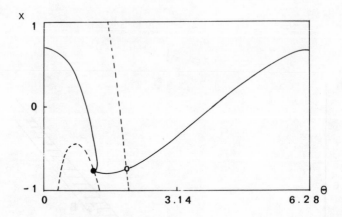

Fig. 6(a). Numerically calculated stable (dashed curve) and unstable (solid curve) manifolds of the saddle for the dissipative standard map (b=0.25, D=0.87, A=0.6). Open circle represents saddle and filled circle represents sink. The dashed curve through the sink is the invariant MSM.

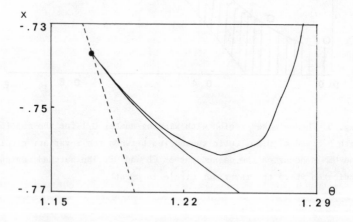

Fig. 6(b). Enlargement of the neighborhood of the sink of Fig. 6(a).

6. TRANSFORMATION OF INVARIANT TORI INTO STRANGE ATTRACTOR

In this section we attempt to illustrate what will happen for an invariant torus in the dissipative system with competing frequencies after the loss of smoothness. As already noted in Sec. 4, in two-dimensional dissipative maps, the tangency between an invariant circle and MSM's will bring about the global destruction of the invariant circle if the map is in quasiperiodic regime. However, the numerical evidence observed by T. Bohr et al.[8] in dissipative standard maps indicates that the transition to chaos through quasiperiodic regimes is unlikely in measure-theory sense. On the other hand, the critical tangency between invariant circle and MSM's in phase-locked regimes does not signal the immediate appearance of a persistent and chaotic motion over a global domain of phase space. In fact, the asymptotic behavior of a phase-locked system is unchanged after the critical tangency, but the approach to periodic sinks will be erratic. Thus, in phase-locked regimes, further bifurcations are needed to carry out the transformation of invariant circle into a strange attractor.

A familiar route to chaos through phase-locked regimes is via period-doubling cascade[14,15]. In the E-A parameter plane of map (5), this route is possible only if the path followed by the system crosses the hyperbola of equal eigenvalues. Another route to chaos is through intermittence, which occurs when the path crosses the saddle-node line above the point of cubic tangency (labelled B in Fig. 5). Our attention is paid to the evolution of the outset of the saddle within the swallow-tailed region between the hyperbola of equal eigenvalues and the saddle-node line where neither period-doubling nor saddle-node bifurcations can occur. The result of computer simulations is shown schematically in Fig. 7. (In drawing this figure we have exaggerated the essential features for the purpose of illustration.)

As seen in the main diagram of Fig. 7, the region between the hyperbola of equal eigenvalues (curve OK) and the saddle-node line (line BH) is divided into eight zones labelled form 1 to 8. The distinctive feature of the outset of the saddle in each of these eight

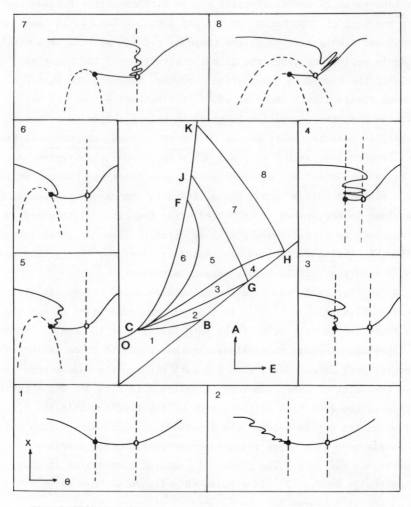

Fig. 7. Bifurcation structure observed in 0/1 phase-locked horn
for dissipative standard maps. The unstable manifold of the saddle
inside eight zones are shown in sub-diagrams as labelled.

zones has been outlined in the correspondingly labelled sub-diagrams on the periphery of the main diagram. Here we describe what is happening on the six boundary arcs inside the swallow-tailed region:

On CB: The outset of the saddle is cubically tangent to non-invariant MSM's. The outset loses smoothness at the sink on passing CB. (Cf. Fig. 3.)

On CG: The outset of the saddle is quadratically tangent to the invariant MSM through the sink (represented by the dashed curve through the filled circle in sub-diagrams of Fig. 7).

On CH: The outset is again quadratically tangent to the invariant MSM through the sink but with the opposite orientation to that occuring on CG. On passing CH both branches of the invariant MSM through the sink pass to negative infinity (as shown in sub-diagrams 5-8).

On CF: The outset is again cubically tangent to non-invariant MSM's but with the opposite orientation to that occuring on CB. On crossing CF from zone 5 the invariant circle recovers smoothness and forms a cusp at the sink.

On JG: The outset is quadratically tangent to the inset (stable manifold) of the saddle (i.e. homoclinic tangency). Above JG the 0/1 phase-locked region overlaps with higher order phase-locked horns so that the period one sink ceases to be a unique attractor.

On KH: Homoclinic tangency again but with the opposite orientation to that occuring on JG. On passing KH both branches of the stable manifold of the saddle pass to negative infinity and a strange attractor (which is formed by the crinkled branch of the outset) suddenly appears.

A closer examination shows that the strange attractor appeared in zone 8 is coexistent with the original attractor (i.e. the sink) and has the stable manifold of the saddle as their basin boundary. On crossing KH from zone 8 to zone 7, the strange attractor seems to collide with the saddle causing the disappearance of both the strange attractor and its basin. This kind of phenomenon has been called crisis or blue sky catastrophe[16,17] to emphasize its discontinuous character. However, the alledged discontinuity occurs only in the asymptotic behavior of the system. From the point of view of the transformation of the outset of the saddle, the bifurcation occuring on KH is really a continuous one.

In conclusion, we have observed a new generic route to chaos through phase-locked regimes in dissipative dynamical systems with two competing frequencies, which can be identified as the inverse route of crisis or blue sky catastrophe. The bifurcation structure shown in Fig. 7 is believed to be true also for all higher order phase-locked horns of a dissipative dynamical system with competing frequencies.

The author would like to thank Prof. B. L. Hao for his helpful collaboration and numerous discussions. He is also grateful to Profs. J. M. Yuan and H. B. Stewart for their valuable comments. This work was supported by the Science Fund of Chinese Academy of Science and a fund from the State Educational Committee of China.

REFERENCES

1. Poincare, H., "les Methodes Nouvelles de la Mecanique Celeste", Vol. 3 (Gautiers-Villars, Paris, 1899).

2. Lorenz, E.N., J. Atmos. Sci. 20, 130 (1963).

3. Smale, S., in "Differential and Combinatorial Topology" (Princeton University Press, Princeton, 1965).

4. Curry, J.H. and Yorke, J.A., in "Lecture Notes in Math.", Vol. 688 (Springer, Berlin, 1978).

5. Aronson, D.G., Chory, M.A., Hall, G.R. and McGehee, R.P., Comm. Math. Phys. 83, 303 (1982).

6. Feigenbaum, M.J., Kadanoff, L.P. and Shenker, S.J., Physica 5D, 370 (1982).

7. Ostlund, S., Rand, D., Sethna, J. and Siggia, E., Physica 8D, 303 (1983).

8. Bohr, T., Bak, P. and Jensen, M.H., Phys. Rev. 30A, 1970 (1984).

9. Bohr, T., Phys. Rev. Lett. 54, 1737 (1985).

10. Gu, Y., "Most Stable Manifolds and Destruction of Tori in Dissipative Dynamical Systems", to be published.

11. Oseledec, V.I., Trans. Moscow Math. Soc. 19, 197 (1968).

12. Gu, Y., Bandy, D.K., Yuan, J.M. and Narducci, L.M., Phys. Rev. 31A, 354 (1985).

13. Henon, M., Comm. Math. Phys. 50, 69 (1976).

14. Belair, J. and Glass, L., Physica 16D, 143 (1985)

15. Mackay, R.S. and Tresser, C., Physica 19D, 206 (1986).

16. Grebogi, C., Ott, E. and Yorke, J.A., Physica 7D, 181 (1983).

17. Abraham, R.H. and Stewart, H.B., Physica 21D, 394 (1986).

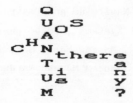

Joseph Ford
School of Physics
Georgia Institute of Technology
Atlanta, Georgia 30332, U.S.A.

"Is quantum chaos merely a new style in Emperor's clothing?"

The Evangelist of Chaos

ABSTRACT

Though not yet widely recognized, there is a
growing body of evidence which strongly indicates that
quantum mechanics does not adequately encompass chaos.
Unfortunately this evidence, while striking, is not yet
sufficient to establish that quantum mechanics is a
theory of limited validity. Thus, we here present four
problems which offer promise of definitively revealing
the lack of chaos in quantum mechanics. Specifically,
we suggest that judicious use of algorithmic complexity
theory can provide rigorous proofs for the following
assertions:

1. The algorithms which compute quantum mechanical
   eigenvalues and eigenfunctions have null
   complexity.
2. The time evolution of the wave function for any
   finite, bounded, conservative quantum system has
   null complexity no matter the smallness of $\hbar > 0$.
3. The time evolution of a specific time dependent
   Hamiltonian, known to be classically chaotic, has
   null complexity not only no matter how small $\hbar > 0$
   but even in the limit as $\hbar \to 0$.
4. An algorithm of null complexity exists for
   obtaining the time evolution of any quantum
   billiard confined within a polygon all of whose
   vertex angles are rational ($m\pi/n$).

Here, the term null complexity means that the problem
can be solved using finite, recursive algorithms.

1. THE PAST

Because quantum mechanics is presumed to contain classical mechanics as a special case and because classical systems are known to exhibit chaos, many investigators regard the existence of quantum chaos as an indisputable reality. But despite the obvious appeal of this view, it nonetheless contains a flaw. Specifically, it provides no definition of quantum chaos. Consequently, over the years each investigator has in turn filled this void with his own definition based upon one or another "erratic" or "irregular" quantum phenomenon which he felt could properly be labelled quantum chaos. As one would expect, in this climate claims and counterclaims have grown like weeds. Moreover, the general acceptance and rapid growth of this subject area have been severely impeded by the lack of consensus regarding the definition of quantum chaos. But before forwarding full blame to this address, recall that no similar controversy occurs at the macroscopic level regarding the existence of classical chaos even though no universally accepted definition exists for it either.[1] Thus, despite the undeniable importance of definition, there are deeper issues involved here.

First, a bit of reflection reveals that, although the founders of quantum mechanics relied even more heavily on the correspondence principle than they did on the uncertainty principle, they nonetheless were invoking or validating the correspondence principle only for systems which were classically integrable. This point was quietly underlined in the teens of this century by Einstein[2] who pointed out that the WKB quantization procedure fails when applied to ergodic systems which form the lowest rung on the ladder of chaos. Nonetheless, by the mid 1920's the founders of quantum mechanics felt supremely confident that no exceptions to their prescription for quantizing dynamical systems existed. Indeed, even now few inquire why the relatively easy and accurate calculation of eigenfunctions and eigenvalues for the helium atom does not directly contradict Poincaré's assertion made some decades ago that the corresponding classical three-body problem is intractably chaotic. But whatever the argument, an unbiased contemporary observer carefully reviewing the evidence would likely conclude that the notion of chaos, whether classical or quantal, has yet to be adequately reconciled with the correspondence principle.

Next, a great deal of confusion regarding quantum chaos has been generated by studies of quantum phenomena which correlate with classical chaos. As an illustrative though hypothetical example, consider an isolated quantum system having a full complement of good quantum numbers which loses

all but one at the precise parameter value for which the
corresponding classical system undergoes the transition
integrable to chaotic. The temptation now becomes
overwhelming to suggest that the loss of good quantum
numbers may be taken as the definition of quantum chaos and,
moreover, that an isolated system which loses its good
quantum numbers (concurrent with the transition to classical
chaos) has made a transition to quantum chaos. But no
matter its appeal, this temptation must be resisted. The
issue of chaos, whether classical or quantal, is much too
complex to be decided by facile or incomplete arguments.
Specifically in the above example, the loss of good quantum
numbers may be shown to be only a necessary but not
sufficient condition for chaos. Indeed, the well-meant
proposal of necessary but not sufficient conditions as
definitions of quantum chaos has been a major problem
afflicting this subject for over a decade. The onset of
extreme sensitivity of energy levels to small outside
perturbation[3] and the transition in the distribution of
nearest neighbor energy level spacings from Poisson to
Wigner-Dyson[4] are but two examples. In summary, although
discoveries of quantum properties which correlate with the
transition to classical chaos[3,4] are most assuredly of
great physical interest,[5] their importance to the notion
of   quantum   chaos,   *per se*,   should   not   be   overblown
Clearly, if we are to avoid these "necessary but not
sufficient" pitfalls, we desperately need a definitive
statement regarding the meaning of quantum chaos. It is to
this issue we now turn.

    A working definition[6] of classical chaos which now
appears in the literature with increasing frequency involves
the notion of exponentially sensitive dependence of final
system state upon initial state-- e. g., positive Liapunov
numbers or the like. This is in fact an excellent working
definition, but why is its attention so sharply focused upon
exponential sensitivity? Why not a large power law
sensitivity or perhaps an exponential raised to an
exponential raised to an exponential sensitivity? The
answer is revealing. Simple exponential sensitivity is
precisely the point at which, in order to maintain constant
calculational accuracy, we must input about as much
information as we get out of our calculations. It is
therefore the point at which our IBM or CRAY computers, no
matter how large their megaflops or magabytes, become
elaborate Xerox machines. But even more important, it is
the point at which algorithmic complexity theory[7] asserts
that our deterministic algorithms are trying to compute
variables which, in fact, are mathematically random. With
this additional insight, it becomes sensible to subsume the
above working definition of classical chaos under the most
general possible theoretical definition: Chaos means

deterministic randomness. Note here that, contrary to popular belief, there is no conflict of meaning between determinism and randomness. Determinism simply means that Newtonian orbits exist and are unique, but since existence-uniqueness theorems are generally nonconstructive, they assert nothing about the character of the orbits they define. Specifically, they do not preclude a deterministic Newtonian orbit from being humanly indistinguishable from a realization of a truly random process. But the attractiveness of chaos as deterministic randomness lies not merely in its lack of internal contradiction; its appeal also rests on man's centuries old intuition that chaos means true unpredictability.[8] But of greatest significance to us here, deterministic randomness is the sole definition of classical chaos which translates into quantum language without change of syllable: in quantum mechanics as well as in classical, chaos means deterministic randomness. In addition to possessing both simplicity and intuitive appeal, this definition provides a clear roadmap which can guide us to the residence of quantum chaos, if such exists. Specifically, the definition directs us to arm ourselves with the tools of algorithmic complexity theory[9] and then search for randomness (or its lack) in the deterministic algorithms which compute quantum eigenvalues and eigenfunctions as well as in the deterministic algorithms, defined by Schrödinger's equation, which compute the time evolution of the wave function Ψ. Equivalently stated, we must establish whether or not any randomness exists in quantum mechanics over and above that contained in the wave function Ψ or probabilities derived from it. The route that research must now follow lies clearly before us. But before turning to the discussion of research particulars which appears in the following section, let us close this section with two final paragraphs of background information.

Recall that, at their beginnings, both quantum mechanics and classical mechanics were derived as analytic expressions for the dynamical order observed in the universe. Indeed, the fundamental equations of motion for both were selected on the basis of agreement with experimental data provided by the harmonic oscillator, the Kepler problem and solar system, the hydrogen atom, and the rigid rotator. Truly then, at their core both classical and quantum mechanics are theories of order; however, the possibility nonetheless remains that a full mathematical investigation of their fundamental equations might reveal circumstances under which chaos can occur. But for classical and quantal theories proven to agree only in their common domain of order, this extrapolation to chaos must be checked with utmost care. Indeed, the types of "mathematical chaos," if any, arising out of these classical and quantal equations of motion need not agree with each other nor with

the chaos observed in nature. These facts have multiple
consequences, but we here emphasize only one: the
mathematical chaos now known to occur in Newtonian dynamics
does not of itself imply the existence of chaos in quantum
mechanics. Nonetheless, the discovery of Newtonian chaos
does permit the hope that chaos may exist in those quantal
systems that are classically chaotic.

But alas, a critical review[10] of the data accumulated
over the years reveals no definitive evidence for the
existence of quantum chaos even in those quantum systems
which are classically chaotic; however, honesty compels the
admission that the evidence is as yet incomplete and that an
impartial jury considering the case against quantum chaos
might well render the Scottish verdict: *NOT PROVED*. Faced
with such an extended hang fire, many investigators have
begun to entertain the obvious question: What serious difficulties, if any, would arise were there no deterministic
chaos in quantum mechanics? As answer, two problems immediately present themselves which are very serious indeed.
First, quantum statistical mechanics justifies its very existence by assuming a randomness in quantum mechanics over and
above that contained in the wave function Ψ; thus if the
underlying quantum mechanics does not in fact contain the
required additional randomness, then both the validity and
the enormous success of quantum statistical mechanics are
without theoretical foundation. Second, if there is no
chaos in contemporary quantum theory and if we continue to
insist that quantum mechanics is our most comprehensive
description of nature, then we thereby forever divorce the
chaotic part of the macroscopic world from its microscopic
description. Only two not equally likely conclusions now
seem defensible. Either quantum mechanics is a true theory
which can eventually be made to expose its deterministic
randomness, or more likely, contemporary quantum mechanics
is a flawed theory which can be modified to include chaos.

2. THE PRESENT

The four problems we now discuss offer the promise of
definitively establishing that quantum mechanics is a flawed
theory which not only fails to exhibit chaos (deterministic
randomness) where many expected it but even where the
correspondence principle requires it. Specifically,
judicious use of algorithmic complexity theory is expected
to reveal that:
  1. The algorithms which compute quantum mechanical
     eigenvalues and eigenfunctions have null
     complexity.
  2. The time evolution of the wave function for any
     finite, bounded, conservative quantum system has
     null complexity no matter the smallness of $\hbar > 0$.

3. The time evolution of a specific time dependent Hamiltonian, known to be classically chaotic, has null complexity no matter the smallness of $\hbar > 0$, indeed even in the limit as $\hbar \rightarrow 0$.

4. An algorithm of null complexity exists for obtaining the time evolution of any quantum billiard confined within a polygon all of whose vertex angles are rational ($m\pi/n$).

In the above, the term null complexity means that the quantity or quantities being computed are not chaotic (random); that is, a computational algorithm exists which requires only a relatively small amount of input information to provide a relatively large amount of output information. In the following as we discuss each item in the above list, we shall omit entering into tedious complexity theory arguments, since in general they are quite long and frequently quite subtle; instead, we shall present selected intuitive arguments which may serve as blueprints for the construction of rigorous complexity theory proofs.

Turning now to Item 1, recall that the procedure for setting up and determining quantum eigenvalues and eigenfunctions is computationally very straightforward; in fact, it is a computable procedure which means that it involves only finite, recursive algorithms. Indeed, one begins in general with well-behaved operators $\hat{A}(\hat{q}, \hat{p})$ generalized from smooth classical functions $A(q, p)$. One then selects a computable basis $\{\Phi_k\}$ for Hilbert space and thence determines matrix elements via the integral $A_{mn} = \int d\tau \, \Phi_n^* \, \hat{A}(\hat{q}, \hat{p}) \, \Phi_m$, which implies that the matrix elements are also computable. Finally, one calculates the eigenfunctions and eigenvalues of the matrix $[A_{mn}]$ using any one of a number of computable diagonalization routines for finite matrices in concert with various computable limit procedures. Thus out of the totality of Hilbert space eigenvalue-eigenfunction equations, most of which have random solutions, quantum mechanics focuses on a subset which is computable and hence not random. In general then, one expects no chaos in quantum eigenvalue-eigenfunction equations. The task now remaining is of course to rigorize these intuitive arguments.

In our later, individual discussions of Items 2 and 3, we shall argue that the unitary time flows of certain quantum systems exhibit no chaos. Thus, it is perhaps worthwhile at the onset to expose a dynamical system whose unitary time flow is chaotic. This exercise will not only emphasize that unitary flows can be chaotic but will also illuminate those characteristics of chaos we might wish to look for as we investigate the properties of unitary flows in quantum mechanics.

Recall now from classical mechanics that the time
evolution or flow of any dynamical system as described by
Hamilton's equations is sympletic,[11] whereas the time flow
of the same dynamical system as described by Liouville's
equation is unitary.[12] Thus, the time evolution of any
chaotic classical system, when transcribed into Liouville
equation language, becomes a chaotic unitary flow. Thus,
for illustrative purposes, let us here discuss the
especially simple, discrete dynamical system called the
Arnol'd cat map[13] which is specified by

$$Q_{n+1} = Q_n + P_n \quad , $$
$$\qquad\qquad\qquad\qquad\qquad\qquad \text{(mod 1)} \qquad (1)$$
$$P_{n+1} = Q_n + 2P_n \quad .$$

Equation (1) describes an area-preserving, fully chaotic[13]
mapping of the unit square or torus upon itself. Now let
$f(Q_0, P_0, 0)$ be the normalized probability density for the
system to reside in state (Q_0, P_0) at time $t = 0$; then
Liouville's equation here takes the form

$$f(Q_n, P_n, n) = f(Q_0, P_0, 0) \quad , \qquad (2)$$

where the Q_n and P_n are related to the Q_0 and P_0 through
Eq. (1). In words, Eq. (2) asserts that the probability
density f at (Q_0, P_0) upon the 0th iteration moves along an
orbit of Eq. (1) arriving at the point (Q_n, P_n) upon the nth
iteration. Using Eq. (2), one may now verify that

$$f(Q_{n+1}, P_{n+1}, n+1) = f(Q_n, P_n, n) \quad . \qquad (3)$$

Thence, because the (mod 1) in Eq. (1) implies that Q and P
are each periodic, one may express both $f(Q_n, P_n, n)$ and
$f(Q_{n+1}, P_{n+1}, n+1)$ as Fourier series. Specifically,

$$f(Q_n, P_n, n) = \Sigma A_{K,\ell}^{[n]} e^{2\pi i (KQ_n + \ell P_n)}, \qquad (4)$$

and

$$f(Q_{n+1}, P_{n+1}, n+1) = \Sigma A_{K,\ell}^{[n+1]} e^{2\pi i (KQ_{n+1} + \ell P_{n+1})}. \qquad (5)$$

One may now invoke Eq. (3) to equate the right sides of
Eqns. (4, 5), use Eq. (1) to express Q_{n+1} and P_{n+1} in
terms of Q_n and P_n and then drop the subscripts on all Q_n
and P_n, relabel dummy summation variables (K, ℓ) as
required, and equate coefficients of $e^{2\pi i (KQ + \ell P)}$ to obtain

$$A_{K+\ell, K+2\ell}^{[n+1]} = A_{K,\ell}^{[n]} \qquad (6)$$

Analysis of Equation (6) paints a rather graphic
picture of the chaos which can occur in a unitary flow.

To perceive this, let us first assume that only one relatively low mode (k, ℓ), k > 0 and ℓ > 0, has nonzero amplitude A at n = 0. Then at n = 1, it is only the mode (k+ℓ, k+2ℓ) which has the nonzero amplitude A; while at n = 2, only the mode (2k+3ℓ, 3k+5ℓ) has this nonzero amplitude. Clearly, the mode numbers for nonzero amplitude will continue to increase indefinitely with iteration number n. The initially smooth probability density $f(Q_0, P_0, 0)$ is thus becoming more and more "wrinkled" as low modes disappear and higher modes appear in the Fourier series for $f(Q_n, P_n, n)$. This is precisely the behavior we would expect to observe in the unitary flow of a dynamical system whose phase space flow is mixing, as is true of Eq. (1). Let us now consider the general situation. To this end, note that the mode number transformation specified by Eq. (6) can be written as the linear mapping k' = k + ℓ, ℓ' = k + 2ℓ. The easily derivable constants of the motion for this mapping are hyperbolas, and iterates of any finite initial mode number may be shown to eventually move out toward infinity at an exponential rate along some hyperbola. Recall now that, due to the requirements of convergence, there is only a finite, though perhaps large, set of Fourier modes having appreciable amplitude for any initial probability density $f(Q_0, P_0, 0)$. Thus, due to the nature of the mode number transformation, all the initial mode numbers of $f(Q_0, P_0, 0)$ are eventually swept toward infinity. Hence, regardless of the initial $f(Q_0, P_0, 0)$, the Fourier series for $f(Q_n, P_n, n)$, provided n is extremely large, has appreciable mode amplitudes only for extremely large mode numbers. As mentioned earlier, this "disappearing" of low modes is the unitary transformation's way of indicating mixing; the "disappearing" of these modes at an exponential rate is its way of indicating chaos.

There are two additional characteristics of the chaos which appear in this unitary flow that are worth mentioning. Although the unitary flow inherits both randomness and "macroscopic" irreversibility (discussed later) directly from the randomness and "macroscopic" irreversibility of Eq. (1), nonetheless two initially close probability densities f_1 and f_2 remain close for all time. Here, if $|f_2(Q_0, P_0, 0) - f_1(Q_0, P_0, 0)| < \epsilon$ where ϵ is small and independent of Q_0 and P_0, then $|f_2(Q_n, P_n, n) - f_1(Q_n, P_n, n)|$ remains less than ϵ for all n independent of Q_n and P_n. This follows from the fact that, although chaos occurs in both f_1 and f_2, the time evolution of iterates Q_n and P_n as well as the exponential "disappearing" of modes is precisely the same for both. Finally, it must be emphasized that the Liouville equation which exhibits chaos is, just like Schrödinger's equation, a linear partial differential equation. Thus, it would appear logically difficult to

attribute any lack of chaos in Schrödinger's equation as
being due solely to its linearity. We now return to the
discussion of items on the problem list.

Item 2 asserts that the time evolution of the wave
function Ψ for a finite particle number, spatially bounded,
conservative quantum system has null complexity. Here the
wave function can always be written

$$\Psi(x, t) = \Sigma \, A_n U_n(x) e^{-iE_n t/\hbar} \, , \qquad (7)$$

where x denotes spatial coordinates and t denotes time, the
sum is over all n, U_n and E_n are energy eigenfunctions and
eigenvalues respectively, the A_n are expansion coefficients,
and ℏ is Planck's constant divided by 2π. Now it is
well-known that the almost periodicity of the time flow
given by Eq. (7) precludes chaos over long time intervals,
but it has frequently been maintained that chaos might
nonetheless occur over laboratory time intervals short with
respect to major almost recurrences. Were this true, there
could then be a smooth blend of quantum chaos into classical
chaos by first taking the limit ℏ → 0 and then the limit
t → ∞. However, this possibility is excluded by a frequently
overlooked, additional property of Eq. (7). For every
fixed x and any ℏ > 0, each term in the sum of Eq. (7) is a
fixed length, rotating vector in the complex plane. The
flow of Eq. (7) is thus nothing more than a simple phase
mixing such as occurs in the spin-echo experiment of solid
state physics or in the reversible oil drop experiment of
couette flow. Such phase mixing can most assuredly mimic
many of the properties of true mixing, including an approach
to apparent equilibrium, but it does not involve true mixing
or chaos as can be easily revealed. In order to make this
last point clear, let us recall that, if sufficient accuracy
is available, then one can always verify the well-known fact
that Hamiltonian orbits are always time reversible. On the
other hand, when only limited calculational or laboratory
precision is available, ordered motion continues to time
reverse back to a relatively distant initial state whereas
chaotic motion does not. Thus in situations where only
limited accuracy obtains, we shall say that ordered motion
exhibits "macroscopic" reversibility whereas chaotic motion
exhibits "macroscopic" irreversibility. Were the time flow
of Eq. (7) deterministically random, it would most certainly
be reversible but not "macroscopically" reversible.
However, phase mixing is, in fact, "macroscopically"
reversible. In consequence, even though the time flow of a
finite, bounded, conservative quantum system mimic
deterministic randomness marvelously well, one need only
expose its "macroscopic" time reversibility to reveal the
fraud. It thus appears that finite, classically chaotic
quantum systems are not themselves chaotic even though they

are expected to be and even though the correspondence
principle apparently requires them to be.

But what of time driven Hamiltonian systems; does chaos
appear in time driven quantum systems which are classically
chaotic? As mentioned in Item 3, we elect to investigate
only one illustrative system, specifically, the Arnol'd cat
map whose classically chaotic Liouville flow was discussed
above; for convenience, we reproduce the mapping equations
here. Although it is perhaps not widely known, the discrete
dynamical system

$$Q_{n+1} = Q_n + P_n \quad , \qquad \text{(mod 1)} \qquad (8)$$
$$P_{n+1} = Q_n + 2P_n \quad ,$$

can be derived[14] from the δ-function kicked, time-
dependent Hamiltonian

$$H = [P^2/2m] + [\epsilon Q^2/2]\sum_n \delta(n - t/T) , \qquad (9)$$

where the sum is over all positive and negative integers.
This time-dependent, classical Hamiltonian system can now
be quantized using standard procedures. Since the classical
Q and P are periodic variables on a torus, the eigenvalues
of $\hat{Q}$ and $\hat{P}$ are both discrete. In consequence, the quantum
motion occurs on a bounded integer grid. Moreover, as
Hannay and Berry have shown,[15] all quantum motion for
this system is periodic. Therefore, the quantum motion has
null complexity, but worse, in the limit as $\hbar \to 0$, the
quantum flow approaches the classical flow on the rationals
which latter also has null complexity. In summary, it
appears that the nonchaotic quantum motion for the Arnol'd
cat map does not asymptote to the fully chaotic classical
motion as required by the correspondence principle. It will
be of considerable interest to establish whether or not this
result can survive careful scrutiny. Indeed, it is quite
difficult not to leap to an immediate final conclusion at
this point, especially since the unitary flow of the quantum
cat map exhibits none of the chaotic properties observed in
its Liouville flow. Regardless, in all these matters it
cannot be over emphasized that the Arnol'd cat map, despite
its seemingly unphysical and nongeneric character, is a
valid theoretical model for testing quantum theory.

Turning now to the final Item 4, let us recall that
Eckhardt and colleagues[16] have already shown that the
motion of classical billiards confined to move in polygons
having all rational (mπ/n) vertex angles is algorithmically
integrable.[16] But in general, when a classical system is
analytically solvable, the corresponding quantum system is
also analytically solvable. Specifically, the key to
obtaining an analytic solution for the classical billiard
problem lay in unfolding the complex billiard orbits within

polygons to form straight lines on one or more flat
surfaces. Therefore, one expects to be able to unfold the
diffractive quantum wave motion in polygons into straight
line patterns on one or more planes as Sommerfeld[17] has
already done for diffraction at a straightedge. A solution
to this problem would have considerable interest for
acoustical engineers, but the interest here centers on the
fact that these classically solvable "rational" billiards
can exhibit an approach to chaos as their polygons tend to
the shape of classical billiard boundaries known to yield
chaos. To what extent will or even can quantally solvable
"rational" billiards mimic this approach to chaos?

Before turning to our final Sec. 3, let us briefly
mention why this review has paid so little attention to
certain topics regarded by many as crucial to the question
of quantum chaos or its lack. Throughout this article, we
have concentrated solely on the issue of intrinsic
quantum chaos. The measurement process has therefore been
neglected because it involves the use of outside classical
devices which in general are themselves chaotic. We have
not sought to discuss any chaotic behavior which might occur
in a quantum system due to an interaction with a classically
chaotic system. However, in regard to the notion that the
measurement process may introduce irreversibility and chaos
into the behavior of a quantum system because of repeated
"collapse of its wave packet," we direct the reader's
attention to a recent paper by van Kampen[18] which
establishes that the measurement process can be accurately
described by Schrödinger's equation and therefore that the
measurement process is fully reversible. In this article,
we do not consider infinite systems, the vacuum
electromagnetic field, or various nonlinear field theories
on the grounds that it is not necessary to exclude the
presence of chaos in every instance where it might occur.
Indeed, to prove that quantum mechanics is flawed, one need
only prove the absence of chaos in various substantive
cases. Turning now to the correspondence principle as it
relates to chaos, some investigators have noted that while
the correspondence principle requires a limit such as $\hbar \to 0$,
a proof of chaos seems to require the limit $t \to \infty$; they then
suggest that care must be taken about the order in which
these limits are taken. Here, there is only a confusion;
randomness is perfectly well-defined for finite sequences or
orbital segments, and the limit $t \to \infty$ is not required in
general. Finally, we have here made no mention of the
exciting and highly relevant work on "chaos" in microwave
driven hydrogen because we have already briefly discussed
this topic in an earlier review.[19]

3.   THE FUTURE

New theories are in part birthed by the limitations and failures of older theories. Each new theory matures, flourishes, and holds sway until its own limitations and failures give rise to yet another theory. In this review, we have sought to initiate such a cycle by enumerating believable assertions whose proofs can definitively reveal the clay feet of contemporary quantum mechanics. But a reigning physical theory can, with the support of its proponents, indefinitely maintain its equilibrium while standing on clay feet. The theories of the late 19th century, for example, continued to thrive even though they were powerless to explain the difference in electrical conductivity between copper and quartz and even though the Raleigh-Jeans formula for black body radiation was totally correct theoretically yet totally wrong experimentally. To a certain extent, a theory loses force only when a tide of evidence runs against it, only when its limitations and failures are sufficiently numerous to shake the faith of even the devout. In this regard, it was not any lack of ability to predict planetary motion that led to the abandonment of the Ptolemaic theory; it was rather the limitations imposed by its *ad hoc* properties which made it totally unacceptable as a general theory of dynamics. Indeed, Newtonian dynamics achieved its dominance over all competitors precisely because it was found to be valid over an enormous range of natural phenomena, but its survival for three hundred years has perhaps depended even more upon its notions of determinism and predictability which serve as the foundation of an appealing theoretical framework for interpreting natural phenomena. Thus, it may add believable flesh to the bare bones of our earlier arguments, if we now describe the appealing theoretical framework of which our arguments are a part.

Over the past two decades, chaos theory has become so preoccupied with extending its theoretical scope and with exporting its methodology to exotic disciplines that little attention has been devoted to addressing foundational issues. Commonly used terms such as *determinism, chaos, unpredictability,* or *randomness* have frequently been treated as primitives which need no definition or as loosely defined terms which mean whatever an investigator wishes them to mean. Moreover, theoretical arguments which seek to clarify foundational issues sometimes omit or simply overlook crucial points. For example, the well-documented sensitive dependence of final state upon initial state which occurs in chaotic Newtonian flows is said to be the source of the unpredictable behavior in these systems. But sensitive dependence is simply an error-growth argument which does not seriously erode the notion of deterministic short-term

predictability for chaotic orbits nor preclude this short-
term predictability from metamorphosing into long-term
predictability as laboratory and computational precision
increase. In short, sensitive dependence only makes
prediction difficult, not impossible. Turning now to
to the issue of randomness in deterministic systems, many
investigators believe that chaos implies randomness, but
few have elevated their beliefs above the intuitive level.
Nonetheless, a well-defined random behavior actually occurs
in chaotic dynamical systems which exhibit Bernoulli
shifts. [13] For these deterministic systems, a time-sequence
of coarse-grained measurements can yield an output which is
indistinguishable from that of a completely random Bernoulli
process. But unfortunately, these considerations leave open
the question of whether it is coarse-graining or an inherent
orbital randomness which is the source of the erratic behavior in these systems.

    To proceed further, we must acknowledge the possi-
bility that chaos involves natural phenomena so complex
that they make meaningful deterministic description not
merely difficult but impossible. If this notion be accepted,
then chaos is seen to imply limitations on the capabilities
of man at least as significant as those implied by $\hbar$ and c.
We have not far to look for verification, for chaos has,
without our notice and certainly without our permission,
quietly ushered Gödel's theorem and its attendant
limitations into the arena of the physical sciences. To
illustrate and verify this heretical sounding statement,
we first discuss a simple physical model which exposes the
presence of Gödel in a chaotic physical system-- the hard
sphere gas. Immediately thereafter, we outline the development of a new world view which is emerging from our
growing understanding of the limitations imposed by chaos.

    As background, let us begin by defining a universal com-
puter as one which can perform every task that any other
computer performs. Then, coming straight to the point, let
us ask if there is any way to determine a *priori* whether,
fed a program, a universal computer will accomplish its assigned task and halt. Turing's halting theorem, [20] a version
of Gödel's theorem, [21] asserts that, in general, the only way
to determine if the computer halts is to run the program and
see. Here prediction is not merely difficult, it is logically impossible. Let us now translate this mathematical
fact into the physical realm. Everyone knows that laboratory computers are physical systems, but perhaps not so well
known is the fact that there are whole categories of
physical systems capable of being used as universal computers, a point which has been emphasized by Wolfram. [22] In
particular, Fredkin and Toffoli[23] have proven the hard
sphere gas in a suitably shaped container to be a universal

computer. Let us use this model to illustrate the
Turing/Gödel theorems as laws of physics.

The following paragraph is not intended to be rigorous
but is rather meant to give the flavor of the issue.
Suppose our hard sphere gas is prepared in an initial state
(q_o, p_o) and is then permitted to evolve according to its
Newtonian evolution operator N. We now ask if the time
evolving gas ever reaches some arbitrarily specified state
(q_f, p_f). Here, preparing the gas in the initial state
(q_o, p_o) "programs" the hard sphere gas computer and the
action of the evolution operator N causes the computer to
run. If the gas finally reaches the given state (q_f, p_f),
the computer has calculated the desired result and is
permitted to halt. But does knowledge of (q_o, p_o) and N
permit us to assert *a priori* whether or not the state
(q_f, p_f) will be reached? Turing/Gödel assert that, in
general, no such prediction can be made. Here again,
prediction is not merely difficult, it is impossible. To
obtain an intuitive understanding of this conclusion, note
that if an algorithm did exist which could predict *a priori*
that state (q_o, p_o) evolves into (q_f, p_f), then this
algorithm could be run by the hard sphere gas itself. In
turn, this means that the hard sphere gas could, as an
isolated logical system, establish its own self-consistency,
a possibility denied by Gödel's theorem. Finally, recall
that the hard sphere gas is known to be a chaotic system,
and it is precisely the dynamical richness of chaos which
causes the hard sphere gas to be unpredictable at the
logically deep level of Gödel's theorem.

Continuing with the notion of predictability, let us
now briefly discuss in physical terms the light that
computational complexity theory[24] sheds on this topic.
Computational complexity theory concerns itself with the
time required to run computer programs. Of particular
interest to us here is its assertion that the integrable
systems of dynamics[25] belong to a category of systems
for which the clock time t_c required to numerically compute
an orbit is some polynomial function P of program size.
Now for integrable systems, the computational algorithm
and initial conditions are fixed. Thus, if we wish to
compute an orbital segment of appreciable size, the bulk of
the computer program will be taken up simply by specification of the system time t_s at which the calculation is to
end. In the computer itself, of course, this crucial datum
takes $\log_2 t_s$ bits. Therefore, computational complexity
provides us with the result: $t_c \sim P(\log_2 t_s)$. Thus,
integrable systems are quite predictable since a computer
can simulate integrable behavior long before the system
itself exhibits this behavior. On the other hand, chaotic
systems belong to that category of systems for which clock

time t_c required to compute a solution is exponential in
program size. As before, program size is very nearly
$\log_2 t_s$ bits. For chaotic systems, complexity theory then
provides the result: $t_c \sim t_s$, and hence, chaotic systems
are seen to be unpredictable because the future behavior of
a chaotic system cannot be foreseen appreciably before the
system actually exhibits that future behavior.

Wolfram[22] provides an illuminating alternative way to
obtain these complexity theory results. He notes that a
universal computer can always simulate step by step the
behavior of any physical system, but for integrable systems
or the like, there are shortcuts which permit the computer
to predict integrable behavior much before it happens.
Note however that this computer cannot determine the outcome
of its own evolution without following it step by step to
the end. But a computer is also a physical system, thus the
time evolution of any physical system which can serve as a
universal computer is computationally irreducible and can
only be simulated step by step to the end. Here as before,
prediction is impossible; there is no quicker way to know
the future of a system which can serve as a universal
computer than to wait for the future to arrive. It is
currently believed that chaotic systems can serve as
universal computers, and hence, they are unpredictable as
a matter of principle.

Finally let us discuss the issues of predictability,
determinism, and randomness from the viewpoint of algorithmic complexity theory which focuses upon finding the
minimum sized algorithm for computing a desired output.
To fix ideas, note that the term *output* may mean a
digit sequence or an eigenvalue sequence, a Newtonian orbit
segment, the time evolution of a wave function Ψ, or the
like. Perhaps of greatest interest to us here is the
algorithmic complexity theory assertion that, if a desired
computer output cannot be obtained from any algorithm whose
bit length is appreciably less than that of the output, the
output is said to be random. It is informationally incompressible; it is its own simplest representation. Such
output must, of necessity, be so lacking in order and so
unpredictable that randomness is indeed a term which leaps
to mind. But much more than intuition is involved here.
Randomness in the algorithmic complexity sense means that
the output so labelled passes every computable test for
randomness.[7] But precisely because randomness is such an
exceedingly subtle and elusive concept, algorithmic complexity theory encounters Gödel and Turing at every stage.
For example, Martin-Lof[7] has proven that the digit strings
for almost all real numbers are random; however, because of
Gödel, no individual digit string can be proven to be
random. The reason can be found in Chaitin's[26] recently

proved, extremely transparent statement of Gödel's theorem:
A one hundred pound theory can no more prove a two hundred
pound theorem than a one hundred pound pregnant woman can
birth a two hundred pound child. Since the information
contained in any infinite, random digit string truly dwarfs
that contained in the totality of all man's logical systems,
proving randomness is here beyond human capability.

But now, what does all this imply for physics? Recall
from Sec. 1 that determinism means existence-uniqueness and
that sensitive dependence in an orbital calculation means
that, of necessity, information in equals information out.
Under this circumstance, algorithmic complexity theory
assures us that chaotic Newtonian orbits are deterministically random. Moreover, such orbits are unpredictable as a
matter of principle. It is of course quite true, as sensitive dependence arguments suggest, that one can use a CRAY
or a CYBER to obtain an approximate short-term or perhaps
even a long-term orbit segment if computer precision is high
enough, but no matter how useful such calculations may be,
nothing is being predicted. Since one cannot obtain the
orbit as output without providing input data having the same
amount of information, the CRAY or CYBER is here merely a
"language" translator which accepts a message in "Sanskrit"
and outputs the same message in "English." The notions of
translation and prediction should not be confused. One
may summarize these points by noting that a chaotic orbit,
in a quite well-defined sense, is its own briefest description and its own fastest computer. These facts not only
imply the physical limitations already stated, but they
also imply additional limitations, as well as some rather
profound conclusions. It is to these matters we now turn.

Let us begin by seeking the source of randomness in
classical dynamics. In view of the fact that the Newtonian
algorithm for computing a chaotic orbit segment is quite
short, what is the required additional input which drives
the bit length of the minimum program up to nearly the size
of orbit segment itself? To answer, first note that, if the
orbit segment has appreciable length, the bits contributed
to the computer program by the short Newtonian algorithm
are negligible. The only thing then left to consider is the
initial conditions. But now the number of bits which must
go into the initial conditions grows linearly with the bit
length of the orbit segment because of the ubiquitous
exponentially sensitive dependence of final state upon
initial state. Thus, at this point, we have that the bit
length of the initial data must approximately equal the bit
length of the orbit segment. However, that fact alone would
not require our program to have this size, were the digit
strings for real numbers informationally compressible. But
alas, according to the Martin-Lof theorem mentioned earlier,

the digit strings for almost all real numbers are random, and they are, therefore, informationally incompressible. In consequence, it is the randomness contained in the digit strings of real numbers which is the ultimate source of randomness in Newtonian dynamics. Specifically, the Newtonian algorithm translates the randomness of real number digit strings directly into the randomness of chaotic orbits.

But now a major flaw in Newtonian dynamics becomes apparent. While randomness is most assuredly a service provided by the real number continuum, the price it extracts for supplying this randomness to deterministic equations is staggeringly high. To understand this point, recall that the digit strings for almost all numbers in the continuum are not computable by any finite, recursive algorithm. Humanly speaking, most real numbers cannot be computed, measured, or even defined; moreover, these defects are inherited by the deterministically chaotic orbits whose initial conditions these numbers specify. Thus, Newtonian dynamics and its accompanying continuum must now be returned to the gods from whom they were stolen, for it is only the gods who possess the logical and computational skills required to cope with them. Following a route now only dimly seen, theorists of the future must abandon the continuum and Newtonian dynamics, replacing them by a theory which removes randomness from its implicit position in the continuum and elevates it to a computable function perhaps like the quantum mechanical wave function Ψ. Let us now turn our attention to a major theoretical flaw in quantum theory.

For centuries, Newtonian dynamics has been viewed as deterministically predictable, whereas contemporary nonlinear dynamics/chaos finds it to be deterministically random. Equally, quantum mechanics has long been regarded as the first fundamental theory of physics to exhibit an inherent, irreducible, uncomputable probability element, whereas we now recognize that Newtonian dynamics preceded quantum mechanics by centuries in this regard. But of much greater significance, Newtonian dynamics is a fully random theory, for, despite the significance of ordered dynamical motion over the centuries, such motion is structurally unstable and hence rare. In consequence, one is led to the remarkably inverted observation that Newtonian dynamics is a fully random theory whereas quantum dynamics is not. Specifically, the Newtonian time evolution for an isolated chaotic system is most assuredly random; yet the time evolution of the Schrödinger probability amplitude Ψ for an isolated quantum system is found to be completely determinate and nonrandom. Finally, quantum mechanics also utilizes the continuum-- the position operator, for example, has a continuous spectrum-- and, in this regard, it is

subject to the same criticisms as Newtonian dynamics.
Specifically, quantum mechanics also claims that it can, in
principle, measure continuum variable digit strings most of
which cannot even be defined. Quantum mechanics is thus
under siege from below and above-- the microscopic level of
the continuum and the macroscopic level of randomness. At
present, it is not certain how long quantum theory can hold
out, for in truth, it is a theory that is not random enough.
Thus far, our discussion has leaned heavily toward theory,
thus let us now wind down this review by discussing briefly
the effects of chaos on experimental measurement.

Newtonian theory has long assumed that measurements on
a system could be made using instruments which do not
essentially modify the quantity being measured because the
system-instrument interaction is either weak or else predictable. These assumptions were obviously made in ignorance
of chaos. Since chaos is ubiquitous at the macroscopic
level, the measuring instruments as well as the system
itself are chaotic. Even a weak interaction between chaotic
systems provides random effects which grow exponentially.
It now becomes clear that the system-instrument interaction
is neither weak nor predictable. In consequence,
measurement at the classical level always introduces a small
but uncontrollable error into the quantity being measured.
Note here that even single variable measurements are subject
to this uncertainty. Thus, chaos implies that no classical
continuum variable can be measured precisely. If this
theoretical argument can be quantified and verified
experimentally, the consequences for Newtonian theory will
be profound. Moreover, since measurement in quantum theory
is almost always performed with macroscopic classical
devices, quantum theory cannot escape the consequences of
chaos either. Further details may be found elsewhere.[27]

This then concludes our discussion of the new theoretical framework which derives from notions related to chaos.
Perhaps the four problems discussed in the earlier sections
can now be seen for what they are-- natural consequences
flowing out of the underlying framework rather than *vice
versa*. Should this review contain some elements of truth,
then the newly recognized limitations on man imposed by
chaos bespeak a new era at the dawning. Reader's who share
our views are most cordially invited to help chart a course
toward the future.

REFERENCES

1.   See, for example, the introduction to the anthology
     Chaos, Edited by Hao Bai-Lin (World Scientific
     Publishing Co., Singapore, 1984).

2. Einstein, A., Verhandlugen der Deutschen Physikalischen Gesellschaft **10**, 82 (1917).

3. Pomphrey, N., J. Phys. B: Atom. Molec. Phys. **7**, 1909 (1974).

4. See, for example, Bohigas, O., Giannoni, M.-J., and Schmit, C., in *Lecture Notes in Physics, Vol. 263*, Edited by Seligman, T. H., and Nishioka, H. (Springer-Verlag, Berlin, 1986).

5. Berry, M. V., "Quantum Chaology," The Bakerian Lecture 1987, to be published by the Royal Society 1987.

6. Devaney, Robert L., *An Introduction to Chaotic Dynamical Systems* (Benjamin-Cummings, Menlo Park, California, 1986).

7. Ford, J., Physics Today **36**, #4, 40 (1983). More technical discussions appear in Zvonkin, A. K., and Levin, L. A., Usp. Mat. Nauk. **25**, 85 (1970); Martin-Lof, P., J. Information and Control **9**, 602 (1966); and Chaitin, Gregory J., *Algorithmic Information Theory* (Cambridge University Press, Cambridge, 1987).

8. As corroborative support for this notion, we quote the following dictionary entry: **CHAOS**... a state of things in which chance is supreme (Webster's New Collegiate Dictionary, G. & C. Merriam Co., **1963**).

9. Ford, J., in *Chaotic Dynamics and Fractals*, edited by Barnsley, Michael F., and Demko, Stephen G. (Academic Press, New York, 1986).

10. See, for example, Ford, J., in *The New Physics*, Edited by Capelin, S., and Davies, P. C. W. (Cambridge University Press, Cambridge, 1987).

11. Consult Abraham, R., and Marsden, J. E., *Foundations of Mechanics* (Benjamin/Cummings, Reading, 1978).

12. Prigogine, I., *Non-Equilibrium Statistical Mechanics* (Interscience, New York, 1962).

13. While Arnol'd and Sinai published an earlier joint paper on hyperbolic maps, the graphical representation of our Eq. (1) hyperbollic map first wore a cat's face in Arnol'd, V. I., and Avez, A., *Ergodic Problems of Classical Mechanics* (Benjamin, New York, 1968), p. 6.

14. Ristow, G. H., "A Quantum Mechanical Investigation of the Arnol'd Cat Map," Master's Thesis, School of Physics, Georgia Institute of Technology, 1987.

15. The first paper to report quantization of the cat map, using a unitary transformation formalism rather than Hamiltonian, was Hannay, J. H., and Berry, M. V., Physica **1D**, 267 (1980).

16. Eckhardt, B., Ford, J., and Vivaldi, F., Physica **13D**, 339 (1984).

17. Sommerfeld, A., *Optics: Lectures on Theoretical Physics, Vol. IV* (Academic Press, New York, 1954).

18. van Kampen, N. G., in *Proceedings of the 1986 Como Conf. on Quantum Chaos*, Edited by Pike, E. R. (Plenum Press, New York, 1987).

19. A review of selected recent theoretical work appears in Ford, J. , Nature **325**, 19, 1 January 1987.
20. Turing's simple proof of the halting theorem is reproduced in Chaitin, G. J. , Adv. Appl. Math. **8**, #2, 119 (1987).
21. Gödel's theorem is presented and discussed by Davis, M. , *Computability and Unsolvability* (Dover, New York, 1958).
22. Wolfram, S. , Phys. Rev. Lett. **54**, 735 (1985).
23. A proof that the hard sphere gas in a suitably chosen container can serve as a universal computer is derived by Fredkin, E. , and Toffli, T. , Int. J. Theor. Phys. **21**, 219 (1982).
24. Garey, M. R. , and Johnson, D. S. , *Computers and Intractability* (Freeman and Co. , New York, (1979).
25. A rather full presentation of the definition of integrability appears in Thirring, W. , *A Course in Mathematical Physics, Vol. I* (Springer-Verlag, New York, 1978).
26. Chaitin, G. J. , Intl. J. Theor. Phys. **22**, 941 (1982).
27. Ford, J. , in *Directions in Chaos*, Edited by Hao Bai-Lin (World Scientific Publishing Co., Singapore, 1987).

QUANTUM MANIFESTATIONS OF CLASSICAL CHAOS: STATISTICS OF SPECTRA

JORGE V. JOSÉ

Department of Physics
Northeastern University, Boston, Massachussets 02115

ABSTRACT

Recent advances in the understanding of the *Quantum Manifestations of Classical Chaos* are briefly reviewed in these lectures. A direct correlation has been found between the *fluctuation* properties in the quantum eigenvalue spectra and the *integrability* properties of corresponding classical models. In the extreme limits of complete integrability and complete chaoticity, the fluctuations in the eigenenergy spectra of time-independent Hamiltonian operators are *quantitatively* described by Poisson and the Gaussian Orthogonal and Unitary ensembles distributions, GOE and GUE respectively. The GOE applies to time reversal invariant Hamiltonian operators while the GUE applies to the time reversal non invariant case. For time-dependent *periodic* Hamiltonians, similar results are obtained for the quasi-energy spectra of the evolution operator. A continuous transition (as $\hbar$ decreases) between Poisson to circular orthogonal ensemble statistics is found to be correlated with a smooth transition from localized to extended state properties of the quasi-energy eigenvectors, respectively. Possible experimental applications of the results obtained so far to different experimental systems are also discussed.

1. INTRODUCTION

Quantum Mechanics is unquestionably one of the most successful physical theories in existence. The information content of the Heisenberg or Schrödinger equations has led to innumerable explanations of existing phenomena and the prediction of new ones. Although plagued from its inception with apparent paradoxes, one can say with some confidence that, as of now, no serious shortcomings nor viable alternatives to the description of the microscopic world have been successfully implemented.

However, quantum mechanics (QM) is not an independent theory, since in order to formulate its problems one needs the corresponding version of the problem in classical mechanics (CM). Often, to define a QM problem one uses the *"quantized"* version of the classical problem. This quantization procedure is not unique, however, and has led to innumerable studies of equivalent quantum problems generated by different classical Hamiltonians. One of the important heuristic pillars of the early formulation of QM was Bohr's correspondence principle, which states that a quantum mechanical system should approach the behavior of its classical counterpart in the limit of vanishing Plank's constant $\hbar$. It is now well known that this simple prescription has problems since, in many cases, the quantum mechanical quantities show non-analytic behavior as $\hbar \to 0$.

In principle, one should be able to derive all possible classical solutions from the more basic solutions of QM. This requirement is fulfilled in most of the problems considered by the forefathers of QM since their systems were classically *integrable* . In the semiclassical limit (SCL) one can obtain the quantum spectra from the classical solutions using the Einstein-Brillouin-Keller (EBK) semiclassical quantization conditions for n-dimensional systems given by the n conditions,[1]

$$\oint_{C_i} \vec{p} \cdot d\vec{q} = 2\pi\hbar(N_i + \alpha_i/2)$$

with α_i the Keller-Maslov index, N_i an integer and $i = 1, 2, 3...n$. To carry out these integrals we need to know the momenta p_i as a function of the q_i's, which implies being able to integrate the classical equations of motion. The contours C_i are defined on the surface of the $n-$dimensional invariant torus. In recent years, a large amount of literature has been devoted to the study of CM systems for which complete integration of the equations of motion is not possible[2]. It is now believed that the set of Hamiltonians that are integrable is of measure zero, *i.e.:* most Hamiltonian systems are *not integrable.* For the class of Hamiltonians which are non-integrable, it is found that although Newton's equations of motion are deterministic, the solutions show very sensitive dependence on initial conditions. This dependence on initial conditions leads to an ensemble of solutions that behave as essentially random or *chaotic.* There are, in fact, dynamical models that by varying a parameter can show a whole spectrum of possible classical solutions that go from being completely integrable to completely chaotic. [2]

Given that there is a large set of classical Hamiltonian systems that are not completely integrable *i.e.:* that present erratic or chaotic behavior, the question immediately arises as to the properties of their quantum mechanical counterparts. It would be nice if we could give simple quantization condition expressions like the one given above, but it is precisely the nonintegrability that makes this an apparent impossibility.

There have been many attempts to answer this question, and from different points of view[3]. The question of being able to find *real* chaotic behavior that is purely random behavior in quantum mechanics, over and above that of the probabilistic interpretation of the wave function has, as of now, been answered in the negative. One often finds in the literature the misnomer *Quantum Chaos* used freely without giving a precise meaning to this term as it has in CM. Of course, if one believes in Bohr's correspondence principle we should be able to see signatures of the classical erratic behavior in the semiclassical limit. However, even fundamental considerations would seem to indicate that there are inherent limitations in QM that preclude this from happening. Specifically, in CM for a chaotic Hamiltonian if we take two initial conditions infinitesimally close to each other after a certain time they will be exponentially apart. In QM there is a physical limitation as to how close two trajectories can be given by Heisenberg's uncertainty principle. So we expect that for a certain time the quantum solution will mimic the classical one, but for sufficiently long times the graininess imposed by the uncertainty principle will limit the similarities between the quantum mechanical and classical solutions[4]

An alternative question one may ask is, if we know that a given classical Hamiltonian is nonintegrable what are the physical properties of its quantum counterpart. Equivalently, what are the *quantum manifestations of classical chaos* $(CMCC)$. This question does not attempt to force the existence of chaos in a quantum system, just to see if there are clear quantum differences between corresponding classical integrable and non-integrable Hamiltonians. Significant amount of work has been done recently trying to answer this question, mostly insofar as the statistical properties of the eigenvalue spectra is concerned both in *time-independent* (TI) and *time-dependent* (TD) Hamiltonians. [5]-[26] In these lectures I will attempt to give a brief summary of some of the main results obtained to date.

The organization of these notes is as follows: We begin by recalling some of the questions one wishes to answer in QM. These are related to the spectral properties of the Hamiltonian for time-independent (TI) problems and the evaluation of the time evolution operator U when the Hamiltonian is time dependent (TD). A particular, but very important class of TD problems has an H that is periodic in time. In this case U is also periodic in time and one can use Floquet's theorem in time to calculate the spectral properties of U within one period, and thus the calculation of the wave function $|\Psi(\vec{X}, t) >$ follows. The spectra of U is known as the *quasi-energy-spectra* (QES). It is important to have in mind specific experimental systems against which the theoretical results could be tested. In section 2., the problems of highly excited hydrogen atoms in strong *static* magnetic or a.c. fields are taken as typical representatives for the TI and TD quantum problems. Since the SCL entails a large number of eigenvalues, it was thought reasonable to use an approach akin to Wigner's **random matrix theory** (RMT) to analyze the statistics of the spectra in this regime. In section 3. we give a brief review of the main ideas and results of RMT, which provides a natural quantitative explanation of the spectral properties obtained from calculations in models corresponding to the extreme classically integrable and chaotic cases. In section 4. we proceed to discuss some of the results obtained from explicit solutions to quantum problems which are known to be either completely integrable or completely chaotic in the TI case. Further analyses of Hamiltonian models that can go from the integrable to nonintegrable regimes by the variation of a parameter lead to spectral properties that can not be easily fitted by the known RMT results. Next in section 5. we go on to discuss the results associated with TD models. We emphasize the statistical properties of the QES associated with TD models with periodic Hamiltonians. Specifically, the study of the Fermi-Ulam acceleration model is discussed in some detail. Finally, we conclude with a critical assessment of the results presented.

2. THE QUANTUM QUESTIONS

2.1 Theory

Given a Hamiltonian operator H, the main goal of QM is to find the wave function $|\Psi(\vec{X}, t) >$, obtained from solving the Schrödinger equation

$$i\hbar \frac{\partial |\Psi(\vec{X}, t) >}{\partial t} = H |\Psi(\vec{X}, t) >, \tag{2.1}$$

plus boundary conditions, from which all observables can in principle be calculated. The Hamiltonian operator has the important property of being hermitian, $H^\dagger = H$ and it is obtained from quantizing the corresponding classical Hamiltonian $h = h(p, q)$. If $h(p, q)$ is integrable, the EBK algorithm permits us to obtain, in principle, the quantum spectra in the

SCL. However, when $h(p, q)$ is not integrable there is no known method to find the corresponding quantum solutions, and in fact such a method may not even exist.

Of importance in QM is the fact that conservative and nonconservative Hamiltonian problems are treated separately. Thus in the TI $(H \neq H(t))$, case the goal is to find the spectral properties of H by solving the equation

$$H\phi_n = E_n \phi_n, \tag{2.2}$$

with the appropriate boundary conditions. The time dependence of the wave function for stationary states can be written at once,

$$|\Psi(\vec{X}, t) >= \sum_n exp(-iE_n t) \phi_n(\vec{X}). \tag{2.3}$$

For bounded Hamiltonian systems the spectrum is discrete; therefore the dependence of $|\Psi(\vec{X}, t) >$ in time is always almost-periodic. This fact, which is directly associated with the first order time derivative of the Schrödinger equation, shows that $|\Psi(\vec{X}, t) >$ can not show sensitive dependence on initial conditions as seen in the dynamic flows of classical problems. Of course, what are measurable are expectation values of observable operators. It is hard to imagine how these averages can show the sensitivity properties of the solutions in the classical limit when $|\Psi(\vec{X}, t) >$ is itself well behaved. Therefore, in this case the only place where we can search for $QMCC$ is in the spectral properties of Hamiltonian operators that have integrable and nonintegrable classical Hamiltonians.

For nonstationary problems, the Hamiltonian depends explicitly on time. The central object of interest in this case is the time evolution operator $U(t, t_0)$ that satisfies the operator equation,

$$i\hbar \frac{\partial U(t, t_0)}{\partial t} = H U(t, t_0), \tag{2.4}$$

with boundary condition $U(t_0, t_0) = \hat{I}$, with $\hat{I}$ the unit matrix. For arbitrary H the evaluation of $U(t, t_0)$ can be done perturbatively, provided there is a small

parameter in the problem. Since we are interested in problems where the interactions can be strong, the perturbative approach can not be used. In this case
the evaluation of $U(t, t_0)$ can be highly nontrivial. An important simplification
arises for Hamiltonians periodic in time, $H(t + \tau) = H(t)$. In this case the time
evolution operator satisfies the properties,

$$U(t + \tau, t_0 + \tau) = U(t, t_0) \rightarrow U(t + N\tau, t_0) = U^N(t + \tau, t_0). \qquad (2.5)$$

Therefore, the problem of finding $|\Psi(\vec{X}, t) >$ when H is periodic in time is reduced
to that of finding the time evolution operator within one period. Moreover, since
$U(t, t_0)$ itself is periodic, Floquet's theorem applies, and thus we can represent
$U(t, t_0)$ in terms of its spectral decomposition,

$$U(t, t_0) = \sum_n |\lambda_n > < \lambda_n| e^{-i\lambda_n(t-t_0)}. \qquad (2.6)$$

Here, $|\lambda_n >$ and $e^{-i\lambda_n}$ are the eigenvectors and eigenvalues of $U(t, t_0)$, respec-
tively. It is more convenient to study the QES defined by the eigenvectors of
$U(t, t_0)$ and the $\lambda_n \epsilon[0, 2\pi]$. Once the QES is known, the wave function can be
obtained from,

$$|\Psi(\vec{X}, t) > = \sum_n |\lambda_n > < \lambda_n |\Psi(\vec{X}, t_0) > e^{-i\lambda_n N}, \qquad (2.7)$$

with $t - t_0 = N\tau$, and $|\Psi(\vec{X}, t_0) >$ is the initial condition in time. The sum over
n includes the possible continuum contribution to the QES.

One has now reduced the TI and the TD problems to studies of spectra.
There are, however, important differences in both cases. For a bounded Hamiltonian in the TI case, the eigenenergies are necessarily discrete and can take values
over the entire negative axis of the real line. In so far as the time dependence
of $|\Psi(\vec{X}, t) >$ is concerned, as mentioned above, it will always be quasi-periodic.
By contrast, the QES eigenvalues are restricted to the interval $[0, 2\pi]$, and since

$U(t, t_0)$ is an infinite-dimensional operator, the QES can be discrete, continuous or singular continuous. In fact, one of the main open questions in the TD case is to try to determine conclusively the nature of the QES. It is clear that, as in the TI case, if the QES is discrete, the evolution of the system will be quasi-periodic. But if the QES has at least a continuous component, the evolution of $|\Psi(\vec{X}, t) >$ will not be recurrent ant perhaps an unpredictable time dependence may arise. A definite answer to this question awaits the discovery of a completely and exactly soluble non-trivial time dependent problem analytically. This is so since most direct evaluations of the QES to date involve numerical calculations which take $U(t, t_0)$ as a finite dimensional matrix. Nevertheless, significant findings have emerged from the study of specific time dependent models which will be discussed in section 6.

2.2 EXPERIMENT

It is clear that the ultimate judge for the validity of the answers to the fundamental questions posed above, is the *experiment*. Thus, it is important to have specific experimental systems where the ideas can be tested and where inspiration can be found. Although one can easily think of many possible examples, some of which will be discussed elsewhere, there are two sets of experiments that can specifically be thought of to define the proper paradigms in this field of research. Both experimental systems involve the hydrogen atom in highly nonlinear regions of parameter values. For the TI case one considers a hydrogen atom in the presence of a *constant* magnetic field B with Hamiltonian,[13)−14)]

$$\mathcal{H} = \frac{P^2}{2} + \frac{-1}{r} + \frac{\gamma}{2} L_Z + \frac{1}{8}\gamma^2[X^2 + Y^2], \qquad (2.8)$$

with L_Z the angular momentum along the Z axis, $r = \sqrt{X^2 + Y^2 + Z^2}$, and $\gamma = \frac{B}{B_c} = 0.42 \times 10^{-5} B \ (Tesla)$. Defining $\eta = \frac{E}{\gamma}$ we can see that if $|\eta| \gg 1$, the Hamiltonian corresponds to the perturbed Coulomb problem that has Rydberg states as solutions. Classically, this regime is integrable and the quantum solutions can be obtained perturbatively. When $\eta \ll 1$ the Coulomb potential is a perturbation on the Landau levels solution and, again, the classical problem is integrable. The interesting region arises when $|\eta| \approx 0$. In this case the classical solutions are found to be **chaotic**. There have been several theoretical studies of this problem that we will mention in the next section.

For the TD case one considers a hydrogen atom with the electron excited to large principal quantum numbers say between $32 \leq n \leq 80$ and in the presence of a time dependent electromagnetic field. The Hamiltonian in atomic units can be taken to be,

$$\mathcal{H} = \frac{P^2}{2} + \frac{-1}{r} + FZ \times cos(\omega t). \tag{2.9}$$

In the 1970's, Bayfield and Koch carried out these type of experiments for the first time and found very high ionization probabilities for $\omega \ll \omega_{resonance}$ as the strength of the field was increased[28]. This strong sensitivity of the ionization probability to the strength of the field is rather puzzling. A semi-quantitative explanation was then proposed by Leopold and Percival from a classical calculation[29]. Recently, more detailed experiments have been performed not only of ionization but also of transition probabilities in the same regimes of parameters[30]-[32]. The calculations of ionization probabilities, *i.e.:* the coupling of the discrete states to the continuum for these highly excited atoms is in several aspects nontrivial theoretically, the difficulties increasing with the dimensionality of the Hamiltonian. Thus experimentalists have tried to produce low-dimensional hydrogen atoms configurations, *i.e.:* where the charge spatial distribution is strongly anisotropic. In particular, by applying a constant electric field to the Rydberg atoms, parallel to the a.c. field, they can mimic the one-dimensional theoretical calculations. In this case the Hamiltonian can be approximated by,[33]-[34]

$$\mathcal{H} = \frac{P^2}{2} + \frac{-1}{r} + FZ \times cos(\omega t) + F_1 Z \approx \frac{P_Z^2}{2} + \frac{-1}{Z} + FZ \times cos(\omega t). \tag{2.10}$$

This "quasi-one-dimensional" hydrogen atom model has been studied extensively theoretically both classically[34] and quantum mechanically. [33]–[36] Some good correspondence between the classical and quantum calculations and the experimental results has been found but the understanding of the physics associated with this problem is not completely understood.

3. RANDOM MATRIX THEORY (RMT)

In the early days of nuclear physics of heavy nuclei, people were confounded by the very large number of resonances obtained from slow neutron capture experiments and the fact that even if an appropriate Hamiltonian to study the problem was known, the prospect of calculating 10^8 levels seemed completely out of the question. Wigner suggested a *statistical ensemble* approach to the nuclear problem. His premise was that since the number of levels can be relatively large, even if one could know with precision where each level sits, the actual measured information could not be easily correlated with a set of precisely calculated resonances. Thus the problem of analyzing the properties of a large set of eigenenergies may be better accomplished by studying their statistical properties[37]. The idea is that of developing a *statistical ensemble theory* in a similar spirit as the Gibbs ensemble theory of classical statistical mechanics. It helps to think of the energy levels as positions of a gas of particles moving along a one-dimensional space for a given set of quantum numbers. There are different questions one can ask to ascertain how much the position of the energy levels are correlated. Of importance in QM is to know if there are *degeneracies* in the spectrum. To answer the degeneracy question, one can look at the local properties of the spectrum given by the frequency function $P(S)$ defined in the previous section. It is also important to find out more detailed properties of the spectrum, like longer range correlation properties among several energy levels. One may ask, given an arbitrary energy interval of length L, what is the probability, $R(n)$, of finding n levels. The probability $R(n)$ can be characterized by its moments. For example, the average number of levels on an interval of length L is given by

$$\overline{n(L)} = \frac{1}{L} \sum_\alpha n(\alpha, L), \tag{3.1}$$

where the bar stands for spectral average, and $n(\alpha, L)$ is the number of levels in an interval of length L starting at α and ending at $\alpha + L$. Furthermore, it is important to know how sharp the mean value $\overline{n}$ is. Thus it is necessary to calculate the variance,

$$\overline{\Sigma^2(L)} = \overline{(n(\alpha, L) - \overline{n(\alpha, L)})^2}. \tag{3.2}$$

If $R(n)$ were to be Gaussian, the mean value and the variance would be enough. However, one can not a priori assume that this is the case, and in practice this seems to be so, and we need to consider higher moments to get a better characterization of $R(n)$:

$$\gamma_1(L) = \frac{\overline{(n(\alpha, L) - \overline{n(\alpha, L)})^3}}{\Sigma^3(L)}, \tag{3.3}$$

known as the skewness of $R(L)$ and

$$\gamma_2(L) = \frac{\overline{(n(\alpha, L) - \overline{n(\alpha, L)})^4}}{\Sigma^4(L)} - 3, \tag{3.4}$$

the kurtosis or excess statistics. All the statistical measures given above go from local to intermediate range for finite L. One should also consider the density of states, $\rho(E)$, which gives a *global* property of the spectrum, and is of significant physical interest.

Given a set of eigenvalues obtained from experiments or from calculations with specific Hamiltonians, one would just need to organize the data in an appropriate way and try to find correlations in the results. It turns out that this is not enough. One needs to have some theoretical guidance to be able to give a cohesive analysis of the data and predict possible specific behaviors for the statistical moments.

Given the fact that in nuclear physics the appropriate Hamiltonian is not known, Wigner suggested to study ensembles of Hamiltonians rather than just one Hamiltonian. The ensembles are defined in a matrix space where all the Hamiltonian members of the ensemble have the same symmetry properties. Take for example an $N \times N$ rotationally invariant Hamiltonian with matrix elements distributed randomly and independently. Each member of the ensemble corresponds to a given realization of a real symmetric matrix with random and independent $\frac{1}{2}N(N+2)$ matrix elements. The spectrum of each matrix in the ensemble is invariant under an orthogonal transformation. The crucial question is what is the specific probability distribution for a given ensemble of H's. Wigner suggested that to construct the ensembles, the only essential input should be the symmetry properties of the Hamiltonians *i.e.:* rotational and translational invariance, or time reversal or non-time reversal invariances.

There are different ways of deriving the specific form for the probability laws for a given ensemble. For pedagogical reasons we follow the line of thought used by Balian[38] based on the *maximum information entropy principle*. From Shannon's information theory, the information content in the probability law $P(H)$ is,

$$I[P(H)] = -\int d\mu(H)\, P(H)\, \ell n\, P(H), \qquad (3.5)$$

where $d\mu(H)$ is the Haar measure of the given matrix space. The specific form of $d\mu(H)$ is of course essential to this discussion. We are interested in minimizing the information content of $P(H)$ or, equivalently, in maximizing the entropy or randomness contained in $P(H)$. The minimization of $I[P(H)]$ is carried out using the variational method that incorporates the constraints imposed on $P(H)$ via the Lagrange method of undetermined parameters. Generally, the constraints can be written as,

$$\int d\mu(H)\, P(H)\mathcal{F}_i = C_i. \qquad (3.6)$$

From the extremum condition $\delta I[P(H)] = 0$, it follows

$$\int d\mu(H) \delta P(H)[\ell n P(H) + 1 - \sum_i \lambda_i \mathcal{F}_i] = 0, \qquad (3.7)$$

where the λ_i are the Lagrange parameters. The general result for $P(H)$ is then,

$$P(H) = e^{-(\sum_i \lambda_i \mathcal{F}_i - 1)}. \qquad (3.8)$$

For this formal result to be useful, one needs to give explicitly the Haar measure for the given matrix space of interest and the constraints imposed on $P(H)$. A constraint on any $P(H)$, so that it represents a well behaved probability law, is the normalization condition,

$$\int d\mu(H) \, P(H) = 1. \qquad (3.9)$$

We can now consider the possible forms for $P(H)$ corresponding to specific symmetry constraints. We start with Hamiltonians that are invariant under orthogonal transformations; this means problems with rotational and time reversal invariance. In this case H is real and the Haar measure is given as,

$$d\mu(H) = 2^{\frac{N(N+1)}{4}} \prod_{1 \le i \le N} dH_{i,i} \prod_{1 \le i < j \le N} dH_{i,j}. \qquad (3.10)$$

The only extra constraint needed to find $P(H)$ is that the strength of the correlations of the matrix elements of H be finite

$$\int d\mu(H) Tr(H^2) \, P(H) = C, \qquad (3.11)$$

with C a finite constant, and thus one arrives at the result,

$$P_N^{GOE}(H) = A_N^{GOE} \, e^{\frac{-Tr(H^2)}{4\sigma^2}}. \qquad (3.12)$$

Here the name GOE has a clear meaning from the explicit expression for the probability law in this case. The constant A_N^{GOE} is just an N dependent constant and the variance σ^2 is obtained from

$$< H_{i,j}^2 >= (1 + \delta_{i,j})\sigma^2 \, and \quad < H_{i,j} >= 0$$

Here the bracket $< \, >$ denotes the average with respect to $P_N^{GOE}(H)$. If time reversal invariance is not satisfied, the Hamiltonian is complex Hermitian and invariant under unitary transformations. The Haar measure is then given by,

$$d\mu(H) = 2^{\frac{N(N+1)}{2}} \prod_{i \, \leq j} dH_{i,i}^{(0)} \prod_{i<j} dH_{i,j}^{(1)}, \tag{3.13}$$

where $H_{i,j} = H_{i,j}^{(0)} + \sqrt{-1}\, H_{i,j}^{(1)}$ with $H_{i,j}^{(0)} = H_{j,i}^{(0)}$ real and $H_{i,j}^{(1)} = -H_{j,i}^{(1)}$. With a constraint equivalent to Eq(4.11)

$$\int d\mu(H) Tr(H^2)\, P(H) = C_1. \tag{3.14}$$

The probability law

$$P_N^{GUE}(H) = A_N^{GUE}\, e^{\frac{-Tr(H^2)}{4\sigma^2}}. \tag{3.15}$$

Here GUE stands for *Gaussian Unitary Ensemble*. There is another ensemble that can be considered where the Hamiltonian is a real quaternion. Since there are no examples with a classically chaotic limit with this symmetry, it will not be discussed here. Using the invariance properties of the probability laws,

$$P(H)d\mu(H) = P(H')d\mu(H'), \tag{3.16}$$

where $H' = A\dagger H\, A$ with A either an orthogonal matrix (GOE) or a unitary matrix (GUE), one can obtain an eigenvalue representation of the probability laws resulting in,

$$P_N^\beta(E_1, E_2, ..., E_N) = C_N^\beta\, e^{-\frac{\sum_i E_i^2}{4\sigma^2}} \prod_{i \leq j} |E_i - E_j|. \tag{3.17}$$

Here $\beta = 1, 2$, correspond to the GOE, and GUE cases, respectively. The function $P_N^\beta(E_1, E_2, ..., E_N)dE_1 dE_2...dE_N$ gives the probability to find the levels $\{E_1, E_2, ..., E_N\}$ in the energy intervals $[E_i, E_+dE_i]$. A crucial property of these probability functions is that they are zero for any two levels that would be equal to each other. This *repulsion* of energy levels is one of the most significant properties of these distribution functions. Given that one can calculate explicitly the $P_N^{GOE}(E_1, E_2, ..., E_N)$ one can then calculate some of the correlation functions we defined above. We can begin by calculating the density of states from $P_N^{GOE}(E_1, E_2, ..., E_N)$,

$$\rho(E) = \int dE_1 dE_2 ... dE_{N-1} \, P_N^{GOE}(E, E_1, E_2, ..., E_{N-1})$$

$$= \frac{\sqrt{4N\sigma^2 - E^2}}{2\pi N\sigma^2}, \tag{3.18}$$

when $|E| \leq 2\sqrt{N\sigma^2}$ and zero when $|E| > 2\sqrt{N\sigma^2}$. This is Wigner's *semicircle law* and gives a global property of the distribution of levels for the *GOE*.

So far we have considered the case of Hamiltonian operators that have their eigenvalue spectra defined in the real line. This is then appropriate to describe time-independent Schrödinger equation problems. For the time dependent problems, we need to consider ensembles of unitary matrices. This case, although for different reasons, was very thoroughly considered by Dyson[39]. We see at once that since the eigenvalues of $\hat{U}$ are constrained to lie in the unit circle, *i.e.*: $\lambda_i \epsilon [0, 2\pi]$, the only constraint on the distribution is to be normalized. It follows immediately from Eq(4.5), given that the Haar measure is properly defined, that $P_N^\beta(\hat{U})$ is a uniform probability density, which means that all members of the ensemble are equally probable. This is reminiscent of the microcanonical ensemble of classical statistical mechanics. For $\beta = 1$ we have the *circular orthogonal ensemble (COE)*, that corresponds to symmetric unitary matrices, invariant under unitary transformations and representing time reversal invariant systems. For systems without time reversal invariance $\beta = 2$, and one has the *circular unitary ensemble (CUE)* for which $\hat{U}$ is not symmetrical but is invariant under arbitrary unitary transformations, which are not necessarily self-dual.

The probability density for the eigenvalues is,

$$R_N^\beta(\lambda_1, \lambda_2, ..., \lambda_N) = A_N^\beta \prod_{1 \leq i < j \leq N} |e^{i\lambda_i} - e^{i\lambda_j}|^\beta. \qquad (3.19)$$

Again we see that the probability of degeneracy of the eigenvalues is zero, as in the Gaussian ensembles. In fact, Dyson introduced the circular ensembles which are better defined mathematically than the Gaussian ensembles, with the idea that the results of both types of ensembles should be the same in the asymptotic limit of a small energy interval with a large number of levels in the Hamiltonian problem. Later on, Dyson and Mehta[40] made this assumption rigorous by specifying the sense and the limits in which the *fluctuations*, (to be defined more specifically in the next section), in both types of ensembles are asymptotically the same.

As already appreciated from the discussion given above the grate advantage of the Gaussian ensembles is that one can explicitly calculate the averages and correlation functions defined through equations (4.1)-(4.4). Specifically it is found that for the local properties,

$$P^{GOE}(S) \approx \frac{\pi}{2} S e^{-\frac{\pi}{4}S^2}. \qquad (3.20)$$

which in the limit $S \to 0$ goes like $\approx \frac{\pi}{2}S$ *i.e.:* energy level repulsion. For the *GUE* case,

$$P^{GUE}(S) \approx \frac{32}{\pi^2} S^2 e^{-\frac{4}{\pi}S^2}, \qquad (3.21)$$

and the $S \to 0$ limit of $P(S)$ goes like $\approx \frac{\pi^2}{3}S^2$, stronger energy level repulsion than in the *GOE* case. To these results We can add those for the purely random separation of eigenvalues,

$$P(S) = e^{-S}, \qquad (3.22)$$

known as the Poisson distribution, plus the distribution for a picket fence or

Harmonic oscillator problem,

$$P_{h.o.}(S) \approx N\delta(S - \frac{2\pi}{N}),\qquad(3.23)$$

With regard to the correlation functions, one gets,

$$\overline{\Sigma^2_{GOE}(L)} = \frac{2}{\pi^2}\ell n L + 0.44,\qquad(3.24)$$

$$\overline{\Sigma^2_{GUE}(L)} = \frac{1}{\pi^2}\ell n L + 0.565,\qquad(3.25)$$

$$\overline{\Sigma^2_{Poisson}(L)} = L,\qquad(3.26)$$

$$\overline{\Sigma^2_{h.o.}(L)} = 0.\qquad(3.27)$$

For the higher order γ correlation functions, only the Poisson case is known to give,

$$\gamma^{1,2}_{Poisson} = \frac{1}{\sqrt{L}},$$

while the harmonic oscillator case gives zero.

From these specific analytic results one arrives at the conclusions that the distribution of energy levels in the Gaussian ensembles, and as a consequence for N large in the circular ensembles, the spectra have the local property of energy level repulsions as well as very specific rigidity properties for the correlation among energy levels. This should be contrasted with the random distribution of levels represented by the Poisson law where there is energy level clustering as well as a correlation between levels that decreases with their separation distance. These very specific properties will be tested against specific set of experimental data and specific numerical calculations on systems that are known to be integrable or nonintegrable.

A test of the predictions of $\tilde{R}MT$ with the resonance spectra measured in different nuclei was attempted for a number of years, but without a conclusive result. To arrive at a conclusive positive comparison between data and theory, one needs a large number of resonances to perform a reliable statistical analysis. It took a number of years for the necessary data to accumulate and it was not until 1982 that Haq et.al., combining the resonances from different nuclei, were able to carry out a thorough statistical analysis of the data. They arrived at the conclusion that the *GOE describes the fluctuations found experimentally.*

It should be stressed that it is the fluctuations in the GOE that correlate with the experimental data. The global behavior for the density of states of the highly excited levels found experimentally has an exponential behavior in E rather than the semicircle law result given in Eq(4.18). In general, the form of $\rho(E)$ depends on the specific physical system under consideration. In order to consider the statistical spectral properties of different systems one should subtract the average behavior of $\rho(E)$ and study only its fluctuations. For the analysis of the experimental as well as the theoretical data, it is more convenient to study the cumulative density of levels $N(E)$ obtained from $\rho(E) = \frac{dN}{dE}$. Usually one can separate N into two contributions,

$$N(E) = N(E)_{av} + \delta N(E), \tag{3.28}$$

where $N(E)_{av}$ is the average behavior of $N(E)$ and $\delta N(E)$ are the fluctuations around the mean value of $N(E)$. The $N(E)_{av}$ is obtained from the best fit to the experimental values for $N(E)$. To be able to compare the fluctuations of the energy spectra for different systems, it is convenient to eliminate the average behavior of $N(E)$, a procedure known as *unfolding* of the spectra. For every energy eigenvalue E_i we carry out the map $E_i \rightarrow x_i$, where $x_i = N(E)_{av}$. Thus we can define the unfolded $\tilde{N}(x)$ as $\tilde{N}_{av}(x(E)) = N_{av}(E)$, given that the average number of levels remains constant under the map. Since $\tilde{N}_{av}(x) = x$, the unfolded average spectrum is a straight line with slope of 45^0, and the density of states is $\rho(x) = 1$. The unfolding of the spectrum is crucial to be able to compare the fluctuation properties of different systems. In the TI case, since one is rarely able to calculate the whole spectrum, the unfolding allows a reliable study of a subset of the spectrum with an average distance between levels of one.

Once the spectrum has been unfolded, one can study the fluctuations about the mean value by using another statistical test introduced by Dyson and Mehta[41] known as the Δ_3 statistics. This is defined by,

$$\Delta_3(L, \alpha) = \frac{1}{L} \min_{A,B} \int\limits_{\alpha}^{\alpha+L} dx[\tilde{N}(x) - Ax - B]^2. \tag{3.29}$$

This is just the least mean square deviation of $\tilde{N}(x)$ from the straight line mean behavior. This statistic is directly proportional to the $\overline{\Sigma^2}$, and thus can be calculated for the Gaussian ensembles as well. The specific theoretical predictions for the averaged $\overline{\Delta_3(L)} = \frac{1}{L}\sum_\alpha \Delta_3(L, \alpha)$, are

$$\overline{\Delta_{3(GOE)}(L)} = \frac{1}{\pi^2}\ell n L - 0.007, \tag{3.30}$$

$$\overline{\Delta_{3(GUE)}(L)} = \frac{1}{2\pi^2}\ell n L + 0.058, \tag{3.31}$$

$$\overline{\Delta_{3(Poisson)}(L)} = \frac{L}{15}, \tag{3.32}$$

$$\overline{\Delta_{3(h.o.)}(L)} = \frac{1}{12}. \tag{3.33}$$

The results given in Eq(4.30) and (4.31) are correct in the asymptotic limit valid for $15 \leq L$.

Notice that in contrast to the TI case, in the TD case the quasi-energy spectrum is constrained to lie between 0 and 2π, and one does not need to unfold the spectrum, since the level density is uniform. Thus, one may surmise that the asymptotic predictions resulting from the Gaussian ensembles may have a bearing on the statistical properties of the QES as well.

It is against these predictions from RMT that we shall compare the results obtained from specific calculations on Hamiltonians that are known to be integrable or nonintegrable in the classical limit.

4. TIME-INDEPENDENT UNIVERSALITY CLASSES

Percival [1] pointed out in 1973 that in the SCL for TI potentials, one should expect that the nature of the spectra of classically integrable (CIS) and classically nonintegrable (CNIS) systems should be different[42]. For integrable systems the quantum states are labeled by n action integrals and, using the correspondence principle, the classical frequencies must correspond to differences in the quantum energies. However, since there are many more classical frequencies that appear randomly distributed in the nonintegrable case , one may surmise that in the SCL the distribution of energy eigenvalues may appear random. He did not specify quantitatively in which way the two types of spectra should be different but called the spectra for CIS *regular* and that for $CNIS$ *irregular*.

Given the fact that the number of eigenstates in the SCL is very large, the motivation, as in nuclear physics, to use statistical methods to analyse the spectrum is clearly justified. What is, of course, completely unexpected is that any of the specific results from the different ensembles discussed in the previous section should have any relevance to the spectral properties of systems which in the classical limit are chaotic. Saslavsky *et. al.* (1974) and Saslavsky (1977) suggested the study of the distribution of energy level separations[43]. Berry and Tabor (1977) provided the first quantitative measure of the local properties of the spectrum in the generic *regular* spectrum[5]. They considered the distribution, $P(S)$, of having nearest neighbor eigenvalues separated by a distance $S = E_{i+1} - E_i$ in the limit where $S \to 0$. By using the EBK torus quantization method they found that in the SCL, where the number of levels increases significantly as $\hbar \to 0$,

$$P(S) = e^{-S}, \tag{3.22}$$

which is characteristic of a Poisson process or random distribution of separations of nearest neighbor energy levels. The harmonic oscillator with incommensurate frequencies was found to be nongeneric, however, with a $P(S) \approx n\delta(S - \frac{2\pi}{n})$ with n the total number of levels.

To find out what the spectral characteristics are for the nonintegrable case, one needs to study specific models both in the TI and TD cases. Below we give a brief account of the different models studied so far. The main conclusion from these studies is that one can clearly see differences in the spectral properties associated to the specific integrability properties of the corresponding classical Hamiltonian. Thus we are led to define *Universality classes* in the spirit, although not yet in the theoretical content, of the theory of critical phenomena.

4.1 Quantum billiards

Since tori no longer exist in the chaotic regions, one can not use the EBK method of quantization to find the quantum solutions in the SCL. Given that for a general classical Hamiltonian the corresponding phase space will have coexisting tori with chaotic regions, it is *a priori* not straightforward to separate the properties of quantum problems with classically chaotic or no-chaotic Hamiltonians. There are very few systems where chaos has been proved to exist rigorously. The classical examples are chaotic billiards like the *Bunimovich* contour or *stadium* as well as the Sinai billiard. Billiards are two-dimensional enclosures that constrain the motion of a free particle.[3]. The particle has elastic collisions with the walls such that depending on the geometry of the domains, the dynamic flows will be either quasi-periodic or chaotic. Examples of integrable billiards are a rectangular box with commensurate or incommensurate sides, a circle, and the ellipse. In these cases, apart from the energy, there is another constant of the motion and the dynamical flows are constrained to lie on the surface of a torus.

For the nonintegrable billiards, however, the only constant of the motion is the energy. The stadium consists of two semicircles separated by straight lines of length δ. For $\delta = 0$, the circle case, the problem is integrable. However, for arbitrary value of $\delta \neq 0$ the flows are mixing. The Sinai billiard is equivalent to the Lorentz model. This consists of a periodic array of circles in a two-dimensional plane. A free particle with a given initial speed and orientation moves in the the lattice of obstacles. Because of the translational invariance of the model, one can replace it by a square with a circular void in the center of radius R. For $R = 0$ the problem is completely integrable, however, for any $R \neq 0$ the dynamic flows are mixing thus chaotic.

A *Quantum Billiard* (QB) problem is defined by the corresponding TI Schrödinger equation

$$\frac{\partial^2 |\Psi(X,Y)>}{\partial X^2} + \frac{\partial^2 |\Psi(X,Y)>}{\partial Y^2} + k^2 |\Psi(X,Y)>= 0, \qquad (4.1)$$

with $k^2 = \frac{2mE}{\hbar^2}$ and Dirichlet boundary conditions for $|\Psi(X,Y)>$, *i.e.*:
$|\Psi(X,Y)>= 0$ in the boundary of the enclosure. Note that this problem is just the Helmholz equation problem for the vibrational modes of an elastic membrane with a given shape. It is important, in solving these problems, to consider the symmetry properties of the billiards that translate into properties of the solutions.

Marcus suggested that avoided degeneracies in quantum systems may be a signature of having chaotic behavior in the classical limit[44]. MacDonald and Kauffman studied the statistics of the spectra of QB, specifically that of the CIS circle and the $CNIS$ stadium[6].. They concentrated mainly on the study of the eigenfunctions, but also carried out a calculation of $P(S)$ in both types of billiards. Their results for the circle were in agreement with the Berry and Tabor result. For the stadium they found some evidence for energy level repulsion, but the statistics of their results was not good enough to give a convincing correspondence with the RMT ensembles predictions. There were other studies of the spectra for the stadium[7]. and Sinai's billiard[8]. that concentrated on the evaluation of $P(S)$.

As mentioned before, even in nuclear physics where the *GOE* was surmised to be the appropriate ensemble to describe the fluctuations of the resonances, a satisfactory comparison between experiment and theory needed several years of painstaking accumulation of the *nuclear ensemble data* to arrive at a statistically significant comparison between theory and experiment[45]. The success from the detailed comparison between *RMT* and the nuclear ensemble data, led Bohigas et.al. to apply the same statistical measures used in the nuclear problem to the spectra of integrable and nonintegrable *QB*. They studied the desymmetrized quantum Sinai's billiard. To unfold the spectrum they used the *SC* Weyl formula,

$$N_{av}(E) \approx \frac{1}{8\pi} \int_{V(x) \leq E} [E - V(x)]dx. \qquad (4.2)$$

To get good statistics they put together the spectra from four different values of the disk radius R, discarding the lowest lying levels, which are representative of the integrable circle. The comparison between the *GOE* prediction and that of $P(S)$ is quite good. Furthermore, they analyzed the longer range correlation functions, and in particular the $\Delta_3(L)$ statistics also shown in Fig. 1

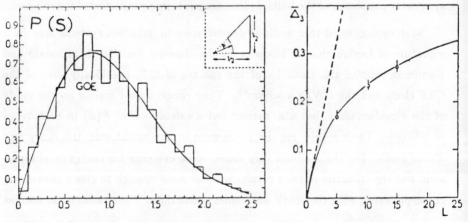

Fig. 1 Energy level separation distribution $P(S)$ and $\Delta_3(L)$ for the desymmetrized Sinai billiard. Taken from ref. 9.

Later they studied the stadium with similar results. These results are significant in that they not only prove that the strongly chaotic billiards have energy level repulsion in their quantum spectra, but also that the short range and medium range properties of their spectra, correspond *quantitatively* to the *GOE* predictions. These results led Bohigas *et.al.* to conjecture that in *generic time reversal invariant chaotic systems the spectral fluctuations are fully described by the GOE.*

To continue the pursuit of the correspondence between *RMT* and quantum systems, Berry and Robnik considered a *QB* without time reversal invariance [46]. Specifically, they studied the conformal map of a disk with a magnetic flux line at the origin. The corresponding Scrhödinger equation is,

$$\{i\frac{\partial}{\partial X} + i\frac{\partial}{\partial Y} + \frac{q}{\hbar}\vec{A}(X,Y)\}^2|\Psi(X,Y) > +k^2|\Psi(X,Y) >= 0, \qquad (4.3)$$

where $\nabla \wedge \vec{A}(X,Y) = \vec{n}_Z\Phi\delta(X,Y)$, and Φ is the magnetic flux. For values of the ratio $F = \frac{\Phi}{\Phi_0}$ (with Φ_0 the quantum of flux), different from integer and half integer, the Hamiltonian is complex and one would expect *GUE* statistics. This is indeed what Berry and Robnick found from their calculations. This is an interesting model since classically the trajectories are not modified by the field, whereas quantum mechanically they get affected by the vector potential as first discussed by Bhom and Aranov.

The stadium with a *homogeneous* magnetic field, was studied by Schmidt *et.al*[47] An asymmetric stadium had to be considered since the stadium is symmetric with respect to the line along its minor and major axis, and thus *GOE* statistics follows. The physics of the problem is the following: when the field is "weak", with respect to the energy of the particle, the orbits of the stadium are modified slightly, remaining chaotic, and thus in the quantum limit the spectra has *GOE* fluctuations. In the limit of a "strong" field the orbits are circular and, when quantized, lead to a Landau levels structure with picked fence spectra. Therefore, the regime of interest is that for intermediate values of the field. A smooth transition from GOE to GUE was found as the field was increased from zero. The results found in this case do not have the statistical significance found in the GOE case, but were certainly not contradictory to the GUE expectation.

It would seem that for all the results discussed above, the conjecture of Bohigas *et.al.* holds. However, studies in triangular billiards, obtained from a compact surface of constant negative curvature known to have chaotic orbits, have led to disquieting results. Schmidt studied the case of such a triangle with inner angles of $\frac{\pi}{8}$, $\frac{\pi}{2}$, and $\frac{\pi}{3}$, that *tiles* the space, and gives results that approach those of Poisson statistics. Nonetheless, if the inner angles are modified slightly, such that the triangle does not *tile* the space, the results are in very good agreement with the GOE statistics. What these results seem to indicate is that, as in the theory of classical dynamical systems, the conjecture may be true for *generic* systems, but that there may be a finite number of *nongeneric* exceptions. More studies are needed to sharpen the general validity of the conjecture for QB.

4.2 Nonlinear oscillators

Although the studies of QB are important since they allow studies of dynamical systems with rigorously known properties, it is important as well to study more realistic physical models. In particular, models in which the size of the chaotic regions changes while changing a control parameter. Seligman *et.al.*[10] and Cerderbaum *et.al.*[11] studied a system of coupled oscillators with scale invariant Hamiltonian,

$$H = \frac{1}{2}\{[P_1^2 + P_2^2] + V_1(X_1) + V_2(X_2) + V_{1,2}(X_1 - X_2)\}, \qquad (4.4)$$

with the nonlinear potentials,

$$V_i(X) = v_i(\alpha_i X^2 + \beta_i X^4 + \gamma_i X^6). \qquad (4.5)$$

Here, v_i, α_i, β_i, and γ_i are constants that when varied lead to a phase space configuration going from purely integrable, to fully chaotic. To differentiate the different amounts of integrability in the problem for a given set of parameters, Seligman *et.al.* used the chaotic volume in phase space as the appropriate measure of chaoticity. Their results are shown in Fig. 2.

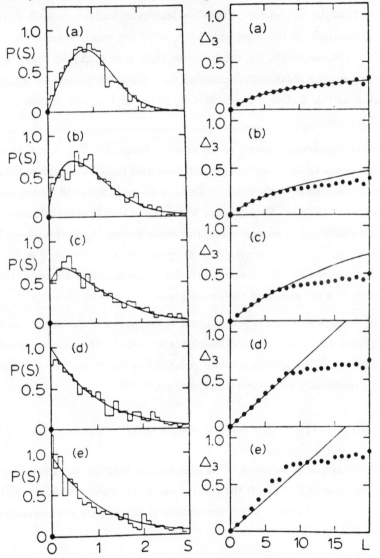

Fig. 2 Fluctuations for the polynomial Hamiltonian given in Eq(4.4). Taken from reference 10.

In Fig. 2 we see that as the chaotic volume goes from ≈ 1 (a) to ≈ 0.0 (e), the distribution of energy level separations goes from fitting the GOE prediction quite well to a Poisson distribution. The rigidity is also shown in the figure. We notice that the fully chaotic limit does lead quantitatively to a GOE $\Delta_3(L)$ result. In contrast, in the integrable limit, while for small values of L, $\Delta_3(L)$ follows the Poisson result, for values larger than $L \approx 6$, $\Delta_3(L)$ shows a *kink*, leading to a result similar to that expected for a picket fence spectra. A physical explanation of this finding was provided by Berry, and we will discuss it at the end of this section.

Another significant property of the results presented in Fig. 2, is that as the chaotic volume is varied, there is a smooth transition from GOE type statistics to Poisson. These results, in the extreme limits of integrability, yield credence to the Bohigas *et.al* conjecture. Nevertheless, it also shows that the results known from RMT are insufficient to accommodate the whole variety of possible integrability properties found in classical systems. Seligman *et.al.* introduced a heuristic model that serves to fit the numerical results in the *intermediate* region[12]. We shall not discuss the details of their model here.

As in the QB problem, it is of interest to consider the statistical properties of the spectra when time reversal invariance is broken. This was accomplished by Seligman *et.al.* by adding a vector potential to the model given in Eq(5.4) and (5.5), specifically, the model considered reads, [48]

$$H = \frac{1}{2}\{[P_1 - aX_2^3]^2 + [P_2 + aX_1^3]^2 + \alpha_1 X_1^6 + \alpha_2 X_2^6 - \alpha_{1,2}(X_1 - X_2)^6\}, \quad (4.6)$$

with a, α_i constants to be varied. This Hamiltonian is antisymmetric under time reversal invariance. For $a = 0$ the spectra has GOE fluctuations, as would be expected. For $a \neq 0$ and $\alpha_{1,2} \neq 0$ the statistical fluctuations in the spectra are those of a GUE.

The physically realizable hydrogen atom in a constant magnetic field given in Eq(2.8), has been studied by Delande and Guy[13] and also by Wintgen and Friedrich[14] In the classical limit the relevant parameter controlling the nature of the flows is $\beta = \frac{\gamma^2}{(-2E)^3}$. The critical value for β above which the trajectories are fully chaotic is ≈ 60.638. For values of β much smaller than β_c, a fair agreement with the Poisson distribution is found as well as the appearance of the *kink* in the $\Delta_3(L)$ statistics. The results are in very close agreement with the GOE analytic results given in the previous section.

The conclusion from all the results discussed in this subsection for *time independent* quantum problems is that: for classically chaotic systems the fluctuations of the spectra follow quantitatively the predictions from RMT. When the Hamiltonian is time reversal invariant the fluctuations are described by the GOE, while for non time reversal invariant systems they are those of GUE. In the integrable case the fluctuations are given by the Poisson distribution, up to a maximum value of L above of which the spectrum becomes rigid. The explanation for why this is so was given by Berry[24] and will be discussed later.

5. TIME-DEPENDENT UNIVERSALITY CLASSES

When the Hamiltonian is independent of time, as discussed above, the time dependence of the wave function is always quasi-periodic. Nonetheless, we have seen that there are clear differences in the spectral properties of systems that classically have different degrees of integrability. For *time-dependent periodic* Hamiltonians, one can calculate the wave function fully from the knowledge of the *quasi-energy* spectra of the time evolution operator (Eq(2.7)). The crucial question is what is the nature of the QES. If it is discrete , the wave function is again a quasi-periodic function of time[49]. If any component of the QES is continuous, the evolution will be nonrecurrent and the possibility of *chaos* in the classical sense could appear. In practice, determining if the QES has a continuous component is highly nontrivial.

In the same spirit as the studies discussed above, it is informative to study quantum versions of simple models that are fully understood classically. Bellow we discuss results from studies in such type of models: We start with the extensively studied periodically-kicked-quantum-rotator model $(PKQRM)$ introduced by Casati et.al.[50], then to the somewhat realistic quasi-one dimensional hydrogen atom model,[33]-[34] to the more recent studies of the Fermi-Ulam-Acceleration-Model $(FUAM)$.[17],[19]-[21]

5.1 Periodically kicked quantum rotator model

The $PKQRM$ is the quantum version of the classical *standard map* model, which has been studied extensively and which shows all the intricacies of the chaotic solutions for systems with two degrees of freedom..[51] The Hamiltonian for the $PKQRM$ is,[50]

$$H = \frac{-\hbar^2}{2I} \frac{\partial^2}{\partial \phi^2} + S \cos\phi \sum_n \delta(t - nT). \tag{5.1}$$

The model represents a particle that is constrained to move in a circle and that gets *kicked* periodically in time T. There are two relevant parameters to describe the physics of the model: $k = \frac{S}{\hbar}$, and $\tau = \frac{\hbar T}{I}$. Classically, when $K = \tau k \gg 1$, the model shows chaotic behavior. The flows in this regime can be described by a diffusion process with diffusion constant $D_{cl} \approx \frac{k^2}{4}$. In the quantum limit, in the regime where $K \gg 1$ together with $k \gg 1$, Casati *et.al.* found that the energy of the $PKQRM$ showed diffusive-like behavior but only up to a maximum time $t^* \approx \frac{k^2}{4}$. For times larger than t^*, the energy of the quantum problem shows quasi-periodic behavior. This important result indicates that the quantum system *behaves* like the chaotic system up to a maximum time after which it stops behaving diffusively. The magnitude of the *brake time* t^* is inversely proportional to $\hbar$ in such a way that the closer we are to the classical limit the longer it takes *to suppress classical chaos.*[4]

Significant progress in the understanding of the quantum solutions of this model emerged from a connection found by Fishman *et.al.* to a tight-binding chain model:[52]

$$T_m u_m + \sum_r W_r u_{m+r} = E u_m. \tag{5.2}$$

The lattice sites m correspond to the different integer values of the angular momentum. The QE eigenfunctions u_m have two different limiting behaviors, depending on the number theoretic nature of τ. If $\tau = 4\pi p/q$, with p and q prime integers, the lattice chain model is periodic and the QE eigenfunctions are extended. That the solutions are extended means that the system will be able to explore all possible values of the angular momentum. This implies that the energy of the model, which is given by $E(t) = \sum_m m^2 |a_n|^2$, grows like t^2. This phenomena occurs for all rational values of τ, and was termed *quantum resonance* by Izrailev and Shepelyanski.[53]

When τ is an irrational number, the onsite energy term in the tight binding model becomes pseudorandom: for the explicit model given in Eq(5.1), $T_m = tan\frac{\{\lambda - \frac{m^2 \tau}{2}\}}{2}$, and the hopping term is $W(\phi) = -tan(\frac{k}{2}cos\phi)$. These types of models, known generically as Anderson models, have been the paradigm in studies of the electronic states in disordered materials.[54] The one-dimensional Anderson model, *with nearest neighbor hopping terms,* has been studied extensively and there are solid mathematical results for its eigen-solutions and spectra: The spectra is pure point and the eigenfunctions are *exponentially localized.*[55] This localization of the quasi-energy eigenfunctions explains the existence of the *brake time* found by Casati *et.al.* in their initial study of the $PKQRM$. Notice, however, that for the $cos\phi$ potential, the hopping term in Eq(5.2) decays exponentially only for values of $k < \pi$. One can then use the results known for the Anderson model rigorously only in the $k < \pi$ regime. There are numerical solutions for larger values of $k(\sim 5)$ that indicate that the states are still localized.[53] Nonetheless, for relatively large values of k say $20,000$, the nature of the eigenfunctions is not rigorously known. The particular choice of the potential of interaction in Eq(5.1), leads to an onsite energy in the tight-binding equation that is somewhat non physical since it has a singularity for $k = \pi$. Shepelyansky has studied another model without this pathology, and with a short range hopping term in Eq(5.2).[56] The problem with this model is that the onsite and off diagonal randomness are correlated. For this model there are no results known in localization theory that one could use. On the other hand, Shepelyansky proposes for his model that the *classical* diffusion constant is directly proportional to the localization length of the quasi-energy eigenfunctions, and thus can be calculated from a purely classical analysis.

From the discussion given in the previous sections one can ask *what are the relations, if any*, between the spectral properties of the QES and those of RMT. Given the connection found by Fishman *et.al.* between the $PKQRM$ and the disordered tight binding model, one can surmise, based on a rigorous mathematical result by Molcanov[57] for the infinite lattice one-dimensional Anderson model, that the distribution of QE level separations should follow a Poisson distribution. This was confirmed explicitly for the $PKQRM$ by Feingold *et.al.* for values of k up to 5.[58]

There were two studies that appeared in print almost simultaneously concentrating on the statistical properties of the QES, that showed behavior consistent RMT predictions.[17),18] In most cases the time evolution operator is of infinite dimension, but one can only analyze numerically finite dimensional matrices. Izrailev considered a finite-dimensional version of the $PKQRM$ with U matrix:[18]

$$U_{mn} = \frac{1}{2N_1 + 1} e^{[-ik\,cos\{\frac{2\pi}{N}m + \phi_0\}]} \sum_{\ell=-N_1}^{N_1} e^{\{\frac{i}{2}\ell^2\}} e^{\{-i\frac{2\pi}{N}\ell(m-n)\}}, \qquad (5.3)$$

where the phase $\phi_0 \epsilon [0, \frac{\pi}{N}]$ and U has periodic boundary conditions. This model has N QE eigenvalues, and in the limit $N \to \infty$ becomes the continuum $PKQRM$ model. For values of $k \approx 20,000$ it was found that the QES does fit the GOE prediction given in Eq(3.20) with good confidence level. Izraleiv also considered a Hamiltonian without time reversal invariance. This consists of adding a contribution $\frac{i}{\sqrt{2}} \frac{\partial}{\partial \phi}$ to the kinetic energy term in Eq(5.1) together with a potential of interaction $V(\phi) \neq V(-\phi)$. For the same value of k, he found a GUE distribution of energy level separations, as expected.

5.2 Quasi-one-dimensional hydrogen atom model

Although the results found in the $PKQRM$ depend on the specific functional dependence of the interaction potential, the localization of the QE eigenfunctions should be a general property of simple quantum models acted on by periodic external perturbations. This has been confirmed in studies of other one-dimensional models. Of interest are the studies carried out in the one-dimensional hydrogen atom model $(ODHAM)$.[51] Of course, the main crucial difference between the $PKQRM$ and the $ODHAM$, is the continuous part of the spectra present in the $ODHAM$. A correct treatment of the continuum component of the spectra has been a major challenge.

Casati *et.al.* have studied the quasi-one-dimensional hydrogen atom model defined in Eq(2.10).[51] They have found that for values of $F \leq F_c$, the angular momentum states are localized, although in the same regime the classical model shows chaotic diffusive behavior in momentum space. For values of the field above F_c, the states are extended indicating the onset of ionization even for subresonant frequencies. The quantum ionization region for $F > F_c$, can be mimicked by the classical solutions in the diffusive or chaotic regime. However, when looking more carefully at the reversibility properties of the classical versus the quantum mechanical solutions, it was found that the quantum solution is inherently more stable numerically than the classical solution. Blumel an Smilansky[36] have presented a study that seems to differ in conclusions from that of Casati *et. al.*. The situation is therefore not conclusive yet.

A more extensive discussion of the $ODHAM$ is given by Casati elsewhere in this proceedings.

5.3 Quantum Fermi-Ulam acceleration model

We introduced a new quantum model to ascertain the generality of the RMT results found in the time independent models.[17] The model is a quantum version of a the Fermi-Ulam accelerator model ($FUAM$). This problem has been studied extensively in the classical limit. The model consists of a free particle moving inside a one-dimensional box. The box has one of its walls fixed and the other one is oscillating periodically in time. The collisions between the particle and the walls are perfectly elastic. This model was first conceived by Fermi when he was trying to explain the existence of high energy cosmic rays. He proposed the model, with the role of the oscillations produced by the effect of interstellar oscillating magnetic fields, to explain why the energy of the charged particles was so large. Unfortunately, due to the existence of the Kolmogorov-Arnold-Moser surfaces, there is a clear upper bound in the amount of energy that the particle can gain and thus the Fermi acceleration does not take place indefinitely. Further extensions and simplifications of the model by Ulam and others helped to clarify the properties of the classical limit of the $FUAM$.

The quantum version of the $FUAM$ is defined by the Schrödinger equation,

$$i\frac{\partial|\Psi(X,t)>}{\partial t} = -\frac{\hbar}{2m}\frac{\partial^2|\Psi(X,t)>}{\partial X^2}, \tag{5.4}$$

where m is the mass of the particle. The spatial coordinate X is in the interval $[0, \ell(t)]$, with $\ell(t)$ the time-dependent length of the box, that is taken to satisfy the symmetry and periodicity relations $\ell(t) = \ell(-t)$, $\ell(t + T) = \ell(t)$, where T is the period of oscillation of the moving wall. The boundary conditions are,

$$|\Psi(X = 0, t) > = |\Psi(X = \ell(t), t) > = 0,$$

$$\int_0^{\ell(t)} <\Psi(X,t)|\Psi(X,t)> dX = 1. \tag{5.5}$$

These time-dependent boundary conditions (b.c.) are what makes the problem non-trivial. The b.c. can be incorporated in the partial differential equation with fixed b.c., leading to a non-Hermitian equation. Instead, what makes the problem almost fully soluble analytically is a specific choice of $\ell(t)$.17) To see how this comes about, we start by expanding $|\Psi(X,t)>$ in terms of the complete set of instantaneous solutions to the rigid box,

$$|\Psi(X,t)> = \sum_n C_n(t) \sqrt{\frac{2}{\ell(t)}} \sin \frac{n\pi X}{\ell(t)} \tag{5.6}$$

Substituting this solution back into Eq(5.4), one gets the equation for the expansion coefficients C_n:

$$i\dot{C}_n = \frac{E_n(t)}{\hbar} C_n - \sum_{m \neq n} [i\frac{\dot{\ell}}{\ell} \frac{2nm}{(n^2 - m^2)} C_m(t)]. \tag{5.7}$$

Here, $E_n(t) = \frac{\pi^2 \hbar^2 n^2}{2m\ell(t)^2}$ are the *instantaneous* "eigenenergies". It is convenient to rewrite Eq(5.7) using the transformation,

$$a_n = e^{-i \int_0^t \frac{E_n(s)}{\hbar} ds} C_n, \tag{5.8}$$

as,

$$i\dot{a}_n = \sum_{m \neq n} [-i\frac{\dot{\ell}}{\ell} e^{-i \int_0^t \frac{2\pi}{T_B}(n^2 - m^2)\frac{\ell_0^2}{\ell^2(s)} ds} \frac{2nm}{(n^2 - m^2)}] a_m(t).$$

$$= \sum_{m \neq n} H_{nm}(t) a_m(t) \tag{5.9}$$

Here ℓ_0 is the box length at $t = 0$, and T_B is the Bohr period that gives the time the particle takes in completing a round trip in the static box in its ground state:

$$T_B = \frac{8m\ell_0^2}{h}. \tag{5.10}$$

Note that the T_B is inversely proportional to h. Thus in the SCL, T_B will be very large. Note that Eq(5.7) has a tight binding equation structure. Of importance is the fact that the hopping terms are neither of short nor long range. Instead, the decay is algebraically slow (for fixed m, say, the hopping term $\sim 1/n$). Since $\ell(t) = \ell(-t)$, the matrix H_{nm} satisfies the symmetry,

$$H_{nm}(t) = H_{mn}(-t). \tag{5.11}$$

The solution to Eq(5.9) can be obtained from the time evolution operator,

$$U_{nm}(-t,0) = AT \, exp\Big[-i \int_0^t H_{mn}(s) \, ds\Big] = U_{mn}(0,t), \tag{5.12}$$

where AT stands for anti-time ordering operator, and the second equality follows from Eq(5.11). Notice that the time evolution operator $U_{nm}(-t,t) = e^{-iF_{nm}}$, with F_{nm} a real symmetric matrix. It is in general quite difficult to calculate U_{nm} for an arbitrary $\ell(t)$ since it involves a time ordering operation. However, from inspection of Eq(5.9), we see that if we take $\ell(t)$ as[17]

$$\ell(t) = \ell_0\sqrt{1 + 2\delta|t/T|}, \tag{5.13}$$

with $|t| \le T/2$, then the ratio $\frac{\dot{\ell}}{\ell} = \pm\delta$, with the $+$ sign for $t > 0$ and the minus sign for $t < 0$. In this case the Hamiltonian for the interval $(-T/2, 0)$ commutes with that for $(0, T/2)$. This choice of $\ell(t)$ represents a significant simplification in the evaluation of U_{nm} since the time evolution operator within a period, $U_{nm}(-T/2, T/2)$, can be evaluated from simple diagonalizations of an $n \times n$ unitary symmetric matrix.

The QES of the quantum $FUAM$ was calculated and was analyzed from the RMT point of view. The two important parameters in the problem are the Bohr period T_B and the amplitude of the wall oscillation δ.

It is found that when changing the values of δ and T_B, such that their product is small, the distribution of energy level separations follows the Poisson law quite well.

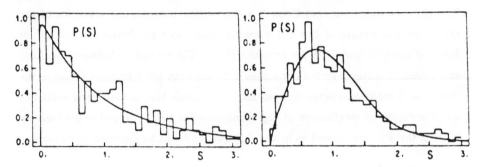

Fig. 3 shows the results for the distribution of energy level separations $P(S)$ for $T_B = 1.57$ and $\delta = 0.2$ (left), and $\delta = 0.5$ and $T_B = 628.31$ (right).

As the values of δ and/or T_B are increased, the distribution $P(S)$ changes over smoothly from Poisson to the GOE behavior. In Fig. 4 we show the results for $\Delta_3(L)$ as a function of δ and T_B. As in the TI case we see that by varying the strength of the interaction one can pass *smoothly* from a Poisson to a COE, or GOE, type of statistic.

The higher order correlations γ_1 and γ_2 for this QES have been analyzed and yield results that quantitatively correspond to the predictions of RMT in the corresponding asymptotic limits.[20]

An important correlation emerges when studying the corresponding QE eigenfunctions. The coefficients a_n give the projection of the time evolution operator onto the angular momentum basis, from which the wave function is obtained via Eq(2.7). When $P(S)$ has the Poisson limit, there is essentially only one a_n that contributes to the wave function. This is understood easily since the perturbation is weak and it can only excite few states around the n-th state.

However, when the statistic is *GOE*, and the coupling is strong there are many a_n's that contribute to the wave function with large amplitudes. In this case the possibility of having resonances between far separated states is significant. Therefore we surmised that there is a connection between the statistics of the *QES* and the nature of the *QE* eigenfunctions, to wit: *Poisson statistics for localized states*[20] and *GOE for extended states*. The transition between localized and extended states is smooth and thus different than the transition found in the theory of disordered electronic systems. The connection between the statistics and the degree of localization of eigenfunctions has been pursued quantitatively both in the *TI* case as well as in the *TD* one, and will be reported elsewhere.

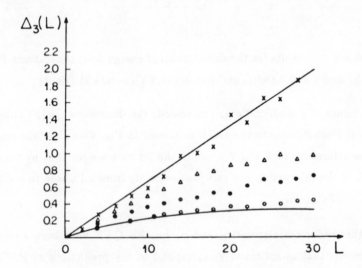

Fig. 4. $\Delta_3(L)$ as a function of T_B and δ. The symbols stand for: X with $\delta = 0.2, and T_B = 1.57$, while $\triangle$, •, $\bigcirc$ correspond to $\delta = 0.5$ and $T_B = 1.25, 50.26, 62.83, 628.31$, respectively.

Further analyses of the *QES* for *FUAM* has been given by Vischer.[21] For the *FUAM* defined by Eq(5.4) and (5.13), he finds agreement with the results given above. However, in his results, as in the *TI* nonlinear oscillator calculations, there is a deviation of the $\frac{L}{15}$ Poisson prediction for large values of L. He also studied a model without time reversal invariance. The Hamiltonian in this case reads,

$$H = \frac{\hat{P}^2}{2m} + \alpha \Big[\frac{X}{\ell(t)}\Big]^n + \hat{P}\,\frac{\beta}{\ell(t)} \tag{5.14}$$

the momentum operator is $\hat{P} = -i\frac{d}{dx}$, and α and β are constants. For $\beta = 0$, this model reduces to the $FUAM$ studied above in the limit $n \to \infty$. The $P(S)$ follows the Poisson distribution for small values of β, δ such that when multiplied by T_B give small numbers. The $\ell(t)$ dependence is chosen as in Eq(5.13) such that the time evolution operator can be diagonalized without having to perform the time ordering operation. Vischer has also generalized the way to find the appropriate time dependence of the Hamiltonian such that the time ordering product is not necessary, thus giving a way to study models not just in one dimension but also in higher dimensions. For larger values of β, say, $\beta = 1$, with $T_B = 230.9$ and $\delta = 0.66$ Vischer finds a GUE QES as one would have expected from RMT ideas.

One of the shortcomings of the analyses of the QES in the $PKQRM$ and the $FUAM$ relates to the truncated nature of the time evolution operators studied. This problem does not exist in spin models where the time evolution operator for a given value of the spin is finite-dimensional. However, since in this case the role of $\hbar$ is played by S, and the SCL corresponds to $S \to \infty$, one needs to consider large matrices as well. Frahm and Mikeska[22] considered the Hamiltonian of a periodically kicked spin:

$$H = A(S^z)^2 - BS^x \sum_n \delta(t - nT). \tag{5.15}$$

A and B are positive constants with $\frac{B}{A} = Sb$, with b the strength of the anisotropy, and $B = \delta^2\sqrt{b}/2$. By varying the coupling constant δ with $S(= 400)$ fixed, they observe a transition from Poisson to GOE as found in the $FUAM$. The same type of transition occurs when fixing $\delta = 1.6$ and varying S from 100 to 400 (see also ref. 23).

Although the results discussed in this section point to the general conclusions arrived at in the TI case, from the theoretical point of view the understanding and limitations of these results is still lacking.

6. CONCLUSIONS

The main conclusion of the work reviewed in these lectures is that there are indeed *quantum manifestation of classical chaos* in the form of specific properties of the spectra. It is the *fluctuation* in the spectra that carry the information about the degree of integrability of the corresponding classical problem. One has been able to identify extreme *universality* classes both in the TI and TD cases, to wit: Integrable $\rightarrow$ Poisson, chaotic $\rightarrow GOE$, for time reversal invariant systems, or $\rightarrow GUE$, for time reversal non-invariant ones. Intermediate regimes between fully integrable and fully chaotic have been found, but a fully developed RMT formulation of the *intermediate regime* is still laking. An important limitation in the region of validity of the universal behavior, which can not be explained by standard RMT, emerged from the specific analyses of integrable systems, which showed a clear deviation from the $\frac{L}{15}$ prediction.[10]

For time-independent Hamiltonians, Berry has provided an understanding for the breakdown as well as the universality, from a study of the $\Delta_3(L)$ statistic evaluated from the Gutzwiller sum over periodic orbit method.[24] An essential difference between integrable and chaotic systems is that in the former case the orbits are stable and occur in families while they are generically isolated and unstable in the later.[59] Berry shows that for $L \ll L_{max}$, where L_{max} is the saturation value after which the universal behavior breaks down, the integrable case leads to Eq(3.32), while for the nonintegrable cases one gets either Eq(3.30) or Eq(3.31) depending on the time reversal properties of the Hamiltonian. In this regime the contributions to $\Delta_3(L)$ come from very long periodic orbits. The difference in the type of orbits in the integrable and chaotic regimes lead to the difference in the asymptotic properties of $\Delta_3(L)$. For smaller orbits, $\Delta_3(L)$ saturates into a nonuniversal value that depends on the system under consideration. For integrable systems $L_{max} \sim \hbar^{(N-1)}$, with N the number of degrees of freedom, while for the nonintegrable case $L_{max} \sim -\ell n(\hbar)$. This result may explain why the "kink" has been seen in the integrable region and not in the nonintegrable one.

With regard to the fluctuations in the eigenvalue spectra of the time independent problems, we now have the beginnings of what may become a full theory. So far RMT has served the purpose of providing the framework against which the eigenvalue spectra of specific models have been tested. It is clear now that RMT needs to be extended to encompass the different types of spectral fluctuation characteristics found in the different limits of integrability. In this regard, there is a suggestion by Berry and Robnik[46] as to how to associate statistics to the different regions of integrability in phase space. This idea needs to be pursued further to prove its validity. Also, there have been other theoretical attempts to prove the universal behavior of the statistics, mainly that of the repulsion of energy levels from rather general considerations. [25],[26] However, some of the assumptions do not seem to be of general validity, and more work needs to be done along those lines.

Although studies of eigenvalue spectra in TI systems have led to solid general results, the same can not be said about the properties of the corresponding eigenfunctions. Several ideas have been proposed as to the nature of the eigenfunctions but they have not been found in specific model calculations. There is a lot of work yet to be done in this respect.

In the time-dependent case, most studies have concentrated on time periodic Hamiltonians. This has the advantage that it is a physically realizable type of Hamiltonian and that the properties of the wave functions can be extracted from a study of the QES. As in the TI case it has been found that the statistics of the QES can also be fitted by the predictions of RMT. For large values of $\hbar$, the

QES shows Poisson statistics while for small values of $\hbar$ one can either have GOE or GUE depending on the time reversibility of the Hamiltonian. Evidence for the "kink" shows up also in the Poisson region. As $\hbar$ is varied, a transition region is found that is very similar to that found in studies of nonlinear coupled oscillators. One should stress the fact that many of these results are obtained for finite-dimensional time development operators. However, there are also calculations for models where U is exactly finite dimensional and the results are qualitatively the same.

These results are much less understood than in the TD case since one does not have the full semiclassical theory needed to calculate the $\Delta_3(L)$ for the QES. This leads us to believe that, either the properties of the spectral fluctuations are truly universal, meaning to say that they are independent of the specific form of the expressions in the semiclassical limit, or one can produce a Gutzwiller-like formulation to calculate the density of QE states. This situation leaves the possibility open that the results found in specific model calculations are not truly universal. This is still an unresolved question.

The situation of the eigenfunctions in the TD case is better understood than in the TI one. It has been found, although not yet fully analyzed, that there is a direct connection between the statistics of eigenvalues and that of the eigen-vectors. Specifically, Poisson $\rightarrow$ localized states, while $GOE \rightarrow$ extended states. There are possible experimental consequences of the existence of these regimes in experiments in quasi-one-dimensional hydrogen atoms acted on by a.c. fields.

All in all, one can say that there has been progress in finding specific signatures of chaotic behavior in quantum systems, but there is still a long road before we can claim that the connections are fully understood. A more extensive review of the material presented here, including experimental applications, will appear elsewhere.[60]

ACKNOWLEDGEMENTS

Special thanks to O. Bohigas, D. Dessner, D. Grempel, M. J. Giannoni, and C. Schmit, T. Seligman for very useful discussions on many aspects of the topics covered in these lectures. Thanks also to W. Vischer for sending a copy of ref. 21 prior to publication. This work was supported in part by NSF Grant DMR-8640360.

REFERENCES

1. I. C. . Percival, J. Phys. **B6**, L229 (1973) and Adv. Chem. Phys. **36**, 1 (1977).

2. For an excelent review see *Regular and stochastic motion* by A.J. Lichtenberg and A.M. Lieberman, Springer-Verlag Publ.,(1983)

3. See for example, M.V.Berry,*Chaotic behavior in deterministic systems.* Les Houches summer school XXXVI, Ed. R. Helleman and G. Joos. (North Holland 1981) and G. Zaslasvskii Phys. Rep.**80**, 157.(1981)

4.B. V. , F. M. Izrailev,D. L. Shepelyansky, *Soviet Scientific Reviews* **2C**, 209 (1981).

5. M.V.Berry and M.Tabor Proc.Roy.Soc.London,Ser **A356**, C375(1977).

6. S.W.McDonald and A.N.Kaufman Phys.Rev.Lett. **42**,1189 (1979).

7. G.Casati *et.al.* Nuovo Cimiento Lett. **28** ,279 (1980).

8. M.V.Berry Ann.Phys.(NY)**131** , 163 (1981).

9. O. Bohigas, M. J. Giannoni, and C. Schmit Phys.Rev.Lett. **52**, 1.(1984); in *Quantum Chaos and Statistical Nuclear Physics*, Cuernavaca, México. Ed. T. .H. Seligman and H. Nishioka (Springer-Verlag 1986). O. Bohigas and M. J. Giannoni in *Mathematical and Computational Methods in Nuclear Physics*, Granada, España. Ed. J. S. Dehesa, J. M. G. Gomez, and A. Polls (Springer-Verlag, Berlin 1983).

10. T. H. Seligman, J. J. Versbaarschot, and M. R. Zirnbauer Phys. Rev. Lett. **53**, 215 (1984); J.Phys. **A18**, 2751. (1985).

11. E. Heller, H. Koppel and L. S. Cerderbaum ibid. **52** 1665.(1984);

12 T. H. Seligman, J. J. Versbaarschot, Phys.Lett. **A108** 183 (1985).

13. D. Delande and J. C. Gay, Pys. Rev. Lett. **57**, 2006 (1986).

14. D. Wingten and H. Friedrich, ibid. **57**, 571 (1986).

15. N. L. Balazs and A. Voros, Physics. Reports **143**, 109 (1986).

16. M.Feingold,S.Fishman, D.R.Grempel and R.E.Prange hys.Rev. **B31**, 6852(1985)

17. J.V. José and R.Cordery Phys. Rev. Lett. **56**, 290 (1986)

18. 11. F.M.Izrailev ibid. **56**, 541 (1986)

19. J.V. José, (to be published)

20. C. Roman and T. Seligman in *Quantum Chaos and Statistical Nuclear Physics*, Cuernavaca, México. Ed. T. .H. Seligman and H. Nishioka (Springer-Verlag 1986).

21. W. Vischer (preprint).

22. H. Frahm and H. J. Mikeska, Z. Phys. **B65**, 249 (1986).

23. M. Kuś, R. Scharf, and F. Haake, ibid **B66**, 129 (1987).

24. M. Berry , Proc. Roy. Soc. **A400**, 229 (1985).

25. P. Pechukas, Phys. Rev. Lett. **51**, 943 (1983).

26. T. Yukawa, ibid. **54**, 1883.

27. E. Bayfield and P. M. Koch ibid. **33**,258(1974).

28. J. G. Leopold and I. C. Percival, ibid **41**, 944 (1978).

29. P. M. Koch and D. R. Mariani ibid. **46**, 1275 (1981);

30. K. A. H. van Leeuwen *et.al.* ibid **55**, 2231 (1985), and references therein.

31. J. Bayfield, "Fundamental Aspects of Quantum Theory", (Notes in Physics, Springer Verlag, 1986).

32. J. E. Bayfield and L. A. Pinnaduwage, Phys. Rev Lett. **54** 313 (1985). J.N. Bardsley at.al. ibid. **56**, 1007 (1986)

33. D. L. Shepelyansky, *Proc. Int. Conf. on Quantum Chaos* Ed. G. Casati (Plenum 1983).

34. R. V. Jensen Phys. Rev.Lett. **49** 1365(1982) and Phys.Rev. A30 386(1984).

35. G. Casati, B. V. Chirikov, D. L. Shepelyansky, and I. Guarneri, Phys. Rev. Lett. **56**, 2437(1986), ibid **57**, 823 (1986). For a review of their work see Phys. Rep. (1987).

36. R. Blümel and U. Smilansky, Phys. Rev. Lett. **58**, 2531 (1987).

37. See the recent updated review by T. A. Brody *et.al.* Rev.Mod.Phys. **53**,385.(1981)

38. R. Balian, Nuovo Cimento **B57**, 183 (1968).

39. F. Dyson, J. Math. Phys. **3**, 140, 157, 166, (1962).

40. F. Dyson, Comm. Math. Phys. **19**, 235 (1970). M. . L. Metha ibid. **20**, 245 (1971).

41. F. Dyson, and M. . L. J. Math. Phys. **4**, 701 (1963).

42. I. C. . Percival, J. Phys. **B6**, L229 (1973).

43. G. M. Zaslavskii, Zh. Exp. Theor. Fiz. **73**, 2089 (1977) [Sov. Phys. JETP, **46**, 1094 (1977)]. Usp. Fiz. Nauk. **129**, 211 (1979). [Sov. Phys. Uspekii **10**, 788 (1979)].

44. R. A. Marcus, in *Nonlinear Dynamics*, Ann. N. Y. Acad. Sci. **357**, 169 (1980).

45. R. U. Haq, A. Pandey and O. Bohigas, Phys. Rev. Lett. **54**, 1617 (1982).

46. M. Berry and M. Robnik, J. Math. Phys. **A17**, 2413 (1984).

47. C. Schmit *et. al.* (to be published).

48. T. H. Seligman, J. J. Versbaarschot, Phys. Lett. **A108**, 183 (1985).

49. T. A. Hogg and B. A. Huberman Phys.Rev.Lett. **48**,711 (1982); Phys.Rev. **A28**, 22.(1983);

50. G. Casati, B. V. Chirikov, F. M. Izrailev, and J. Ford in, *Stochastic behavior in classical and quantum Hamiltonian Systems*, ed G. Casati and J. Ford (Lecture Notes in physics **Vol 93**, 334.(1979).(Springer, NY).

51.

52. S. Fishman, D. R. Grempel and R. E. Prange Phys.Rev.Lett. **49**, 509 (1982). D.R.Grempel and R.E.Prange and S.Fishman Phys. Rev. **A 29**, 1639 (1984)

51. G. Casati, B. Chirikov, D. L. Shepelyanskii, and I. Guarneri, Phys. Rep. **154**, 77 (1987).

53. F. M. Izrailev and D. L. Shepelyanski, Dokl.Akad.Nauk **SSSR 249**,1103 (1979). [Sov.Phys.Dokl. **24**, 996 (1979)];

54. P. W. Anderson Phys. Rev. **109**, 263 (1958).

55. R. Herendon and P. Erdös, Adv. Phys. **31**, 65 (1982).

56. P. L. Shepelyansky, Phys. Rev. Lett. **56**, 677 (1986).

57. A. Molcanov Comm.Math.Phys. **78**, 429(1981)

58. M. Feingold, and S. Fishman, Physics **25D**, 181 (1987).

59. J. H. Hannay and A. M. Ozorio de Almeida, J. Phys **A17**, 3429 (1984).

60. J. V. José. (in preparation).

THE THERMODYNAMICS OF FRACTALS

Tomas Bohr

The Niels Bohr Institute
University of Copenhagen

Tamás Tél

Institute for Theoretical Physics
Eötvös University
Budapest

Introduction.

Statistical mechanics, as developed by J.W.Gibbs in the last century, is a theory of great beauty and coherence. By basing it on the concept of *ensembles* Gibbs freed the theory from the ambiguities related to the introduction of probabilities in deterministic systems, and, at the same time, he created a framework which lends itself easily to generalizations - most importantly the subsequent development of quantum statistical mechanics. The advantage of statistical mechanics over many other fields of science is that the number of degrees of freedom is usually enormously large. This means that the "law of large numbers" singles out certain members of the ensemble as "typical" for given external conditions, and their properties are then described by *thermodynamical functions* which contain most of the relevant information about macroscopic systems.

Recently there has been a strong interest in the socalled "thermodynamical description of fractals". A *fractal* is an object having structure on all length scales described by power laws with noninteger exponents.[1] In the analysis of fractals one introduces *partitions* which approximate the structure with a certain accuracy. Regarding the elements of the partitions as an "ensemble" a "statistical mechanics" can now be defined and properties like dimensions and entropies, which describe the fractal, emerge in the limit of infinitely fine partitions in the same way as the thermodynamic limit emerges from usual statistical mechanics.

This formalism goes back to the pioneering work of Sinai[2], Ruelle[3] and Bowen[4] in the seventies, but - until recently - it hasn't attracted much attention among physicists because of the lack of concrete applications and the rather technical presentations available. Within the last few years the situation has changed. The introduction of the concept of "multifractals"[5-9] has given new impulses to the thermodynamical description of fractals both experimentally and theoretically in the study of dynamical systems[9-28] and in other areas where fractals occur, notably the study of aggregates and percolation[29].

The following article is intended as an introduction to certain aspects of this ramified subject. We shall restrict our attention to fractal sets (Cantor sets) generated by chaotic motion of a dynamical system. Due to the existence of natural, hierarchical constructions for such sets the thermodynamical formalism is comparatively well understood and in order to keep the context as simple as possible only fractals embedded in one dimension will be studied, i.e. fractals generated by a one-dimensional map. These sets are either *strange attractors* [30] which characterize the asymptotic motion of generic points, or they can be characterized as *strange repellers* [31-32] which characterize the transient behaviour. In fact, the simplest systems on which the statistical theories can be applied are chaotic repellers of one dimensional maps, and in this article we shall concentrate on such repellers and on the attractors obtained in certain limits. The layout is as follows: Chapter 1 contains general definitions of the type of Cantor sets, which we shall look at, and the different forms of thermodynamics to be considered. Chapter 2 describes the thermodynamics on hyperbolic sets. Here a considerable body of theory is available, and in particular we shall show that the different thermodynamical formalisms are all connected in that case. In chapter 3 we look at a specific example of a non-hyperbolic system at a limiting, completely chaotic point, which is exactly soluble. It introduces the concept of phase transitions and gives hints of more general connections between the thermodynamical functions. Finally in chapter 4 we conclude by discussing results for more general non-hyperbolic systems, concentrating on the general case of "completely chaotic maps".

1. THE THERMODYNAMICAL FORMALISM.

The kind of fractal structures that we shall primarily discuss are generated by a one-dimensional map $x' = f(x)$. Imagine that f maps the unit interval, I, unimodally[33] on an interval which is *larger* as shown in fig.1.1. Thus $f(0) = f(1) = 0$ and f has a single maximum where $f(x) > 1$. The points that are mapped outside the unit interval "escape" and never return again. If there are no attractive periodic orbits almost all initial conditions will eventually lead to escape. There is, however, a Cantor set, the *repeller* $\Lambda \subset I$ which will remain forever and we can construct this Cantor set by looking at the set of points, Λ_n, which remain for n iterates, $n = 1,2,...$ and let $n \to \infty$. The easiest way to obtain the elements of Λ_n is by iterating the map *backwards:* if we denote the two branches of the inverse map f^{-1} by h_0 and h_1 respectively, Λ_n consists of the 2^n intervals obtained by mapping I backwards through any sequence of h_i's, of length n, where each i is either 0 or 1. These small intervals will be called *cylinders*, [4,17] they are disjoint and within each cylinder $f^n(x)$, the n-fold iterate of f, is monotonic.

The construction of this Cantor set is completely analogous to Cantors original example where the middle 1/3 is removed at each stage. Going from Λ_n to Λ_{n+1} a part of each cylinder is indeed removed, but the fraction is different for the different cylinders, depending on the derivative $|f'|$. Such Cantor sets are often referred to as "Cookie cutters". A particularly interesting case is obtained in the limit where the function maps the interval, I, exactly onto itself, since I may then be a chaotic attractor. The map is then said to be "Smale complete"[33] or to be in a fully developed chaotic state and in this case the cylinders fill out I completely. The formalism which we shall develop in this chapter can be applied to such limiting cases without problems, although, as we shall discuss in chapter 3 and 4, the theorems of chapter 2, based on hyperbolicity, break down.

1.1. The Spectrum of Characteristic Exponents.

The points on the Cantor set Λ will remain forever in I and do therefore have well-defined characteristic exponents. If the motion is ergodic almost all points have the same characteristic exponent, but in fact Λ can be decomposed into a set of interwoven Cantor sets each consisting of points with some specific characteristic exponent. Given the cylinders of, say, Λ_n it is very easy to find the different characteristic exponents: each little cylinder expands to the unit interval, I, in precisely n iterates so the length, Δ of the cylinder is related to its characteristic exponent, λ, by

$$\Delta = e^{-n\lambda}. \tag{1.1.1}$$

The fundamental quantity describing this set is the *entropy* $S(\lambda)$, which expresses how many cylinders have given λ (or equivalently, given length). More precisely the meaningfull quantities are the *growth rates* of those numbers so that $e^{nS(\lambda)} d\lambda$ is the number of cylinders with characteristic exponent in an interval of size $d\lambda$ around λ. In stead of focussing on a specific value of λ it is usually simpler to use a "Canonical formalism" and introduce a *partition function* $Z_n(\beta)$ as a sum over all cylinders, $I_j^{(n)}$, on a given level[12,17]:

$$Z_n(\beta) = \sum_j \Delta(I_j^{(n)})^\beta. \tag{1.1.2}$$

The parameter β is the analog of inverse temperature in thermodynamics, but one should note that we are interested in the whole interval $-\infty < \beta < \infty$ so "temperatures" can be both positive and negative. The growth rate of the partition function defines the *pressure* $P(\beta)$, i.e.

$$P(\beta) = \lim_{n \to \infty} \frac{1}{n} \log Z_n(\beta). \tag{1.1.3}$$

In fact, thermodynamically speaking, $P(\beta)$ is more naturally analog to the *free energy density, F(T)* (times $-\beta$), but since the originators use the word pressure for (1.1.3) we shall stick to it. One value, namely $P(1)$ has a simple physical meaning: it is simply $-\kappa$, where κ is the *escape rate* of points in $I^{31,32}$. The escape rate can be determined by distributing N_0 points uniformly on I and counting the number N_n of survivors after n step. The ratio N_n/N_0 decays exponentially as $N_n/N_0 \sim e^{-n\kappa}$. According to the construction of the cylinders, this ratio is the total length of Λ_n so

$$\sum_i \Delta(I_j^{(n)}) \sim e^{-n\kappa} \tag{1.1.4}$$

from which the relation to $P(1)$ immediately follows. For an attractor, of course, $\kappa = 0$. If we can compute the pressure for all β, the entropy follows by a Legendre transform: The sum over cylinders in (1.1.2) can be replaced by an integral over characteristic exponents by using (1.1.1) and the fact that the number of cylinders of given λ is $\exp(nS(\lambda))$. Thus

$$Z_n(\beta) \approx \int e^{nS(\lambda) - n\beta\lambda} d\lambda \tag{1.1.5}$$

and as $n \to \infty$ the saddle-point dominates the integral, which defines the *thermodynamic* value $\lambda = \lambda(\beta)$ through

$$S'(\lambda(\beta)) = \beta \qquad (1.1.6)$$

and since the growth rate of (1.1.5) is $P(\beta)$ we get the ralation

$$S(\lambda(\beta)) = P(\beta) + \lambda\beta. \qquad (1.1.7)$$

In the language of thermodynamics $\lambda(\beta)$ is analog to the internal energy and (1.1.7) becomes the familiar relation $S = (E-F)/T$. The function $S(\lambda)$ usually looks as shown in fig.1.2. It is non-zero on an interval $[\lambda_1,\lambda_2]$ and has a single maximum. The value of S at the maximum is $log2$ since there are 2^n cylinders in Λ_n (note, from (1.1.6) that $S(\lambda)$ is maximal for $\beta = 0$). The *Hausdorff dimension* of the "sub Cantor set" of points with characteristic exponent λ will be called $D(\lambda)$. Obviously this set is covered by $N = e^{nS(\lambda)}$ intervals of length $\Delta = e^{-n\lambda}$ so we find[17].

$$D(\lambda) = \lim_{n \to \infty} -\frac{logN}{log\Delta} = \frac{S(\lambda)}{\lambda} \qquad (1.1.8)$$

The maximal dimension of such subsets is the Hausdorff dimension D_0 of the whole set. Thus $D_0 = D(\lambda_0)$, where λ_0 maximizes $D(\lambda)$. According to (1.1.8) this means that $S(\lambda_0) = \lambda_0 S'(\lambda_0)$. In terms of the pressure function this translates into the well-known relation[2] $P(D_0) = 0$.

Another way of defining a spectrum of characteristic exponents for a repellor or an attractor is via the (unstable) *periodic points* [34-36] i.e. points satisfying

$$f^n(x^*) = x^* \qquad (1.1.9)$$

Note that any fixed point or cycle of a length that devides into n satisfies (1.1.9). The stability of such a cycle point is determined by the derivative $|f^{n\prime}(x^*)|$ and a natural thermodynamics can be formed by

$$Z_{fix,n}(\beta) = \sum_{x^* \in fixf^n} e^{-\beta log|f^{n\prime}(x^*)|} \qquad (1.1.10)$$

whose growth rate defines a "Free energy"

$$\beta F_{fix}(\beta) = -\lim_{n \to \infty} \frac{1}{n} logZ_{fix,n}(\beta) \qquad (1.1.11)$$

in analogy with the pressure $P(\beta)$, except for the sign and a factor of β conventional in thermodynamics. In general F_{fix} and P are not necessarily related, but for *hyperbolic systems* they are, as we shall see in chapter 2. This formalism is often easy to implement and extends easily to higher dimensional systems. It has gained additional actuality after the recent appearance of algorithms that locate periodic trajectories in experimental time series[37].

1.2. Cantor Sets with Measure: Generalized Entropies and the $f(\alpha)$ Spectrum.

The notion of a *measure* doesn't seem directly to apply to the Cantor set generated above. One can think of all the different cylinders of Λ as "equally probable". This is not the case if one wants to describe e.g. the motion on a set generated by (forward) iteration of a map $x' = f(x)$. If f is a map *on* the interval (i.e. such that all points remain in it) each segment of the interval can be assigned a measure telling how often it is visited when the map is iterated. This is called[33] the *natural* measure, μ. As we shall see a natural measure also exists for chaotic repellers.

The natural measure defining the probabilities of each element of the partition (in our case the cylinders) is seldomly known in an analytic way i.e. there is no simple connection between the lengths, Δ_i, of the cylinders and the probabilities $p_j = \mu(I_j^{(n)})$. For hyperbolic systems such a relation does exist as will be explained in the next chapter and in that case, therefore, the different "thermodynamic" quantities introduced in this chapter are all related.

Beside the natural measure one can define many other invariant measures. The generalized entropies, with respect to a measure ν, are defined as follows. For the set of n-cylinders with probabilities p_j given by some invariant measure ν the partition function

$$Z_{\nu,n}(q) = \sum_j p_j{}^q \tag{1.2.1}$$

defines the Renyi entropies[38]. The generalized entropies[30,39] are the growth rates of these Renyi entropies

$$h_q(\nu) = \lim_{n \to \infty} \frac{1}{1-q} \frac{1}{n} \log Z_{\nu,n}(q) \tag{1.2.2}$$

Here we have used the fact that the partition function defined by the cylinders becomes infinitely fine everywhere, it is a "generator"[30]. In terms of the "Free energy"

$$qF_\nu(q) = -\lim_{n \to \infty} \frac{1}{n} \log Z_{\nu,n}(q) \qquad (1.2.3)$$

we get

$$h_q(\nu) = \frac{qF_\nu(q)}{q-1} \qquad (1.2.4)$$

The denominator $q-1$ secures the correct limit as $q \to 1$:

$$h_1(\nu) = \lim_{q \to 1} h_q(\nu) = \lim_{n \to \infty} -\frac{1}{n} \sum_j p_j \, \log p_j \qquad (1.2.5)$$

The case where the measure ν is the natural measure μ is of particular importance and to avoid confusion we use K for the corresponding generalized entropies i.e. $K_q = h_q(\mu)$. The entropy K_1 is called the *metric entropy* or "Kolmogorov - Sinai invariant" and describes the rate at which the chaotic dynamics generates "information".

The generalized entropies, K_q contain information about the dynamics of the system, e.g. how far into the future we can predict it with a given initial knowledge. The so-called $f(\alpha)$ *spectrum* [8] contains, in contrast, metric-geometric information. It generalizes the Hausdorff dimension and is related to the generalized dimensions[6], D_q. In every point, x, (or family of cylinders containing x) we define the *pointwise dimension* or *crowding index* by looking at the scaling of the probability with the length of the intervals. If the pointwise dimension is α it means that

$$p_j = \nu(I_j^{(n)}) \sim (\Delta(I_j^{(n)}))^\alpha \qquad (1.2.6)$$

where ν is some measure, usually the natural one. The function $f(\alpha)$ then describes the "density" of cylinders with that α. More precisely, $f(\alpha)$ is the Hausdorff dimension of the set of points with pointwise dimension α analogos to the function $D(\lambda)$ defined above. As shown in ref.8 this function can be obtained from a generating function, Γ, whose terms are like the *ratio* of two Boltzmann factors:

$$\Gamma_n(q,\tau) = \sum_i \frac{p_i^q}{\Delta_i^\tau} \qquad (1.2.7)$$

For each q, $\tau(q)$ is found by requiring that $\Gamma = \lim_{n \to \infty} \Gamma_n$ is finite. For large n (1.2.7) can be written as an integral and

$$\Gamma(q,\tau) \approx \lim_{\Delta \to 0} \int d\alpha \Delta^{\alpha q - \tau} \Delta^{-f(\alpha)} \tag{1.2.8}$$

The condition for this to remain finite can be found by saddle point arguments analogous to (1.1.5), and one obtains[8]

$$f(\alpha) = \alpha q - \tau(q) \tag{1.2.9}$$

where

$$\alpha(q) = \tau'(q) \tag{1.2.10}$$

The generalized dimensions, D_q are then simply

$$D_q = \frac{\tau(q)}{1-q}. \tag{1.2.11}$$

In special cases where something is known about the relation between the measure (p_i) and length scales (Δ_i) there is a direct thermodynamical formalism for $f(\alpha)$ analogous to section 1.1. In particular[12,15], if the 2^n intervals Δ_i of the repellor of section 1.1 are given equal probability (2^{-n}) it is easy to see that the generating function becomes

$$\Gamma = 2^{-nq} \sum \Delta_j^{-\tau} \tag{1.2.12}$$

For this to remain finite as $n \to \infty$ demands that $P(-\tau) = q \log 2$, which implicitely defines $\tau(q)$. The relation (1.1.1) together with $p_j = 2^{-n}$ gives us by definition

$$\alpha = \frac{\log 2}{\lambda} \tag{1.2.13}$$

and further, for this λ

$$f(\alpha) = S(\lambda)/\lambda = D(\lambda) \tag{1.2.14}$$

A relation between $f(\alpha)$ and $D(\lambda)$ might have been expected, since they are both Hausdorff dimensions of a subset of the Cantor set.

This formalism has been succesfully applied to fractals appearing at the borderline of chaos for which a Renormalization Group exists[11,13,19]. Those fractals can be looked upon as cookie cutters generated by a map (f in section 1.1) which is related to the "universal function" coming from the Renormalization Group[17]. The map (e.g. unimodal map or circle map), which is becoming chaotic, maps cylinders onto each other precisely at this point and therefore the "balanced" measure $p_j = 2^{-n}$ is the dynamically relevant one.

1.3. Measure and Density: The Generalized Frobenius-Perron Equation.

One might ask whether there is a "natural" way to assign a measure to a cookie cutter. If we regard the function f generating the Cantor set as a formal device there is of course no a priori reason to choose one measure in stead of another, but if we want the measure to represent probabilities with respect to the dynamics generated by f, this is not the case. It is true that (almost) all points will eventually escape so the measure must be concentrated on the invariant Cantor set, Λ. But if we start out close enough to Λ we can stay for as many iterates as desired and this maps out a measure[32].

To find analytic methods of describing this measure let us begin by reminding ourselves how invariant measures are constructed for an *attractor*.

The natural measure μ of a one dimensional strange attractor can be described by its *density*, $\rho(x)$, such that the measure of the set A is

$$\mu(A) = \int_A \rho(x)dx \qquad (1.3.1)$$

The condition for this measure to be *invariant* is the Frobenius-Perron equation[4,33]

$$\rho(y) = \sum_{x \in f^{-1}(y)} \frac{\rho(x)}{|f'(x)|} \qquad (1.3.2)$$

as we shall se in detail in section (2.2). One can also regard the Frobenius-Perron equation as an iterative equation

$$\rho_{k+1}(y) = \sum_{x \in f^{-1}(y)} \frac{\rho_k(x)}{|f'(x)|} \qquad (1.3.3)$$

and for an attractor the natural measure can be found by iterating the uniform density $\rho_0(x) = const$. For a repellor the result of iterating (1.3.3) would be $\rho(x) = 0$: all measure eventually escapes so the probability of being in the interval, I, becomes zero. One can, however, compensate for this by inserting a factor, R, into (1.3.3):

$$\rho_{k+1}(y) = R \sum_{x \in f^{-1}(y)} \frac{\rho_k(x)}{|f'(x)|}$$

For one and only one value of R, iteration of this equation, starting from some smooth ρ_0, converges to a finite, positive $\rho(x)$, and this value turns out to be related[40] to the escape rate: $R = e^{\kappa}$.

Independently of the the problem of how to find the natural measure for repellers also other ways have been found to generalize the Frobenius-Perron equation (1.3.3). It is particularly appealing, from a thermodynamical view point, to introduce a parameter β and for each value of β try to solve

$$Q_{k+1,\beta}(y) = e^{\beta F(\beta)} \sum_{x \in f^{-1}(y)} \frac{Q_{k,\beta}(x)}{|f'(x)|^{\beta}} \tag{1.3.4}$$

again starting from a smooth, positive initial function. This process turns out to define a unique $F(\beta)$ such that the limiting $Q_{\beta}(x) = \lim\limits_{n \to \infty} Q_{n,\beta}(x)$ becomes finite and positive and independent of the initial condition for both attractors and repellers[18]. An important special case is obtained when $\beta = D_0$, the Hausdorff dimension of the set. It is shown that $F(D_0) = 0$, which provides a useful method for calculating the dimension[41]. In general $e^{-\beta F(\beta)}$ is the largest *eigenvalue* of (1.3.4) and the generalized density Q_{β} is the corresponding eigenfunction.

In order to see the connection between this formalism and statistical mechanics, let us consider the n ($\gg 1$) fold iterated density $Q_{n,\beta}$. Since the limiting value is unique we can start with the uniform density and we find

$$Q_{n,\beta}(y) = e^{n\beta F(\beta)} \sum_{x \in f^{-n}(y)} e^{-\beta \log |f^{n'}(x)|} \tag{1.3.5}$$

The sum plays the role of a partition function over the ensemble of all *preimages* of y under f^n. Since $Q_{n,\beta}$ remains finite, the growth rate of the partition function is exactly $-\beta F(\beta)$. Indeed the function $F(\beta)$ turns out to be closely connected to the pressure defined in section 1.1. In fact we shall show in the next chapter that

$$\beta F(\beta) = -P(\beta) \tag{1.3.6}$$

for hyperbolic sytems and in the last chapter that this holds at least in some non-hyperbolic system -in particular the socalled "completely chaotic cases" studied in the last two chapters. Note that the ensemble (1.3.5) depends on y, whereas the growth rate does not. Further, it is interesting to note that in this formalism a new quantity, the generalized measure $Q_\beta(x)$, appears which characterizes the chaotic system. Its interpretation in terms of *Gibbs measures* will be given in the next chapter.

2. THERMODYNAMICS OF HYPERBOLIC SYSTEMS.

A one-dimensional map, $f(x)$, is called *hyperbolic* if the slope is everywhere finite and larger than one. More generally it is sufficient to require that $1 < |f^{k\prime}(x)| < \infty$ for sufficiently large k on a partition of the attractor, or, for a repellor, over the cylinders. The "Cookie cutters" of section 1.1 satisfy this condition with certain assumption about the map, f [33], and in this chapter we shall always assume those conditions to be fulfilled and explore the strong consequences of hyperbolicity basically following refs.4 and 17. In the next chapters simple non-hyperbolic cases will be discussed.

2.1. Gibbs Measures.

For hyperbolic systems there exists a particularly attractive class of invarant measures, the so-called Gibbs measures. The *invariance* of a measure ν means that

$$\int g(x)d\nu = \int g(f(x))d\nu \tag{2.1.1}$$

for any real function $g(x)$ on I. One should note that $\int g(x)d\nu$ is a short hand notation for the limit as $n \to \infty$ of $\sum_j g(x_j)\nu(I_j^{(n)})$, where x_j is any point inside the cylinder $I_j^{(n)}$.

A Gibbs measure can be defined uniquely for any (nice) real function ϕ on I. In fact we shall be interested in particular choices of ϕ, but the derivations become clearer if we don't specify ϕ until necessary. If we introduce the notation

$$S_n \phi(x) = \phi(x) + \phi(f(x)) + \cdots + \phi(f^{n-1}(x)) \tag{2.1.2}$$

then, for any n, any cylinder, $I_j^{(n)}$, and any $x \in I_j^{(n)}$ the Gibbs measure satisfies

$$\nu_\phi(I_j^{(n)}) = c_j^{(n)} e^{-nP_\phi + S_n \phi(x)} \tag{2.1.3}$$

the *Gibbs condition*. Here the prefactor is bounded between positive, n-independent numbers, say $c_j^{(n)} \in [c_1, c_2]$ and P_ϕ normalizes the measure:

$$P_\phi = \lim_{n \to \infty} \frac{1}{n} \log \sum_j e^{S_n \phi(x_j)} \tag{2.1.4}$$

where $x_j \in I_j^{(n)}$ is chosen such that $S_n \phi(x)$ is maximal.

The "principle of bounded variation" plays an important rôle: In a hyperbolic system all cylinders become exponentially small and hence the variation of ϕ (which, for our purposes can be assumed smooth) is also exponentially small. Thus, if we assume the existence of a positive number b less than 1 and a positive c such that $|\phi(x)-\phi(y)|<cb^n$ for all n and x,y in any $I_j^{(n)}$, then, since $\phi(f^{n-m}(x))$ and $\phi(f^{n-m}(y))$, $m=1,...,n$ belong to the same m-cylinder we get

$$|S_n\phi(x)-S_n\phi(y)|\le|\phi(x)-\phi(y)|+...+|\phi(f^{n-1}(x))-\phi(f^{n-1}(y))| \qquad (2.1.5)$$

which is bounded *independently* of n, namely by $d=c/(1-b)$. This means that growth rates like (1.1.3) are independent of the precise choice of x_j within each cylinder - any point can be chosen with the same result. The corrections are of order $\dfrac{1}{n}$ and vanish in the thermodynamical limit.

We shall now prove an important *variational principle* for the Gibbs measures[2,4,42]. Let P_ϕ be defined by (2.1.4) and $h(v)=h_1(v)$ be the metric entropy (1.2.5) for an invariant measure v. Then, in general

$$h(v)+\int\phi\,dv\le P_\phi \qquad (2.1.6)$$

and the equality is attained precisely for $v=v_\phi$, the Gibbs measure. We first show that (2.1.6) is true for any invariant measure v: From the invariance property (2.1.1) follows, in particular, that

$$\int S_n\phi(x)dv = n\int\phi(x)dv \qquad (2.1.7)$$

Combining this with (1.2.5) we can write the left hand side as

$$h(v)+\int\phi\,dv = \lim_{n\to\infty}\frac{1}{n}\sum_j v(I_j^{(n)})\left[-\log(v(I_j^{(n)})+S_n\phi(x_j)\right] \qquad (2.1.8)$$

where $x_j\in I_j^{(n)}$. Now denote $v(I_j^{(n)})$ as p_j. Then we can certainly bound (2.1.7) by maximizing over all, positive p_j (not necessarily coming from an invariant measure) with the condition $\sum_j p_j = 1$. The result is

$$p_j = e^{S_n\phi(x_j)-nP_\phi} \qquad (2.1.9)$$

Note the similarity between this procedure and the derivation of the canonical distribution in standard statistical mechanics. When this expression is inserted into (2.1.8) we obtain precisely P_ϕ.

We now show that the Gibbs measure satisfies

$$h(v_\phi) + \int \phi \, dv_\phi = P_\phi \qquad (2.1.10)$$

We look at

$$v(I_j^{(n)})(-\log v(I_j^{(n)}) + S_n \phi(x_j))$$

for one cylinder $I_j^{(n)}$ containing x_j and $v = v_\phi$. Now using the bounded variation (2.1.5) with constant d, and (2.1.4) we can bound this below by $v(I_j^{(n)})(nP_\phi - \log c_1 - d)$. We now sum over cylinders

$$h(v) + \int \phi \, dv = \lim_{n \to \infty} \frac{1}{n} \sum_j v(I_j^{(n)})(-\log(v(I_j^{(n)}) + S_n \phi(x_j))$$

$$\geq \lim_{n \to \infty} \frac{1}{n} \sum_j v(I_j^{(n)})(nP_\phi - \log c_1 - d) = P_\phi$$

Combining this with (2.1.6) gives (2.1.10). For any given ϕ we can create a family of Gibbs measures by introducing a multiplicative parameter β and looking at the Gibbs measures $v_{\beta\phi}$ for the function $\beta\phi(x)$. We then obtain a function $P_\phi(\beta) = P_{\beta\phi}$. By inserting into the variational principle (2.1.10) we see that the derivative of $P(\beta)$ is

$$P_\phi'(\beta) = \int \phi \, dv_{\beta\phi} \qquad (2.1.11)$$

which we shall use later. Now we have a complete analogy with statistical physics for arbitrary ϕ. The quantities $-nP_\phi(\beta)$ and $-nP'_\phi(\beta)$ play the role of a free energy and of internal energy, respectively. The Boltzmann factor (2.1.3) shows that the energy of a microstate is $S_n \phi(x_j)$, where x_j is a point in the cylinder $I_j^{(n)}$. Equation (2.1.11) tells us that the internal energy is in fact the average over the ensemble. The total entropy is given by $nh(v_{\beta\phi})$ and (2.1.6) then shows that other invariant measures do not correspond to equilibrium states.

2.2. Thermodynamics on the Sinai-Bowen-Ruelle (SRB) Measure.

The physically most interesting Gibbs measure is found when

$$\phi = -\log |f'| \qquad (2.2.1)$$

For hyperbolic attractors and repellers the Gibbs measure $\mu = \nu_\phi$ for this particular ϕ is the *natural measure* and is named the Sinai-Bowen-Ruelle (SRB) -measure after its discoverers. It is *ergodic* i.e.

$$\lim_{n \to \infty} \frac{1}{n} \sum_{i=1}^{n} g(f^i(x)) = \int g(x) d\mu \qquad (2.2.2)$$

for any continuos function g on I and almost all points x on the invariant set. Further, any *smooth* invariant measure must be the SRB measure. From now on we shall always assume that ϕ is chosen according to (2.2.1). Then (2.1.2) aquires a special meaning:

$$S_n\phi(x) = -\log|f'(x)| - \log|f'(f(x))| - \cdots - \log|f^{n-1}{}'(x)| \qquad (2.2.3)$$

$$= -log|f'(x)....f'(f^{n-1}(x))| = -\log|f^n{}'(x)|$$

which means that $-\frac{1}{n}S_n\phi$ converges to the characteristic exponent, λ. Thus the same reasoning that led to (1.1.1) now gives

$$\Delta(I_j^{(n)}) = e^{-n\lambda} = e^{S_n\phi(x_j)} \qquad (2.2.4)$$

where x_j lies in cylinder $I_j^{(n)}$ with length $\Delta(I_j^{(n)})$. The Gibbs condition (2.1.3) can now be written

$$\nu(I_j^{(n)}) = c_j^{(n)} e^{-nP_\phi} \Delta(I_j^{(n)}) \qquad (2.2.5)$$

As in last section we now look at the Gibbs measures $\nu_{\beta\phi}$ built on (2.2.1) denoted shortly as ν_β. Here

$$\nu_\beta(I_j^{(n)}) = c_j^{(n)}(\beta) e^{-nP_\phi(\beta)} (\Delta(I_j^{(n)}))^\beta \qquad (2.2.6)$$

where the c's are bounded in some interval $[c_1, c_2]$.

If we now sum (2.2.6) over all cyliders, the normalization condition is

$$1 = \sum_j \nu_\beta(I_j^{(n)}) = e^{-nP_\phi(\beta)} \sum_j c_j^{(n)}(\beta)(\Delta(I_j^{(n)}))^\beta$$

The sum is bounded between $c_1 \sum (\Delta(I_j^{(n)}))^\beta$ and $c_2 \sum (\Delta(I_j^{(n)}))^\beta$ both of which have *growth rates* $P(\beta)$, defined by (1.1.2-3) and therefore the sum itself must also grow as

$\exp(nP(\beta))$. This shows that $P_\phi(\beta)$ with ϕ given by (2.2.1) is equal to the *pressure* $P(\beta)$ defined in section 1.1, i.e.[17]

$$P_\phi(\beta) = P(\beta) \tag{2.2.7}$$

The Legendre transform of $P(\beta)$ is $S(\lambda)$ - the growth rate of cylinders with characteristic exponent λ. Thus

$$S(\lambda) = P(\beta) + \lambda\beta \tag{2.2.8}$$

where

$$\lambda(\beta) = -P'(\beta) \tag{2.2.9}$$

Now, using (2.1.11)

$$\lambda(\beta) = -\int \phi \, d\nu_\beta \tag{2.2.10}$$

and from the variational principle (2.1.10) we get

$$h(\nu_\beta) = P(\beta) - \beta \int \phi \, d\nu_\beta = P(\beta) - \beta P'(\beta) = S(\lambda) \tag{2.2.11}$$

Thus[17] $S(\lambda)$ is the *metric entropy of the Gibbs measure* ν_β.

The entropy spectrum based on ν_β is given by (1.2.1-2). Using the Gibbs condition (2.2.6) and the same reasoning that led to (2.2.7) we get

$$Z_{\nu_\beta,n}(q) = e^{-nqP(\beta)} \sum_j (\Delta(I_j^{(n)}))^{\beta q} = e^{-nqP(\beta)} e^{nP(\beta q)} \tag{2.2.12}$$

which shows that

$$h_q(\nu_\beta) = \frac{P(\beta q) - qP(\beta)}{1-q} \tag{2.2.13}$$

Note that in (2.2.12) we have equated expression with the same growth rate being always interested in the limit $n \to \infty$. In the limit $q \to 1$ we recover (2.2.11) and the generalized entropies for the SRB measure appear for $\beta = 1$.

The $f(\alpha)$ spectrum can also be found. To stress the fact that it is built on the measure ν_β we shall use tha notation $f_\beta(\alpha)$. As before we set $p_j = \nu_\beta(I_j^{(n)})$ and substitute into (1.2.7) using the Gibbs condition (2.2.6). Again we neglect the constants $c_j^{(n)}$ since we only want growth rates and we find

$$\Gamma_{n,\beta}(q,\tau) = e^{-nqP(\beta)} \sum_j \frac{\Delta(I_j^{(n)})^{q\beta}}{\Delta(I_j^{(n)})^\tau} = e^{-nqP(\beta)} e^{nP(q\beta-\tau)} \tag{2.2.14}$$

The condition that Γ_β remains finite selects a $\tau_\beta(q)$ which, via (1.2.17) determines the spectrum of generalized dimensions. This $\tau_\beta(q)$ is determined implicitly by

$$qP(\beta) = P(q\beta-\tau) \tag{2.2.15}$$

A more explicit relation follows by defining the pointwise dimension, α, through $p_j \sim \Delta(I_j^{(n)})^\alpha$. From (2.2.4-6) we then immediately obtain the relation

$$\alpha = \beta + \frac{P(\beta)}{\lambda} \tag{2.2.16}$$

between α and λ and finally

$$f_\beta(\alpha) = \frac{S(\lambda)}{\lambda} = D(\lambda) \tag{2.2.17}$$

Therefore, as (1.2.11) based on equal measure of all cylinders, $f(\alpha)$ is the same as $D(\lambda)$, but the relation between α and λ is different and depends on β.

In particular we find for the SRB measure

$$\alpha = 1 + \frac{P(1)}{\lambda} = 1 - \frac{\kappa}{\lambda} \tag{2.2.18}$$

where κ is the *escape rate*. The most probable value of λ is the Liapunov exponent $\bar{\lambda}$, and the corresponding α is the information dimension, D_I. Hence

$$\kappa = \bar{\lambda}(1-D_I) \tag{2.2.19}$$

which relates escape rate, Liapunov exponent and information dimension for a hyperbolic system[31,32]. For an attractor $\kappa = 0$ so $\alpha = 1$, which shows that a one-dimensional, hyperbolic strange attractor is really not so strange.

Next we show that the pressure in a hyperbolic system can be found via the *periodic points* as in (1.1.9-11). This can now be seen as a simple consequence of the bounded variation (2.1.5). For a Cookie cutter f^n is monotonic in each cylinder and maps each of them to the entire interval, I. Thus each cylinder contains precisely one fixed point of

f^n, and the exponent of (1.1.10) is precisely $-\beta S_n \phi(x)$ evaluated at those points. Now (2.1.5) ensures that growth rates are unchanged if we choose different points in the cylinders, so the growth rates of $Z_{fix,n}(\beta)$ and $Z_n(\beta)$ must be identical i.e.

$$\beta F_{fix}(\beta) = -P(\beta) \tag{2.2.20}$$

The invariance property (2.1.1) of the Gibbs measures leads to a generalized Frobenius-Perron equation - in fact that is how they are usually constructed (see ref.4). Let us look for Gibbs measure with a smooth *density*. By the density we mean that the Gibbs condition (2.2.6) can be written

$$\nu_\beta(I_j^{(n)}) = \rho_\beta(x)e^{-nP(\beta)}(\Delta(I_j^{(n)}))^\beta = \rho_\beta(x)e^{-nP(\beta)}|f^{n\prime}(x)|^\beta \tag{2.2.21}$$

where $\rho_\beta(x)$ is a smooth function. Note that $\beta = 1$ leads to a normal density, which shows that a smooth measure must necessarily be SRB. That implies also that the spectrum of characteristic exponents given by $S(\lambda)$ will also tell us the probability of finding given "finite time" Liapunov exponents[10] in, say, a time series from an experiment (which of course probes the natural measure). Now, invariance means that the two cylinders on level $n+1$, say $I_k^{(n+1)}$ and $I_l^{(n+1)}$, that are mapped by f onto $I_j^{(n)}$ must have the same measure as the latter: $\nu_\beta(I_k^{(n+1)}) + \nu_\beta(I_l^{(n+1)}) = \nu_\beta(I_j^{(n)})$. When we insert (2.2.19) we get

$$\rho_\beta(y) = e^{-P(\beta)}\left[\frac{\rho_\beta(x_1)}{|f'(x_1)|^\beta} + \frac{\rho_\beta(x_2)}{|f'(x_2)|^\beta}\right] \tag{2.2.22}$$

where $y \in I_j^{(n)}$ and $y = f(x_1) = f(x_2)$. This is precisely the "generalized Frobenius-Perron equation" introduced in section (1.3) and the derivation shows that the function F of (1.3.4) must be related to the pressure by (1.3.5) and that the solution, $Q_\beta(x)$, is precisely the density of the Gibbs measure. Thus we have, finally,

$$P(\beta) = -\beta F_{fix}(\beta) = -\beta F(\beta) \tag{2.2.23}$$

and all the formalisms of chapter 1 are equivalent.

3. THERMODYNAMICS FOR THE LOGISTIC MAP AT THE "ULAM - VON NEUMANN POINT."

In this chapter we shall treat a simple example in detail: the limiting "Cookie cutter" Cantor set generated by the logistic map $f(x) = \mu x(1-x)$ with $\mu = 4$ (the "Ulam - von Neumann" point[33]). This is the simplest example of a *non-hyperbolic chaotic attractor*. It is exactly soluble due to the conjugacy of f to a trivial map, but since the conjugating function has zero slope in the ends of the interval the results are not trivial. The thermodynamic functions display a phase transition[17,20-22,43] and the $Q_\beta(x)$ densities, which can be calculated explicitly, show interesting change of behaviour.

3.1. The Thermodynamic Functions.

To construct the "Cookie cutter" for the map

$$f(x) = 4x(1-x) \tag{3.1.1}$$

we use the fact that f is conjugate to the "tent map"

$$t(w) = 1 - |2w - 1| \tag{3.1.2}$$

as

$$f(x) = 4x(1-x) = h \circ t \circ h^{-1}(x) \tag{3.1.3}$$

where

$$h(w) = \sin^2 \frac{\pi}{2} w \tag{3.1.4}$$

The cylinders of the tent map at a given level, n, are simply the 2^n intervals $C_j = [\frac{j-1}{2^n}, \frac{j}{2^n}]$, where $j=1,\dots,2^n$, and since f^n is conjugate to t^n, i.e. satisfies

$$f^n(x) = h \circ t^n \circ h^{-1}(x) \tag{3.1.5}$$

the cylinders, D_j, of the logistic map, are the images

$$D_j = [h(\frac{j-1}{2^n}), h(\frac{j}{2^n})] = [\sin^2(\frac{\pi}{2} \frac{j-1}{2^n}), \sin^2(\frac{\pi}{2} \frac{j}{2^n})] \tag{3.1.6}$$

with lengths

$$\Delta_j = \sin^2\left[\frac{\pi}{2}\frac{j}{2^n}\right] - \sin^2\left[\frac{\pi}{2}\frac{j-1}{2^n}\right] \tag{3.1.7}$$

If j is large the difference can be approximated by the derivative and $\Delta \sim 2^{-n}$. That is the case for most j (almost all, in fact). If, however,

$$j \sim 2^{\sigma n} \tag{3.1.8}$$

with $0 < \sigma < 1$, we find

$$\Delta_j \approx 2^{\sigma n - 2n} \tag{3.1.9}$$

Now, using (1.1.1): $\Delta \sim e^{-\lambda n}$, we can find the spectrum of characteristic exponents $S(\lambda)$. First, (3.1.9) and (1.1.1) gives

$$\lambda(\sigma) = (2-\sigma)log2 \tag{3.1.10}$$

which lies in the interval $[log2, 2log2]$. Second, the number of cylinders in an interval $d\sigma$ around σ is proportional to $n2^{n\sigma}d\sigma$ so

$$S(\lambda) = \sigma log2 = 2log2 - \lambda \tag{3.1.11}$$

The function $S(\lambda)$ is shown in fig 3.1a and has a shape which is very different from the shape of $S(\lambda)$ for a hyperbolic system. Since $S'(\lambda) = -1$ for all λ the pressure cannot be found from (1.1.5) and (1.1.6). We must go back to the definitions (1.1.2-4). When we insert the known $S(\lambda)$ into (1.1.5) we get

$$Z_n(\beta) = \int e^{(2log2 - \lambda - \lambda\beta)n} d\lambda \tag{3.1.12}$$

which means that

$$P(\beta) = 2log2 + \lim_{n\to\infty}\frac{1}{n}\log\int_{log2}^{2log2} e^{-\lambda(1+\beta)n} d\lambda \tag{3.1.13}$$

In this case the saddle-point method doesn't work: the maximal value of the exponential is obtained at one of the two *endpoints* of the interval depending on the value of β. Thus[17]

$$P(\beta) = \begin{cases} (1-\beta)log2 & \beta > -1 \\ -2\beta log2 & \beta < -1 \end{cases} \tag{3.1.14}$$

which is shown in fig3.1b and we see that the system undergoes a first order phase transition at $\beta = -1$. The "disordered" phase is $\beta > -1$ and here almost all cylinders contribute equivalently. In the "condensed" phase $\beta < -1$ only the cylinders very close to the edge contribute. It is easy to see that $\beta F_{fix}(\beta) = -P(\beta)$ also in this case. In fact the slope at all fixed points of f^n is 2^n except at the origin where the slope is 4^n (see (3.2.4) below) and, again, for $\beta < -1$ that one cylinder dominates the partition function.

Because of the conjugacy to the tent map the natural measure can simply be carried over and this shows that all cylinders have equal measure. That is very different from the Gibbs condition (2.2.6), but in fact it simply means that the constant $c_2 \to \infty$, i.e. that the density in (2.2.21) *diverges*. The logistic map is very special in that the divergence in the density $\rho(x)$, which is determined by the order of the critical point precisely cancels the "extra compression" of the edge cylinders, which is determined by the slope of f at the origin. Thus the generalized entropies (1.2.2) are trivial i.e.

$$K_q = \log 2 \tag{3.1.15}$$

for *all* q. Further in this special case $f(\alpha)$ can be found from (1.2.12) giving a straight line from (1/2,0) to (1,1) as it should in any system with a square root divergence of the invariant density[8].

3.2. Solution of the Generalized Frobenius-Perron Equation.

We shall now calculate the $Q_\beta(x)$ functions by solving the generalized Fronbenius-Perron equation (1.3.4). We shall explicitly see that (1.3.6) holds, i.e. $\beta F(\beta) = -P(\beta)$. By inserting this value of $F(\beta)$ into (1.3.4) we shall se that $Q_\beta(x)$ can be found iteratively and is positive and finite (except for possible divergences at the ends.)

We then have to solve

$$Q_\beta(y) = e^{-P(\beta)} \sum_{x \in f^{-1}(y)} \frac{Q_\beta(x)}{|f'(x)|^\beta} \tag{3.2.1}$$

with $f(x)$ given by (3.1.1) and $P(\beta)$ given by (3.1.14). We solve this by iteration, as described in section 1.3, starting from a uniform density. Thus

$$Q_\beta(y) = \lim_{n \to \infty} e^{-nP(\beta)} \sum_{x \in f^{-n}(y)} \frac{1}{|f^{n'}(x)|^\beta} \tag{3.2.2}$$

Using (3.1.5) we can now express f^n as

$$f^n(x) = \sin^2(2^n(arcsin\sqrt{x} - \frac{\pi}{2}\frac{j}{2^n}))$$ (3.2.3)

which, for any even j, is valid when $x \in D_j \cup D_{j+1}$. The derivative which appears in (3.2.2) can now be expressed as

$$|f^{n\prime}(x)| = 2^n \left[\frac{y(1-y)}{x(1-x)}\right]^{1/2}$$ (3.2.4)

where $y = f^n(x)$. In each cylinder D_j there is one $x = x_j(y) \in f^{-n}(y)$ found by inverting (3.2.3). That is, for j even,

$$x_j(y) = \sin^2(\frac{\pi j - 2z(y)}{2^{n+1}})$$ (3.2.5)

and

$$x_{j+1}(y) = \sin^2(\frac{\pi j + 2z(y)}{2^{n+1}})$$ (3.2.6)

where we have defined $z = arcsin\sqrt{y}$. The sum (3.2.2) can now be rewritten as

$$Q_\beta(y) = \lim_{n\to\infty} e^{-nP(\beta)} 2^{1-(n+1)\beta} (y(1-y))^{-\beta/2} F_{N,\beta}(z)$$ (3.2.7)

where $N = 2^{n-1}$ and

$$F_{N,\beta}(z) = \sin^\beta\frac{z}{N} + \sum_{k=1}^{\frac{N}{2}-1}\left[\sin^\beta(\frac{\pi k + z}{N}) + \sin^\beta(\frac{\pi k - z}{N})\right] + \cos^\beta\frac{z}{N}$$ (3.2.8)

When the explicit form of the pressure function $P(\beta)$ ((3.1.14)) is inserted this becomes

$$Q_\beta(y) = \lim_{N\to\infty} \begin{cases} (y(1-y))^{-\beta/2} 2^{-\beta} N^{-1} F_{N,\beta}(z) & \beta > -1 \\ (y(1-y))^{-\beta/2} 2 N^\beta F_{N,\beta}(z) & \beta < -1 \end{cases}$$ (3.2.9)

The expression $F_{N,\beta}(z)$ looks like a Riemann sum. In fact

$$N^{-1}F_{N,\beta}(z) \rightarrow \frac{2}{\pi} \int_0^{\pi/2} \sin^\beta w \ dw \qquad (3.2.10)$$

independently of z provided the integral converges. This is the case when $\beta > -1$ and then

$$Q_\beta(y) = (y(1-y))^{-\beta/2} c(\beta) \qquad (3.2.11)$$

where $c(\beta)$ is a constant.

For $\beta < -1$, $N^{-1}F_{N,\beta}$ diverges for $N \rightarrow \infty$ and to obtain $Q_\beta(y)$ we must look closer at $N^{-|\beta|}F_{N,\beta}(z)$. We do this by subtracting the divergent part. That is, we look at

$$G_{N,\beta}(z) = N^{-|\beta|} \left[F_{N,\beta}(z) - (\frac{N}{z})^{|\beta|} - \sum_{k=1}^{N/2} (\frac{N}{\pi k + z})^{|\beta|} + (\frac{N}{\pi k - z})^{|\beta|} \right] \qquad (3.2.12)$$

which can be rearranged as

$$G_{N,\beta}(z) = N^{-|\beta|} \sum_{k=1}^{N/2} \sin^{-|\beta|}(\frac{\pi k + z}{N}) + \sin^{-|\beta|}(\frac{\pi k - z}{N}) - (\frac{N}{\pi k + z})^{|\beta|} - (\frac{N}{\pi k - z})^{|\beta|}$$

Now we have a Riemann sum for a *convergent* integral:

$$G_{N,\beta}(z) \rightarrow N^{-|\beta|+1} \frac{2}{\pi} \int_0^{\pi/2} (sin^{-|\beta|}(w) - w^{-|\beta|}) \ dw \qquad (3.2.13)$$

which approaches 0 for $N \rightarrow \infty$ when $\beta < -1$. Thus

$$Q_\beta(y) = c(\beta) (y(1-y))^{|\beta|/2} T_\beta(z) \qquad (3.2.14)$$

where

$$T_\beta(z) = \frac{1}{z^{|\beta|}} + \sum_{k=1}^\infty (\frac{1}{\pi k + z})^{|\beta|} + (\frac{1}{\pi k - z})^{|\beta|} \qquad (3.2.15)$$

Here, again, $c(\beta)$ is a constant and $z = arcsin \sqrt{y}$.

For $\beta = -2$ the result is particularly simple. Using

$$(\sin(z))^{-2} = \sum_{k=-\infty}^\infty (\frac{1}{\pi k + z})^2 \qquad (3.2.16)$$

we find

$$Q_{-2}(y) \propto 1-y \qquad (3.2.17)$$

We can check that $Q_\beta(x)$ defined by (3.2.14) really do satisfy the Frobenius-Perron equation (3.2.1) for $\beta < -1$. Inserting (3.2.14) into (3.2.1) gives

$$T_\beta(\frac{z}{2}) + T_\beta(\frac{\pi-z}{2}) = 2^{|\beta|} T_\beta(z) \qquad (3.2.18)$$

which is satisfied by $T_\beta(x)$ defined by (3.2.15)

It is interesting to note that the functions $Q_\beta(x)$ are symmetric in the "disordered" phase, whereas this symmetry is broken in the ordered phase. Fig 3.2 shows $Q_\beta(x)$ for different β. Notice that Q_β given by (3.2.9) does not remain finite as $\beta \to -1^-$, which may also be interpreted as a sign of a phase transition.

4. THE THERMODYNAMICS OF "COMPLETE CHAOS".

In the last chapter we treated the special case of the logistic map at the completely chaotic point. We shall now look into the more general case of completely chaotic, smooth maps of the interval. The main issues will be to determine which parts of the theory of chapter 2 that remain valid, i.e. the relation between different thermodynamical formalisms, and how close the general case is to the logistic one - in particular whether the phase transitions found in that case (chapter 3) will remain. The answers to those questions will be incomplete and partly conjectural, reflecting our present understanding and emphasizing the need of further work.

The fractals to be studied in this chapter are limiting cookie cutters generated by a unimodal map that maps the interval, I, *onto* itself. If the map is smooth such a system is *non-hyperbolic* since the attractor (the whole of I) contains the critical point x_c, where $f'(x_c)=0$. The cylinders are still well-defined since every point in I has two preimages inside I. In the case of the logistic map (3.1.1) the existence of a function (3.1.4) conjugating it to a map with trivial thermodynamics made it possible to solve everything exactly. In the general case this is of course not possible and one has to resort to numerical computations, using techniques borrowed from statistical mechanics. These techniques will not be discussed in any detail here, but relevant references will be given.

One simplifying aspect should be noted from the outset. All the maps discussed here have nice, smooth natural measures. That is, the density is smooth everywhere except around the edges where singularities develop. In general any forward image of the critical point must have a singularity (as will be discussed in section 4.2 below), but in these special cases the only images are the two edges since the left edge is a fixed point. The nature of these singularities depend only on the *order of the critical point* and therefore the $f(\alpha)$ spectrum (with respect to the natural measure) are all identical.This is, as we shall see, not true for the entropy function or the pressure of section 1, and this difference goes beyond simply reflecting that sizes of characteristic exponents must change when the maps change slopes, since indeed the existence of phase transitions depends on the particular map. The differences comes from the fact that the cylinders carry different measures. Consequently the K_q spectrum is not q-independent. Furthermore the distribution of characteristic exponents given by $S(\lambda)$ is not the same as the distribution with respect to the natural measure[10,44,45]. It can be shown, however, that the thermodynamics based on the cylinders, periodic orbits and preimages are equivalent for completely chaotic maps.

4.1. A Family of Quadratic Maps.

To analyze a more general case of "completely chaotic" maps the limiting cookie cutters for a family of maps, f_γ, was analyzed in ref.24. They chose the maps

$$f_\gamma(x) = Ax(1-x)(1+\gamma x) \tag{4.1.1}$$

where A is adjusted so the maximal value of f_γ is 1. The maps can be thought of as non-symmetric "perturbations" of the logistic map ($\gamma=0$), they all map the unit interval onto itself and all have quadratic maxima. The cookie cutters of the maps (4.1.1) for a sequence of γ-values were studied and an attempt was made to determine the structure of the entropy functions, whether they showed phase transitions and how the different phases could be described.

To do this one has to overcome two difficulties. First, one has to compute pressure functions with high accuracy (without, of course, being able to take the limit $n \to \infty$) and second, one has to find a good way of locating phase transitions and describing the structure of a given phase. Although we shall not describe any details here we shall briefly indicate the relevant concepts and refer the reader to the literature.

The first difficulty is most efficiently overcome by the *transfer matrix technique* [12] in (almost) complete analogy with statistical mechanics. The elements[11,14] of the transfer matrix are constructed from the *scaling function* [46], which expresses how the length of cylinders scale when we go from Λ_n to Λ_{n+1}. The reason why the analogy to statistical mechanics is not quite complete is that in that case the transfer matrix relates, say, two columns containing the same number of spins (see e.g. ref.47), whereas in our case Λ_{n+1} contains twice as many cylinders as Λ_n. The advantage of this formalism is that the pressure simply emerges as the largest eigenvalue of the transfer matrix which can be evaluated very accurately.

The second difficulty was overcome in ref.24 by introducing the concept of an *order parameter*. This is done by using the formal analogy between the thermodynamical formalism and statistical mechanics for an Ising spin chain[12,20,25]. Every cylinder has a unique, binary address which simply expresses the symbolic dynamics (with respect to the critical point) of the points inside as the map is iterated. But the "energies" (i.e. $e^{-\beta \log \Delta}$) entering the partition function (1.1.2) can then be thought of as those of an appropriate Ising model, where each spin (binary digit) can be up or down. Out of this analogy an order parameter can be constructed simply from the local spin averages (although care has to be taken in defining the spins appropriately). It should be noted that the ensuing Ising models have very complicated interactions, far from the standard nearest neighbor case. Indeed to have phase transitions in a one-dimensional spin system the interactions must be infinite ranged, which turns out to be intimately connected with the existence of a critical point for the map, f, inside I.

The resulting $S(\lambda)$ spectra for the maps (4.1.1) depend strongly upon γ and particularly upon its sign. Fig.4.1 represents the results for a negative and a positive γ. The figure is not drawn to fit any particular value of γ, but is an idealization showing the essential features as given in ref.24. To understand the results, let us first remind ourselves of the behaviour for the logistic map ($\gamma=0$) as given by fig.3.1. There the entropy is positive (and linear) between $\log s_1$ and $\log s_0$, where $s_0=4$ is the slope at the origin and $s_1=2$ is the slope at the fixed point inside I (or, in this case, for any cycle away from the origin). There is a phase transition at $\beta_c=-1$, which is the slope of the entropy at the right hand edge.

For the map (4.1.1) with $\gamma<0$, the slope s_0 at the origin is larger than 4 and it still represents the maximal characteristic exponent: $\lambda_{max}=\log s_0$. The minimal λ is still close to $\log s_1$, which, in this case, is less than 2. The phase transition remains: indeed, the whole decreasing part of the entropy seems to be a straight line going from ($log2,log2$) to ($0,logs_0$). Thus the phase transition takes place at

$$\beta_c = - \frac{\log 2}{\log s_0 - \log 2} = - \frac{\log 2}{\log s_0/2} \qquad (4.1.2)$$

In fact[48], if the idealized entropy of fig.4.1a really represents the correct asymptotic structure (which can be hard to extract unambiguosly from the data) the discontinuity of the slope at $\lambda=log2$ should imply another transition at large, positive $\beta=\beta_1$ in order of magnitude given by (4.1.2) with s_0 replaced by s_1. According to the Ehrenfest classification the latter transition should be 2.order: Since the the entropy is continuos at β_1, the pressure function should only have a discontinuity in the second derivative. This is not inconsistent with fig.4d of ref.24, where the order parameter seems to grow up continuosly for $\beta>\beta_1$, but further work is need to clarify the situation.

For $\gamma>0$ the situation is different. Here $s_0<4$ and $s_1>2$ so the λ-interval shrinks. Now, for all "complete" cases the total length of cylinders must equal 1 (the whole of I) for any n. Therefore the interval [$\lambda_{min},\lambda_{max}$] must contain log2 and λ_{min} can no longer be given by $\log s_1$. There is, however, another important scaling due to the critical point. The cylinders on level n+1 are obtained from those on level n by iterating the map (4.1.1) once backwards. In general it is scaled down by the slope of f_γ, but if the cylinder passes close to the critical point, where f' is zero, it is not scaled down by a factor but becomes proportional to its square root. This can only happen if the cylinder is very close to the origin so the new relevant scaling is $\sqrt{s_0}$. Indeed $\lambda_{min}=1/2\log s_0$, at least within a few percent.

If γ is not too big (up to $\gamma\approx1$) the phase transition at $\beta<0$ remains, this time with $\beta_c>1$. In this case probably the entropy is a straight line only in a finite segment of the decreasing ($\beta<0$) part as shown in fig.4.1b. The transition at positive β has dissappeared. For greater values of γ no transition is observed at all. It disappears around (probably slighly before) the point where s_0 and s_1 become identical and presumably the critical point has $\beta_c=-\infty$, i.e. takes place at zero temperature.

4.2. Analysis Based on the Functional Equation

The study of the eigenfunctions of the generalized Frobenius-Perron equation (1.3.4) provides additional information about the thermodynamics and phase transition of completely chaotic maps[49]. For the sake of simplicity we keep on working with quadratic maps i.e. we assume that the function $f(x)$ behaves as $f(x)=1-a(x-x_c)^2$ around the critical point x_c. We perform a *singularity* analysis of the eigenfunctions Q_β and study how they arise.

Let us write the equation in the form

$$S_{k+1,\beta}(y) = R(\beta) \sum_{x \in f^{-1}(y)} \frac{S_{k,\beta}(x)}{|f'(x)|^\beta}. \tag{4.2.1}$$

As initial functions we now allow also ones having singularities at least at one of the endpoints of the interval. $1/R(\beta)$ is the corresponding eigenvalue ensuring convergence to a finite limiting $S_\beta(x)$. We are interested in $S_{k+1,\beta}$ around the two endpoints of the interval. For $y \to 1$

$$S_{k+1,\beta}(y) = \frac{R(\beta)S_{k,\beta}(x_c)}{2^{\beta-1}a^{\beta/2}}(1-y)^{-\beta/2} \tag{4.2.2}$$

since both preimages of y coincide in this limit. Consequently, $S_{k+1,\beta}$ is always singular at the right corner. For $y \to 0$ the preimages of y can be written as y/s_0 and $1-y/s_1$ where s_0 and s_1 are the slopes of $f(x)$ at the origin and at $x = 1$, respectively. Thus

$$S_{k+1,\beta}(y) = R(\beta) \left[S_{k,\beta}(y/s_0)s_0^{-\beta} + S_{k,\beta}(1-y/s_1)s_1^{-\beta} \right] \tag{4.2.3}$$

The singularity of $S_{k+1,\beta}(y)$ at $y = 0$ depends on the singularity of the previous iterate at the two corners.

Consider first functions $S(x)$ having the same singularity, of order $-\beta/2$, at both ends. An example is $(x(1-x))^{-\beta/2}$. We call these functions of type I. Note that the eigenfunctions $Q_\beta(x)$ for $\beta > -1$ of the logistic case are exactly of this type. When chosing S_0 from this class, $S_{1,\beta}$ and all $S_{k,\beta}$ will be member of the class, too, since both terms on the right hand side of (4.2.3) have the same singularity for $y \to 0$. The requirement of having a finite S_β selects a prefactor

$$R(\beta) = e^{\beta F_1(\beta)}. \tag{4.2.4}$$

The reciprocal value of it is the largest eigenvalue of (4.2.1) for functions of class I. Note that $F_1(\beta)$ is not necessarily $F(\beta)$ of (1.3.4) since the latter is defined for smooth initial functions. (The relation betwee $F_1(\beta)$ and the generalized entropies is discussed in ref.59.) The spectrum $F_1(\beta)$ can easily be obtained numerically by iterating (4.2.1) with $(x(1-x))^{-\beta/2}$ as initial function and requiring a finite limiting value of Q.

Let us now turn to functions of type II, defined by having a singularity of order $-\beta/2$ at the right corner, but being finite at $x = 0$, like e.g. $(1-x)^{-\beta/2}$. The eigenfunctions of the logistic case are of this type for $\beta < -1$. By starting with such an S_0 the first iterate for $\beta > 0$ will be of type I since the first term is then negligible on the right hand side of (4.2.3). For $\beta < 0$, however, the first term dominates, and S_1 and its further iterates will remain inside class II. This class for $\beta < 0$ possesses its own eigenvalue. Applying now (4.2.3) on the limiting function S_β, one immediately sees that the prefactor must be

$$R(\beta) = e^{\beta F_0(\beta)} = s_0^\beta. \tag{4.2.5}$$

We can now turn to the original problem of smooth initial functions Q_0. It follows from (4.2.2) that $Q_{1,\beta} \sim (1-y)^{-\beta/2}$ for $y \to 1$. In the range $\beta > 0$ the same singularity appears also at the left corner and Q_β will be of type I. For $\beta < 0$, however, $Q_{2,\beta}$ will be of type II since $Q_{1,\beta}(1-y/s_1)$ vanishes for $y \to 0$. In this range, therefore, besides the eigenvalues characterizing class I, also $e^{-\beta F_0(\beta)}$ is to be taken into account. The actual fate of $Q_{k,\beta}$ is determined by the *largest* eigenvalue. If $s_0^{-\beta} > e^{-\beta F_1(\beta)}$, $Q_{k,\beta}$ remains in class II, otherwise $Q_{k,\beta}(0)$ decreases in course of the iteration as one sees from (4.2.3). The limiting function will be then of type I.

In other words, the free energy $\beta F(\beta)$ defined by (1.3.4) for smooth initial functions appears as a minimum of βF_0 and βF_1:

$$\beta F(\beta) = \min \left[\beta F_0(\beta), \beta F_1(\beta) \right]. \qquad (4.2.6)$$

This is in complete analogy with thermodynamics, where a first order phase transition occurs when the minimum of the free energy crosses over from one branch onto another. In our case F_0 does not exist for positive β. A phase transition takes place only if the branch of F_0 intersects that of F_1 at some negative β_c. For $\beta < \beta_c$ $F(\beta) = \log s_0$. The slope at the origin determines the complete thermodynamics in this range, so in these models the "condensed phase" is trivial, like a $T=0$ phase in thermodynamics, and nothing varies with β.

As an illustrative example we take the map

$$f(x) = 1 - (1-\varepsilon)(1-2x)^2 - \varepsilon(1-2x)^4 \qquad (4.2.7)$$

with $\varepsilon = -0.263$. On fig4.2 numerical approximations to $\beta F(\beta)$, $\beta F_0(\beta)$ and $\beta F_1(\beta)$ are shown -the latter two as thin lines. One can see clearly how the full free energy emerges as the lowest of those two curves, and that this method gives a rather precise way of locating the phase transition point, which in this case is $\beta_c \approx -0.869$. The asymptotic behaviour is given by $F_1(\infty) = 0.54$ and $F_1(-\infty) = 0.76$ ($F_0 = 1.08$).

As mentioned above, for the completely chaotic cases the pressure, $P(\beta)$ and the free energy, $F(\beta)$, coming from the Frobenius-Perron equation carry the same information and are still related by (1.3.6) although the map is not hyperbolic. The reason for this can be best understood from the representation (1.3.5). As long as the points x (preimages of y) avoid the critical point the "Boltzmann factors" $e^{-\beta \log |f^{n'}(x)|}$ are nothing but the corresponding Δ^β of eqn. (1.1.2) just as argued in section 2.2. Now, for x close to the critical point this breaks down, but this happens (for our complete maps) only at $y=0$ and $y=1$. In these points $Q_\beta(y)$ can diverge so $F(\beta)$ is no longer related to the growth rate of the partition function in (1.3.5). If a y-*independent* prefactor $e^{\beta F(\beta)}$ exists for (1.3.5) it must therefore be given by $e^{-P(\beta)}$.

It is worth briefly mentioning the case of more general maps with a maximum of order $z \neq 2$ for which $f(x) = 1-a\,|x-x_c|^z$. The most important change occurs for $z < 1$, i.e, if the maximum is a cusp. Then the critical point and the "condensed phase" lie in the region $\beta > 0$. Another interesting property is that the K_q spectrum may also exhibit phase transition, if the map is intermittent[23]. Thus, the following picture emerges.

For complete nonhyperbolic maps there are three essentially different thermo-dynamic formalisms providing different spectra:

- The spectrum of characteristic exponents, periodic orbits and preimages (the latter defined by (1.3.5)). This seems to be the most sensitive one. A phase transition may or may not appear in it. This defines the pressure function and the densities $Q_\beta(x)$. The Legendre transform $S(\lambda)$ gives the distribution of characteristic exponents with respect to Lebesgue measure (i.e. usual length) not with respect to the natural measure, which has singularities.

- The spectrum of generalized dimensions or $f(\alpha)$ is a rigid spectrum. A phase transition always occurs in it and the spectrum can be expressed in terms of a single parameter, the order of maximum z.

- The spectrum of generalized entropies is not quite as well understood at present. In the case of intermittent maps a singularity appears which is connected with a real time slowing down. In that case the transition is independent of z.

4.3. Final Remarks

Functions mapping an interval *into* itself define incomplete maps since in such cases not all possible (binary) sequences can be realized by trajectories. The thermodynamic description of incomplete maps is most hopeful at parameter values where *finite* symbol sequences (and their permutations) are forbidden only. This means that a so-called local grammar exists and a kind of topological universality holds. In such cases cylinders are well defined, the convergence of (1.3.1) is established and different spectra can be worked out. Maps at general parameter values can be approached through such cases just like irrational numbers through rationals.

The simplest cases are the so-called Misiurewicz points[33], where the critical point maps onto an unstable periodic orbit. The simplest one is actually the "complete" case studied above, where the critical point after two iterations falls onto the unstable fixed point in the origin. The next simplest possibility is to have the critical point fall onto the unstable fixed point inside the interval. That requires three iterates and happens for the logistic map $f(x)=\mu x\,(1-x)$ as $\mu=3.6786$, the situation which is shown in fig.4.3. Fig4.3b shows the second iterate and the boxes show that the dynamics is actually no more complicated than the complete case. Within each box (i.e. between two iterates of the critical point) we have a small, completely chaotic map. The thermodynamic functions in each box must be identical since they map onto each other by f (and the critical points are mapped onto each other). If we look at a sequence of Misiurewicz points, where the critical point falls onto an unstable 2^{n-1}- cycle we get 2^n little boxes and within each we have the same thermodynamical functions. Aside from rescaling the function within the boxes are given by a "universal chaos function"[50], which is well represented by the expansion (4.2.7). Thus fig.4.2 shows the "universal" pressure function (with a minus sign).

It should be noted that the $f\,(\alpha)$-spectrum is the same for all these cases. Since the map is complete in every little box the $f\,(\alpha)$-spectrum is still a straight line between $(0,1/2)$ and $(1,1)$. Although any "chaotic point" can be reached by a sequence of (generalized) Misiurewicz points this does *not* imply that they all have that same spectrum. Indeed these $f\,(\alpha)$ curves represent the asymptotic scaling reached only for distances much less than the distances between images of the critical point. For a general chaotic point these images are ergodic, filling out the whole attractor, so this length-scale can never be reached.

In closing we would like to mention some recent work which goes beyond the scope of this article, but which might interest the reader. A severe limitation of our exposition has come from the constraint to one-dimensional problems. Recently a lot of work has been done in higher dimensional systems, specifically the Hénon map[26-28,51-53] and maps of the annulus[54] to which we would like to draw attention. A different issue, the thermodynamics of mode-locking, (see the article on circle maps by Bak et.al. in this volume) is dealt with in ref.55 and refs.56-57 discuss relations between the different thermodynamical formalisms of the same type as those discussed in the preceeding chapters. Further, in ref.58, the spectra of fractal aggregates have been modelled in terms of Julia sets. There it turns out that the "balanced measure" (1.2.12) precisely gives the "harmonic measure" of the aggregate showing where the growth takes place.

Aknowledgements.

We are thankful for valuable discussions and collaboration with A.Csordas, Z.Kovacs, P.Cvitanovic, M.H.Jensen, I.Procaccia, D.Rand and P.Szépfalusy. One of us (T.T.) is indebted to Itamar Procaccia for the kind hospitality at the Weizmann Institute, where part of the manuscript was completed. This work was partially supported by the Hungarian Academy of Sciences (Grant. No. AKA 28-3-161, OTKA 819).

REFERENCES.

1   B.B.Mandelbrot, "The Fractal Geometry of Nature" (Freeman, San Fransisco, 1982).

2   Ya.G.Sinai, Usp.Mat.Nauk 27 , 21 (1972) [Russ.Math.Surveys 166 , 21 (1972)]

E.B. Vul, Ya.G. Sinai, and K.M. Khanin, Usp.Mat.Nauk. 39 , 3 (1984) [Russ.Mat.Surveys 39 , 1 (1984)].

3   D. Ruelle, "Statistical Mechanics, Thermodynamic Formalism" (Addison-Wesley, Reading 1978).

4   R. Bowen, Lecture notes in Math. 470 (Springer, N.Y. 1975)

5   B.B. Mandelbrot, J.Fluid Mech. 62 , 331 (1974).

6   H.G. Hentchel and I. Procaccia, Physica 8D ,435 (1983).

P.Grassberger, Phys.Lett 97A , 227 (1983). Phys.Lett. 107A ,101 (1985).

7   U. Frisch and G. Parisi, in "Turbulence and Predictability in Geophysical Fluid Dynamics and Climate Dynamics", International School of Physics 'Enrico Fermi', Course LXXXVIII, eds. M. Ghil, R. Benzi, and G. Parisi (North-Holland, New York, 1985), p.84;

R. Benzi, G. Paladin, G. Parisi, and A. Vulpiani, J.Phys.A 17 , 352 (1984).

R.Badii and A.Politi, J.Stat.Phys. 40 , 725 (1985)

8   T.C. Halsey, M.H. Jensen, L.P. Kadanoff, I. Procaccia, and B.I. Shraiman, Phys.Rev.B 33 , 1141 (1986)

9   M.H. Jensen, L.P. Kadanoff, A. Libchaber, I. Procaccia, and J. Stavans, Phys.Rev.Lett. 55 , 2798 (1985).

E.G.Gwinn and R.M.Westervelt, Phys.Rev.Lett. 59 , 157 (1987)

10 J.-P. Eckmann and I. Procaccia, Phys.Rev.A 34 , 659 (1986).

11 M.J. Feigenbaum, M.H. Jensen, and I. Procaccia, Phys.Rev.Lett. 56 , 1503 (1986).

12 M.J. Feigenbaum, J.Stat.Phys. 46 , 919 and 925 (1987)

13 L.P.Kadanoff, J.Stat.Phys. 43 ,395 (1986).

D.Bensimon,M.H.Jensen and L.P.Kadanoff, Phys.Rev.A 33 , 3622 (1986)

14   M.H. Jensen, L.P. Kadanoff, and I. Procaccia, Phys.Rev.A 36 , 1409 (1987)

15   P.Collet, J.Lebowitz and A.Porzio, J.Stat.Phys. 47 , 609 (1987)

16   H.Fujisaka and M.Inoue, Prog.Theor.Phys. 77 , 1334 (1987)

17   T. Bohr and D. Rand, Physica 25D , 387 (1987).

     D.Rand, The singularity spectrum for hyperbolic Cantor sets and attractors, preprint 1986.

18   T. Tél, Phys.Rev.A 36 , 2507 (1987)

19   P. Cvitanovic, in Proceedings of the Workshop in Condensed Matter Physics. Trieste, Italy 1986.

20   D. Katzen and I. Procaccia, Phys.Rev.Lett. 58 , 1169 (1987).

21   M. Kohmoto, Phys.Rev.A to be published.

22   M. Duong-van, Phase transition in the logistic map, preprint 1987.

23   P. Szépfalusy, T. Tél, A. Csordas and Z. Kovacs, Phys.Rev.A 36 , 3525 (1987)

24   T.Bohr and M.H.Jensen, Phys.Rev.A 36 , 4904 (1987)

25   E. Aurell, Phys.Rev.A 35 , 4016 (1987).

26   G.H.Gunaratne and I.Procaccia, Phys.Rev.Lett. 59 , 1377 (1987)

27   C.Grebogi, E.Ott and J.Yorke, Phys.Rev.A 36 , 3522 (1987)

28   P.Grassberger, R.Badii and A.Politi, Scaling laws for invariant measures on hyperbolic and non-hyperbolic attractors. Preprint 1987.

29   For a recent collection of articles see "Fractals in Physics", ed. by L.Petroniero and E.Tosatti. (North Holland, Amsterdam, 1986)

30   J.P.Eckmann and D.Ruelle, Rev.Mod.Phys. 67 , 617 (1985)

31   L.P.Kadanoff and C.Tang, Proc.Natl.Acad.Sci.USA 81 , 1276 (1984)

32   H.Kantz and P.Grassberger, Physica 17D , 75 (1986)

33   P. Collet and J.-P. Eckmann, "Iterated Map of the Interval". (Birkhäuser, Boston, 1980)

34   T.Kai and K.Tomita, Prog.Theor.Phys. 64 , 1532 (1980)

35   Y. Takahashi and Y.Oono, Prog.Theor.Phys. 71 , 851 (1984)

36   R. Benzi, G. Paladin, G. Parisi, and A. Vulpiani, J.Phys.A 18 , 2157 (1985).

37 D.Auerbach, P.Cvitanovic, J.P.Eckmann, G.Gunaratne and I.Procaccia, Phys.Rev.Lett. 58, 2387 (1987)

38 A. Renyi, "Probability Theory", (North Holland, Amsterdam, 1970)

39 P.Grassberger and I.Procaccia, Phys.Rev.A 28, 2591 (1983)

40 T.Tél, Phys.Rev.A 36, 1502 (1987)

41 P.Szépfalusy and T.Tél, Phys.Rev.A 34, 2520 (1986)

42 P.Walters, Amer.J.Math. 97, 937 (1975)

43 E. Ott, W.D. Withers and J.A. Yorke, J.Stat.Phys. 36, 687 (1984)

44 G.Paladin, L.Peliti and A.Vulpiani, J.Phys.A 19, L991 (1986)

45 P.Szépfalusy and T.Tél, Phys.Rev.A 35, 477 (1987)

46 M.J. Feigenbaum, Comm.Math.Phys. 77, 65 (1980).

47 T.D.Schultz, D.C.Mattis and E.H.Lieb, Rev.Mod.Phys. 36, 856 (1964)

48 T.Bohr, P.Cvitanovic and M.H.Jensen, unpublished.

49 I.Procaccia and T.Tél, Scaling properties of multifractals: a functional equation approach, Weizmann Institute preprint, 1988.

50 S.-J.Chang and J.Wright, Phys.Rev.A 23, 1419 (1981).
G.Györgyi and P.Szépfalusy, J.Stat.Phys. 34, 451 (1984).

51 A.Arneodo, G.Grasseau and E.J.Kostelich, Phys.Lett.A 124, 426 (1987)

52 P.Cvitanovic, G.H.Gunaratne and I.Procaccia, Topological and metric properties of Hénon type strange attractors. Preprint 1987.

53 M.H.Jensen, comment on "The organization of chaos". NORDITA preprint 1987.

54 G.H.Gunaratne, M.H.Jensen and I.Procaccia, Nonlinearity 1(1988)

55 R.Artuso, P.Cvitanovic and B.Kenny, Phase transitions on strange irrational sets. Niels Bohr Institute preprint 1987.

56 D.Bessis, G.Paladin, G.Turchetti and S.Vaienti, Generalized dimensions, entropies and Lyapunov exponents from the pressure function for strange sets. Preprint 1987.

57 S.Vaienti, Generalized spectra for the dimensions of strange sets. Preprint 1987.

58 M.H.Jensen, T.Bohr and P. Cvitanovic, Fractal aggregates and Julia sets. Preprint 1987.

I.Procaccia, private communication.

59   A.Csordas and P.Szépfalusy, Generalized entropy decay rates of 1-d maps.   Preprint 1987.

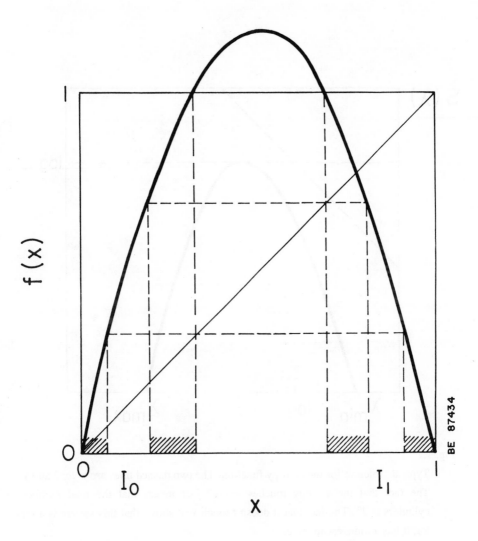

Fig. 1.1

The construction of a "Cookie cutter" Cantor set. The hatched intervals are the cylinder for $n=2$, i.e. the points that remain inside I for at least 2 iterates.

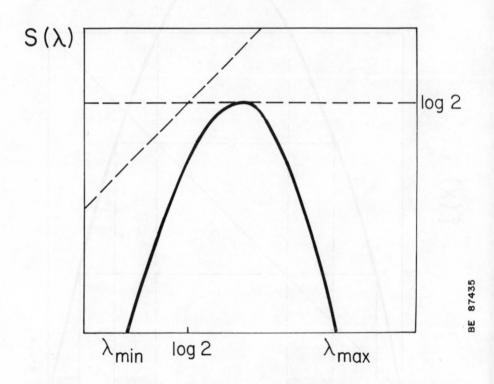

Fig. 1.2

Typical structure for the entropy function. The two dashed lines are $y=log2$ and $y=x$. The fact that the entropy touches $y=log2$ just means that the total number of cylinders is 2^n. The fact that it doesn't touch $y=x$ shows that this system is a repellor, it has a finite escape rate.

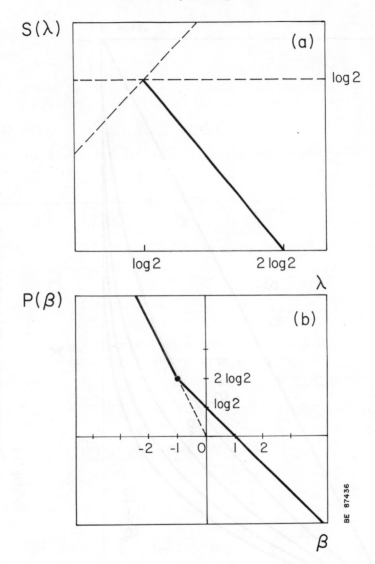

Fig. 3.1

Thermodynamical functions for the logistic map at the "Ulam -von Neumann" point. The function $P(\beta)$ is analog to the free energy so the discontinuity in its slope implies a first order transition.

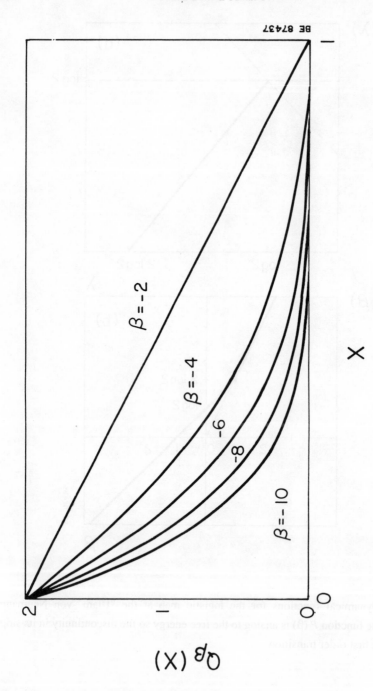

Fig. 3.2   The invariant densities $Q_\beta(x)$ (3.2.14) for different β

Fig. 4.1

Sketch of the map (4.1.1) (left) and the associated entropy curve (right) for a positive (top) and a negative (bottom) value of γ.

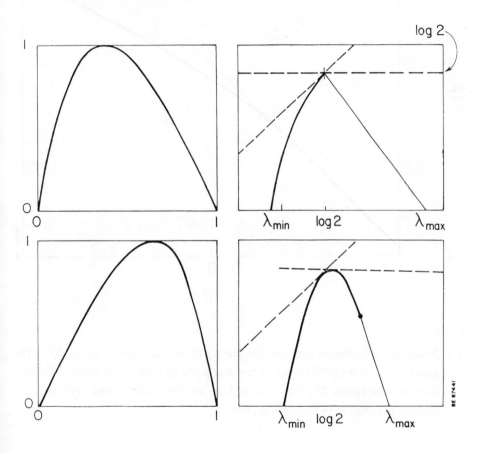

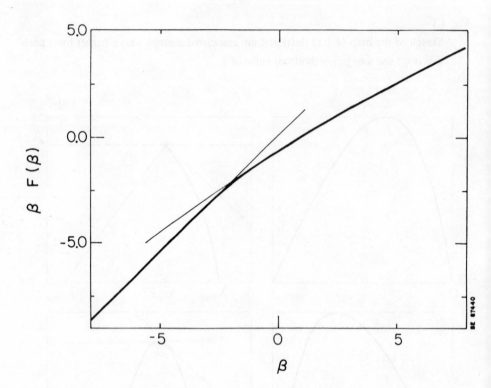

Fig. 4.2

Numerical approximation to the free energy of the "universal map" (4.2.7). The heavy curve shows $\beta F(\beta)$ found by iterating a non-singular initial density. The thin lines are the curves βF_0 (the straight line) and βF_1 which come from iterating singular densities as explained in the text.

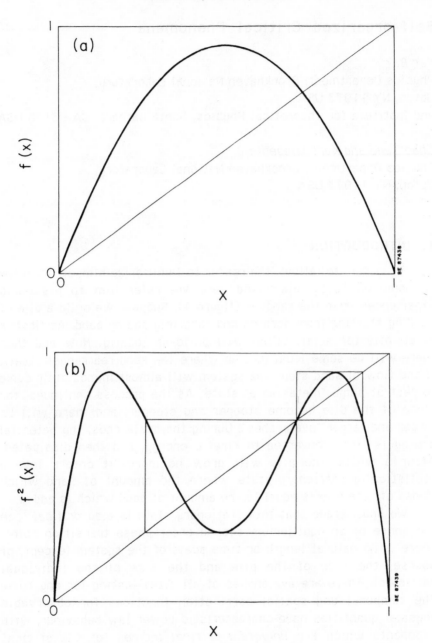

Fig. 4.3

The first Misiurewicz point. Within the boxes f^2 behaves like a "complete map".

Self-organized Critical Phenomena

Per Bak
Physics Department, Brookhaven National Laboratory,
Upton, NY 11973 USA
and Institute for Theoretical Physics, Santa Barbara, CA 93106 USA

Chao Tang and Kurt Wiesenfeld
Physics Department, Brookhaven National Laboratory,
Upton, NY 11973 USA

1. INTRODUCTION

In order to allow the reader to familiarize himself with the phenomenon to be discussed here we refer him to his early experiences from the sandbox (figure 1). Suppose we build a pile of sand by starting from scratch and randomly adding sand. At first, a small pile (or small piles) will build-up locally. Now and then there will be some local motion where the sand rearranges itself. If the flow of sand stops the system will almost immediately come to rest at some "metastable" state. As the process continues, the slope of the pile become steeper and steeper, and there will be larger and larger "avalanches". During the avalanches, the potential energy is first converted to kinetic energy, and then dissipated. After a while, the pile will grow no more: it enters into a statistically *stationary* state where the amount of sand which slides off the pile is equal to the amount of sand which is added.

We shall argue that this stationary state is also *critical* [1], in the sense of an equilibrium second order phase transition point. There is no natural length or time scale of the system (except, of course, the size of the pile and the size of the individual particles). There are avalanches of all sizes lasting for any time. The temporal and spatial correlation functions for observable physical quantities have characteristic power law behaviour, with exponents which are *universal critical indices* for the critical point. For instance, the size distribution of avalanches is given by a power law, in analogy to the distribution of magnetic clusters at

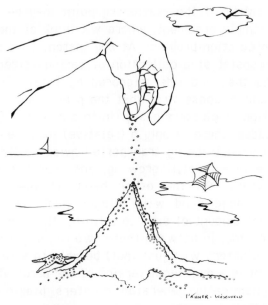

Figure 1. The formation of a sandpile. After a while, the pile reaches its critical state, and the slope will grow no further. We predict that the flow of sand falling off the edges will exhibit 1/f noise.

a second order phase transition, or the distribution of clusters of connected bonds at the percolation transition. The critical indices are related by *scaling relations* which can be verified (or falsified) experimentally.

The picture above is probably very general for *extended non-equilibrium dissipative dynamical systems*, in particular for transport systems. Transport is usually described in terms of some partial differential equation such as the Boltzman equation or a diffusion equation. Such equations ignore the microscopic details of the process (which for many purposes is perfectly legitimate). For instance, the possibility of metastable states is ruled out. Our picture is different. We imagine a situation where the substance being transported can "get stuck" in some static state, until the local force or pressure (the slope of the sandpile) exceeds a critical value. Then the particle moves for some time, until it gets stuck somewhere else, and so on. For visualization, think of the traffic on the "Long Island Expressway".

The criticality of the stationary point may be understood by means of a simple argument, where we think of the process as a dynamical percolation problem. At any instant we can think of the sandpile to consist of a collection of various-sized "clusters", a cluster being the sandslide triggered by locally adding a small amount of sand. Suppose first that the pile is subcritical, so that the distribution of clusters has a finite size cut-off. In this case, as sand is added, there is only a (relative) local rearrangement of the sand; no sand is being transported away to the boundary (at infinity). The pile is still growing, and thus has not reached a stationary state. On the other hand, if the clusters grow indefinitely, added sand will trigger huge avalanches which transport great amount of sand to infinity; the pile will therefore collapse over time. In between these two regimes lies the critical point, where clusters will just (not) be able to spread to infinity. This statement is a dynamical version of the percolation threshold, with the dynamically generated clusters playing the role of connected lattice clusters.

Low dimensional non-equilibrium dynamical systems also undergo "phase transitions" (bifurcations, mode-locking, intermittancy etc.) where the properties of attractors change. The critical point for these systems, and for the phase transitions in equilibrium statistical mechanics mentioned above, can be reached only by fine-tuning a parameter (e.g.) temperature, and so may occur only accidentally in nature. In contrast, the critical phenomena described above requires no fine-tuning: it is *self-organized.* Thus, we expect it to be much more widespread in nature where in general no experimental collaborator is available to turn the knobs. Indeed, we argue that the self-organized criticality may be invoked as an explanation for a number of phenomena where spatial and/or temporal power-law behaviour is known to occur, but no general mechanism has been identified: 1/f noise[2], fractal structures[3], and (maybe) turbulence.

The dynamics of the critical state has a specific temporal fingerprint, namely "flicker noise", in which the power spectrum $S(f)$ scales as $1/f^\beta$ at low frequencies, with β near one. The exponent β is a bona fide critical index. Flicker noise is characterized by correlations extended over a wide range of time scales, a clear indication of some sort of cooperative effect. Flicker noise has been observed, for example, in the light from quasars[4], the intensity of sunspots[5], the current through resistors[6], the sand flow in an hour glass[7], the flow of rivers such as the Nile[2], and even stock exchange price indices[8]. Another signature of the criticality is spatial self-similarity. It has been pointed out that nature is full of self-similar "fractal" structures, though the physical reason for this is not understood[9]. Turbulence is a phenomenon where self-similarity is believed to occur in both space and time.

The notion of self-organized criticality introduced here is unrelated to the concept of self-organization introduced by Prigogine[10] and others[11], which refers to the observation that some physical problems with many degrees of freedom can be effectively described by a few: the "slaving principle", expressed mathematically as the "central manifold theorem". Indeed, the behaviour of some real systems has been very successfully described in terms of low dimensional models; most notable are the transitions to chaos through period doubling[12] and mode-locking[13] which have been quantitatively explained by the study of 1D discrete maps. This success has led to some hope (unjustified, we believe) that "chaotic phenomena" such as turbulence may eventually be identified as a low-dimensional phenomena. The self-organized critical point is intrinsically a many-body phenomenon, which can not even in principle be reduced to low dimensions, but is most conveniently described in terms of its behaviour under spatial and temporal scale transformations. Thus, if this description applies to the situations above, then turbulence is not low-dimensional "chaos".

Specifically, we consider dissipative dynamical systems with local interacting degrees of freedom. It is essential that the system has many metastable states (There is evidence that a high multiplicity of attractors is a common - if not generic - characteristic of locally coupled dissipative systems). Although the details of the local states are not important to the general theory, for the sake of clarity we focus attention on specific models. We choose the simplest possible models rather than wholly realistic and therefore complex models of actual physical systems. Besides our expectation that the overall qualitative features are captured in this way, it is certainly possible that quantitative properties (such as scaling exponents) may apply to more realistic situations, since the system operates at a critical point where universality may apply. The philosophy is analogous to that of equilibrium statistical physics where results are based on Ising models (and Heisenberg models, etc.) which have only the symmetry in common with real systems. Our "Ising models" are discrete cellular automata, which are simple to study numerically. The metastability is a consequence of the discreteness. Our philosophy is very different from recent studies of turbulence where the actual differential equation (Navier-Stokes equation) appears as the continuum limit of discrete automatons. The continuum limit of our automaton is a diffusion equation describing none of the interesting physics.

2. MINIMAL STABILITY

Figure 2 shows a sandpile of length N. The boundary conditions are such that particles can be transported through the system and leave it at the right hand side only. We may think of this arrangement as half of a symmetric sandpile with both ends open. The idea is to simulate a situation resembling some of the systems having $1/f$ noise. In the hour glass, sand is transported through the system; in quasars (and in the case of sunspots) light is transported through an open ended system, and in the case of rivers, water is transported. The numbers z_n represent height differences $z_n \equiv h(n)-h(n+1)$ between successive positions along the sandpile. The dynamics is very simple. From the figure one sees that sand is added at the n'th position by letting

$$z_n \rightarrow z_n+1,$$
$$z_{n-1} \rightarrow z_{n-1}-1. \qquad (2.1)$$

When the height difference becomes higher than a fixed critical value z_c, one unit of sand tumbles to the lower level, i. e.

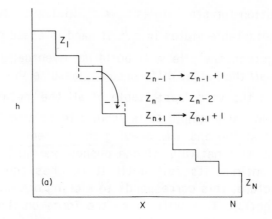

Figure 2. One dimensional "sandpile automaton". The state of the system is specified by an array of integers representing the height difference between neighboring plateaus.

$$z_n \to z_n-2,$$
$$z_{n\pm1} \to z_{n\pm1}+1, \quad \text{for } z_n > z_c. \tag{2.2}$$

Closed and open boundary conditions are used for the left and right boundaries, respectively.

Equation (2.2) is a nonlinear discretized diffusion equation (nonlinear because of the threshold condition). The process continues until all the z_n are below threshold, at which point another grain of sand is added (at a random site) via Eq.(2.1). The model is a *cellular automaton* where the state of the discrete variable z_n at time t+1 depends on the state of the variable and its neighbors at time t.

Alternatively, the system can be thought of as an array of damped pendula in a gravitational field, coupled by torsion springs: The heights h(n) are the winding numbers of the springs, and the z_n are the spring forces on the pendula. When z_n exceeds the critical value z_c, so that the spring force exceeds the gravitational force, the pendulum rotates one revolution, leading precisely to the dynamics Eq.(2.2).

The condition for stability is $z_n < z_c$, $(n=1,2, \ldots ,N)$ so the total number of metastable states is z_c^N. If sand is added randomly from an empty system, the pile will build up, eventually reaching the point where *all* the height differences z_n assume the critical value, $z_n = z_c$. This is the very *least* stable of all the metastable states. Any additional sand simply falls from site to site (left to right) and falls off at the end $n=N$, leaving the system in the minimally stable state. Alternatively, if one pushes one unit downwards it will also continue its fall until it reaches the edge. In the pendulum picture, this corresponds to kicking one pendulum in the forward direction. This will cause the force on the two nearest pendula to exceed the critical value and the perturbation will propagate by a domino effect until it hits the two ends of the array. At the end of this process the forces are back to their original values and all pendula have rotated one period. In other words, despite the strong dissipation, the effect of a small local perturbation is communicated throughout the system, but the system is robust with respect to noise insofar as it returns to the globally minimally stable state. If units are added randomly, the resulting sand-flow is also random white noise, i.e. with power spectrum $1/f^0$. As we shall see in the next section, the robustness of the minimally stable state is lost in two and higher dimensions.

The dynamical selection principle leading to the least stable stationary state is quite independent of how the sandpile is built up. Instead of building the pile by adding sand, we might start with a flat sand surface and slowly raise the left end of the bar. Or, we could randomly add "slope", $z_n \to z_n + 1$, and let the system obey the dynamics Eq.(2.2). This would represent the dynamics of a system with a random distribution of critical height differences and a uniformly increasing slope. We could also start with a very unstable state, $z_n > z_c$ for all n, and let the system relax. In all these cases the minimally stable state will be reached even if the boundary conditions are such that the sand cannot leave the bar, i.e. closed boundary conditions at both ends.

In one dimension, the minimally stable state is critical in the restricted sense that any small perturbation can just propagate infinitely through the system, while any lowering of the slope will prevent this. This is analogous to some other 1D critical phenomena, such as percolation where at the percolation threshold particles can just percolate to infinity. Also, like other 1D systems the critical state has no spatial structure, and correlation functions are trivial. In the next section we shall see that in higher dimensions the critical states and their dynamics are dramatically different.

3. SELF-ORGANIZED CRITICALITY.

We consider a situation in two dimensions where particles are flowing in an axial direction. The geometry is illustrated in figure 3. Particles can leave the system at one side only. Figure 4 shows part of the two-dimensional lattice on which the model is defined. Columns of particles of heights h_1, h_2, h_3, and h_4 are associated with the bonds around the lattice points. At each site $1 \leq (x,y) \leq N$ a discrete variable $z(x,y)$ is defined which is the local slope

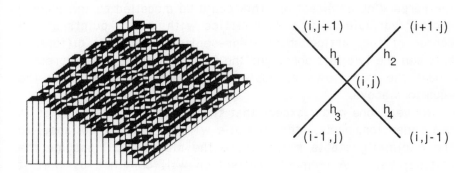

Figure 3 (left). Two dimensional sandpile with open edge perpendicular to flow. The pile is critical. Figure 4 (right). Arrangement of columns at bonds of 2D lattice.

$h_1 + h_2 - h_3 - h_4$. The dynamics are simple: if the slope exceeds a critical value z_c then two grains tumble in the diagonal direction from 1 and 2 to 3 and 4. The corresponding equations are straightforward generalizations of the rules (2.1) and (2.2) for the one-dimensional model,

$$z(x,y) \rightarrow z(x,y) - 4,$$
$$z(x,y\pm1) \rightarrow z(x,y\pm1) + 1,$$
$$z(x\pm1,y) \rightarrow z(x\pm1,y) + 1, \quad \text{for } z(x,y) > z_c. \tag{3.1}$$

Adding particles is described by

$$z(x-1,y) \rightarrow z(x-1,y) -1$$
$$\text{or } z(x,y-1) \rightarrow z(x,y-1) -1,$$
$$z(x,y) \rightarrow z(x,y) +1. \tag{3.2}$$

Thus, we describe a flow in the diagonal direction from the wall to the edge in figure 3. For numerical convenience we have chosen periodic boundary conditions in the perpendicular direction. Of course, in a realistic model, particles will also be transported in the perpendicular direction. This could be modelled by introducing another variable z' on a dual lattice with lattice points at the center of the lattice above, representing the height difference between the two left bonds and the two right bonds. In the present model the two flows decouple, so we shall consider only the equations above.

Naively, one might expect that the situation is the same as in one dimension, namely that the pile will build up (or collapse) to the minimally stable state where the slopes z all assume the critical value. A moments reflection will convince us that it cannot be so. Suppose we push two units of sand downwards in the diagonal direction by applying the rule (3.1). This will render the surrounding sites unstable ($z > z_c$), and the noise will spread to the neighbors, then *their* neighbors, in a chain reaction, *ever amplifying* since the sites are generally connected with more than two minimally stable sites, and the perturbation eventually propagates throughout the entire lattice. The minimally stable state is thus unstable with respect to small fluctuations and cannot represent an attracting fixed point for the dynamics. As the system further evolves, more and more more-than-minimally stable domains will be generated, and these states will impede the motion of the "noise". <u>The system will become stable precisely at the point when the network of minimally stable clusters has been broken down to the level where the noise signal cannot be communicated through infinite distances. At this point there will be no length scale, and consequently no time scale.</u> We remind the reader of the general discussion in the introduction. Hence one might expect that the system approaches, through a self-organizing process, a critical state with power law correlation function for physically observable quantities. In analogy with the discussion for the one dimensional case, the slope (or "pressure") will build up to the point where stationarity is obtained: *this is assured by the self-organized critical state*, but not the minimally stable state. The slope of the critical state is reduced compared to the slope of the minimally stable state.

Suppose that we perturb the critical state locally, by adding one unit, or by locally changing the slope. We expect the perturbation to grow over all length scales. That is, a given perturbation can lead to anything from a shift of a single unit to an avalanche. Consequently, *local* (in space and time) input fluctuations lead to *correlated* space-time output fluctuations. The lack of a characteristic length scale in the cluster structure leads directly to a lack of characteristic time scale for the output fluctuations.

Thus, armed with confidence arising from the speculations above, we start from scratch and build the sandpile by adding particles randomly according to the rule (3.2). After each particle is added, the system is allowed to relax (if need be) according to (3.1). After an initial transient period, the system indeed reaches a statistically stationary state with clusters of all sizes.

Figure 3 actually shows a snapshot of the critical state. The heights of the columns represent the values of the bonds as described above. Figure 5 shows typical clusters, each generated by adding single particles. Note their irregular "fractal" shapes, reflecting the spatial self-similarity at the critical points (actually, the Hausdorff dimension of the clusters is a critical

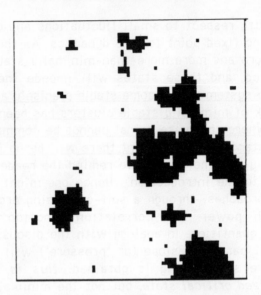

Figure 5. Typical domain structures for a 75×75 array. Each cluster is triggered by a single perturbation.

index of the self-organized critical point). For each cluster we monitor only the flow f(t) of sand that falls off the edges of the box. Most of the time, the relaxation following a single kick results in no sand falling off the edge, though of course on average 1 unit of sand falls off for every unit added, once steady state has been achieved. Similarly, the duration, T, of the sand flow off edge may assume any integer value limited only by the size of the system. To illustrate the criticality (and for later use) we have calculated the distribution of lifetimes, D(T), weighted by the average response s/T where s is the total flow of particles caused by a single event. Figure 6 shows that this quantity for a system of size 75×75 has a *power law behaviour* for a decade or so, as expected for a system at criticality,

$$D(T) \approx T^{-\alpha}, \alpha \approx 1.0 \ . \tag{3.3}$$

We are interested in the response F(t) to a situation where the system is locally perturbed randomly in space and time, so that the

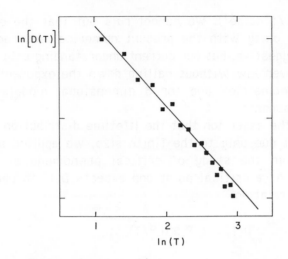

Figure 6. Lifetime distribution of the sand flow for a system of size 75x75. The straight line has a slope -1.05.

flow is a superposition of the events above, acting concurrently and independently. We want to calculate the power spectrum S(f) of the quantity F(t) defined by

$$S(f) = \int <F(t_0+t)F(t_0)>\exp(2\pi i f t)dt , \tag{3.4}$$

where < > represents an average over all times t_0. In fact, the power law distribution of lifetimes, Eq.(3.3), leads to a power law for $S(f)$[14]

$$S(f) = \int (TD(T)/(1+(fT)^2))dT$$

$$\approx f^{-2+\alpha} \equiv f^{-\beta} , \beta \approx 1.0 \tag{3.5}$$

(To convince himself that this description has some connection with reality, we suggest that the reader studies visually the fluctuating flow of an hour glass and notes the pulses of widely different time scales. The hour glass is a dynamical system known to exhibit $1/f$ noise[7]). We cannot rule out that the exponent is identical to unity with the present numerical accuracy. This is certainly suggestive, but our current understanding only tells us to expect a power law without nailing down the exponent. Actually, for other geometries and for 3 dimensional models we found different exponents.

To test the assertion that the lifetime distribution is scaling, with cut-off due only to the finite size, we applied a technique familiar from the study of critical phenomena of 2nd order transitions. At a critical point one expects D(T) to obey a finite size scaling relation

$$D(T) = T^{-\alpha} \mathcal{F}(L^{\sigma}/T), \qquad (3.6)$$

where $\mathcal{F}$ is a crossover scaling function of the single scaled variable L^{σ}/T, and σ is a dynamical critical exponent. Figure 7 shows the product $D(T)T^{\alpha}$ vs. L^{σ}/T for $\sigma = 0.75$: the points for different L all fall on the same curve $\mathcal{F}(x)$ to within numerical

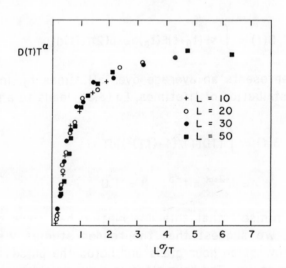

Figure 7. Finite size scaling plot.

accuracy.

The nature of the boundary conditions is essential to the nature (though not to the existence) of the critical state, since the dynamics and the physical situation is largely defined by the properties at the boundaries, for instance whether material is being transported in or out. We have also performed simulations with closed boundary conditions where particles cannot leave the boundary. Starting from a flat surface, z=0, the slope or "pressure" is increased by one unit at a random position (x,y)

$$z(x,y) \rightarrow z(x,y) + 1 . \tag{3.7}$$

Then z is increased by one at another point and so on. When z eventually exceeds the critical value z_c somewhere, the system evolves according to (3.1) until it becomes stable again, creating a cluster of size s in the process. After a while the system arrives at a (statistically) stationary state with clusters of all sizes up to the size of the system. The process described here simulates a

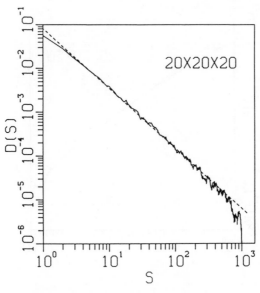

Figure 8. Distribution of cluster sizes for closed system at criticality for 20×20×20 array computed as described in the text. The data have been coarse grained. The dashed straight line has a slope -1.37.

situation where the slope increases gradually and takes the system to the critical point. This has a lot of similarity with "turbulence" where energy is fed into the system in a long wavelength mode. We find, as before, that energy is dissipated on all length scales and all time scales. The process (3.1) represents dissipation of one unit of energy at the position **x** at time t, and the total number of "slidings" at time t represents the instantaneous dissipation rate. The model can be viewed as a "toy" model for turbulence.

Figure 8 shows the distribution of cluster sizes D(s) measured in a 3D system of size 20×20×20 after the system arrived at the critical state. The log-log plot follows a pretty respectable straight line, with slope -1.37, i.e $D(s) \approx s^{-\tau}$, $\tau \approx 1.37$. In 2D simulations on a 50×50 array one again finds a power law, but now $\tau \approx 1.0$. Figure 9 shows the distribution of lifetimes of the clusters weighted again by the average response. This quantity also has power law behaviour, $D(T) \approx T^{-\alpha}$, $\alpha \approx 0.43$ for D=2; $\alpha \approx 0.92$ for D = 3, translating into the "1/f" spectrum

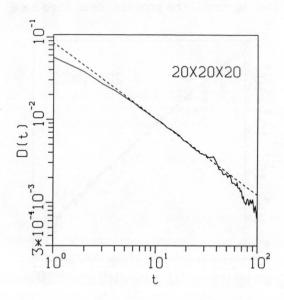

Figure 9. Distribution of lifetimes corresponding to figure 8. The exponent $\alpha \approx 0.92$ yields a "1/f noise" spectrum $f^{-1.08}$.

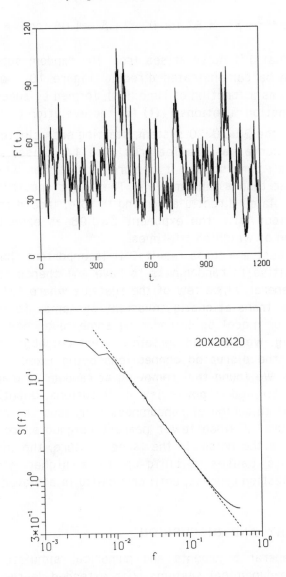

Figure 10. Above: F(t) generated by superimposing randomly the response represented by the individual clusters for a 20×20×20 array. Note the fluctuations on a wide range of time scales. Below: Power spectrum S(f) of the function F(t).

$$S(f) \approx f^{-\beta} , \; \beta \approx 1.57 \text{ for } D = 2 \; ; \; \beta \approx 1.08 \text{ for } D = 3. \qquad (3.8)$$

The fact that $1/f$ noise arises from the random superposition of events can be demonstrated directly. Figure 10 shows the total dissipation as a function of time, $F(t)$, formed by superimposing the time dissipation functions $F_{cl}(t)$ for the evolution of the individual clusters in the $20 \times 20 \times 20$ system, starting each cluster at a random time. The curve has the features of a $1/f$ noise, i.e. there are events on all time scales. It is more regular than white noise, and less regular than a random walk. The power spectrum $S(f)$ of the curve, calculated directly according to (3.4) indeed shows a power law behaviour with the exponent $\beta \approx 0.98$ as expected from the distribution of weighted lifetimes.

How robust are these results? It is important that the results are insensitive to randomness to have any chance to explain $1/f$ noise in general, since few of the systems where $1/f$ noise occurs are simple translationally invariant systems. To test this, we modified our model by introducing some "quenched randomness". Specifically, we removed certain nearest neighbor connections in the array, the disrupted connections being fixed throughout the simulation. We found that removing at random as many as 25% of the bonds still led to power law distributions. Quantitatively, the presence of this kind of randomness increases the mean value of the "pressure" z since fewer nearest neighbors are available to communicate the noise. In the "sand" picture, the introduction of impediments causes a build-up to a higher slope, by the self-organization process, until criticality is achieved again.

4. SUMMARY AND DISCUSSION

Our general arguments and numerical simulations show that dissipative dynamical systems with extended degrees of freedom can evolve towards a self-organized critical state, with spatial and temporal power law scaling behavior. The frequency spectrum is "$1/f$ noise" or "flicker noise" with a power law spectrum $S(f) \approx f^{-\beta}$. Thus, in this picture "$1/f$ noise" is not noise but reflects the generic dynamics of extended dynamical systems. We found values of β tolerably close to one (and certainly between 0 and 2). It remains to be seen to what extent systems can be grouped into universality classes within which the exponents are the same, depending on symmetry, dimension, and so on. We strongly suspect that the criticality discovered here cannot depend on the local details of the models, in analogy with equilibrium 2nd order phase transitions.

Moreover, we conclude that "1/f noise" is intimately related to the underlying spatial organization. This can be tested directly, for instance by measuring the frequency cutoff vs. the system size. In retrospect, it is hard to see how 1/f noise, with long temporal correlations, could possibly occur without long range spatial correlations, except by "fine-tuning" models with few degrees of freedom[15].

We believe that the concept of self organized criticality can be taken much further and might be the underlying concept for temporal and spatial scaling in dissipative nonequilibrium systems. One of our models (with closed boundary conditions) could be considered a "toy" model of generalized turbulence, with dissipation correlated on all length scales. Of course, there is no direct connection with (for instance) the Navier-Stokes equation, where the metastability is due to the storage of kinetic energy in vortices, not potential energy as in the models discussed here. Nevertheless, there is a one-to-one connection for the phenomenology used to describe the two situations, and our simulations may provide an entrance into the problem of dynamical scale invariance.

It would be desirable to have analytical results on our cellular automaton to support the general arguments and numerical simulations. There is reason not to be overly optimistic, since the models, despite their simplicity, combine all of the evils of criticality of equilibrium statistical mechanics systems with those of low dimensional dynamical systems. Even much simpler 1D automatons are believed to be "computationally irreducible"[16], i.e. there is no faster method to study the models than actually to simulate them. Nevertheless, analytical efforts are encouraged.

The authors thank Elaine Wiesenfeld for preparing figure 1. This work was supported by the Division of Materials Science, U. S.

Department of Energy, under contract DE-AC02-76CH00016, and by
the Institute for Theoretical Physics, NSF grant No PHY82-17853
supplemented by N.A.S.A.

REFERENCES

1. Bak, P., Tang, C. and Wiesenfeld, K., Phys. Rev. Lett. 59, 381
   (1987), and Phys. Rev. A (in press).
2. For reviews on 1/f noise, see Press, W. H., Comm. Mod. Phys.
   C7,103 (1978); Dutta, P. and Horn, P. M., Rev. Mod. Phys. 53, 497
   (1981).
3. Mandelbrot, B., "The Fractal Geometry of Nature" (W. H. Freeman,
   San Fransisco, 1982).
4. Nolan P. L. et al., Astrophys. J. 246 494-501 (1981); Lawrence, A.,
   Watson, M. G., Pounds, K. A. and Elvis, M., Nature 325, 694 (1987);
   McHardy, I. and Czerny, B., *ibid* 696 (1987).
5. Mandelbrot, B. B. and Wallis, J. R., Water Resources Res. 5, 321
   (1969).
6. see, e.g., Voss, R. F. and Clarke, J., Phys. Rev. B13, 556 (1976).
7. Schick, K. L. and Verveen, A. A., Nature 251, 599 (1974).
8. Mandelbrot, B. B. and Van Ness, J. W., SIAM Review 10, 422 (1968).
9. Kadanoff, L. P., "Fractals: Where is the Physics", Physics Today 39,
   3 (1986).
10. Prigogine, I., "From being to Becoming" (Freeman, San Fransisco,
    1979).
11. Haken, H., Rev. Mod. Phys. 47, 67 (1975).
12. Feigenbaum, M. J., J. Stat. Phys. 19, 25 (1979).
13. Jensen, M. H., Bak, P. and Bohr, T., Phys. Rev. A 30, 1960 (1984),
    and this volume.
14. van der Ziel, A., Physica 16, 359 (1950).
15. Manneville, P., Journal de Physique, 41, 1235 (1980).
16. Wolfram, S., Phys. Scripta T9, 170 (1985).

Discrete Fluids

by Brosl Hasslacher

PART I
BACKGROUND FOR LATTICE GAS AUTOMATA

The question is simple: Find a minimal logic structure and devise a dynamics for it that is powerful enough to simulate complex systems. Break this up into a series of sharper and more elementary pictures. We begin by setting up a collection of very simple finite-state machines with, for simplicity, binary values. Connect them so that given a state for each of them, the next state of each machine depends only on its immediate environment. In other words, the state of any machine will depend only on the states of machines in some small neighborhood around it. This builds in the constraint that we only want to consider local dynamics.

We will need rules to define how states combine in a neighborhood to uniquely fix the state of every machine, but these can be quite simple. The natural space on which to put all this is a lattice, with elementary, few-bit, finite-state machines placed at the vertices. The rules for updating this array of small machines can be done concurrently in one clock step, that is, in parallel.

An extreme case of complexity is physical systems with many degrees of freedom. These systems are ordinarily described by field theories in a continuum for which the equations of motion are highly nonlinear partial differential equations. Fluid dynamics is an example, and we will use it as a theoretical paradigm for many "large" physical systems. Because of the high degree of nonlinearity, analytic solutions to the field equations for such systems are known only in special cases. The standard way to study such models is either to perform experiments or simulate them on computers of the usual digital type.

Suppose a cellular space existed that evolved to a solution of a fluid system with given boundary conditions. Suppose also that we ask for the simplest possible such space that captured at least the qualitative and topological aspects of a solution. Later, one can worry about spaces that agree quantitatively with ordinary simulations. The problem is threefold: Find the least complex set of rules for updating the space; the simplest geometry for a neighborhood; and a method of analysis for the collective modes and time evolution of such a system.

At first sight, modeling the dynamics of large systems by cellular spaces seems far too difficult to attempt. The general problem of a so-called "inverse compiler"—given a partial differential system, find the rules and interconnection geometry that give a solution—would probably use up a non-polynomial function of computing resources and so be impractical if not impossible. Nevertheless cellular spaces have been actively studied in recent years. Their modern name is cellular automata, and specific instances of them have simulated interesting nonlinear systems. But until recently there was no example of a cellular automaton that simulated a large physical system, even in a rough, qualitative way.

Knowing that special cases of cellular automata are capable of arbitrarily complex behavior is encouraging, but not very useful to a physicist. The important phenomenon in large physical systems is not arbitrarily complex behavior, but the collective motion that develops as the system evolves, typically with a characteristic size of many elementary length scales. The problem is to simulate such phenomena and, by using simulations, to try to understand the origins of collective behavior from as many points of view as possible. Fluid dynamics is filled with examples of collective behavior—shocks, instabilities, vortices, vortex streets, vortex sheets, turbulence, to list a few. Any deterministic cellular-automaton model that attempts to describe non-equilibrium fluid dynamics must contain in it an iterative mechanism for developing collective motion. Knowing this and using some very basic physics, we will construct a cellular automaton with the appropriate geometry and updating rules for fluid behavior. It will also be the simplest such model. The methods we use to do this are very conservative from the viewpoint of recent work on cellular automata, but rather drastic compared to the approaches of standard mathematical physics. Presently there is a large gap between these two viewpoints. The simulation of fluid dynamics by cellular automata shows that there are other complementary and powerful ways to model phenomena that would normally be the exclusive domain of partial differential equations.

Fluid dynamics is an especially good large system for a cellular automaton formulation because there are two rich and complementary ways to picture fluid motion. The kinetic picture (many simple atomic elements colliding rapidly with simple interactions) coincides with our intuitive picture of dynamics on a cellular space. Later we will exploit this analogy to construct a discrete model.

The other and older way of approaching flow phenomena is through the partial differential equations that describe collective motions in dissipative fluids—the Navier-Stokes equations. These can be derived without any reference to an underlying atomic picture. The derivation relies on the idea of the continuum; it is simpler to grasp than the kinetic picture and mathematically cleaner. Because the continuum argument leads to the correct functional form of the Navier-Stokes equations, we spend some time describing why it works. The continuum view of fluids will be called "coming down from above," and the microphysical view "coming up from below."   In the intersection of these two very different descriptions, we can trap the essential elements of a cellular-automaton model that leads to the Navier-Stokes equations.

Coming down from Above—
The Continuum Description

The notion of a smooth flow of some quantity arises naturally from a continuum description. A flow has physical conservation laws built-in, at least conservation of mass and momentum. With a few additional remarks one can include conservation of energy. The basic strategy for deriving the Euler and Navier-Stokes equations of fluid dynamics is to imbed these conservation laws into statements about special cases of the generalized Stokes theorem. We use the usual Gauss and Stokes theorems, depending on dimension, and apply them to small surfaces and volumes that are still large enough to ignore an underlying microworld. The equations of fluid dynamics are derived with no reference to a ball-bearing picture of an underlying atomic world, but only with a serene reliance on the idea of a smooth flow in a continuum with some of Newton's laws added to connect to the observed world. As a model (for it is not a theory), the Navier-Stokes equations are a good example of how concepts derived from the intuition of daily experience can be remarkably successful in building effective phenomenological models of very complex phenomena. It is useful to go through the continuum derivation of the Euler and Navier-Stokes equations presented in "The Continuum Argument" for several reasons: First, the reasoning is short and clear; second, the concepts introduced such as the momentum flux tensor, will appear pervasively when we pass to discrete theories of fluids; third, we learn how few ingredients are really necessary to build a fluid model and so mark out that which is essential—the role of conservation laws.

It is clear from its derivation that the Euler equation describing inviscid flows is essentially a geometrical equation. The extension to the full Navier-Stokes equations, for flows with dissipation, contains only a minimal reference to an underlying fluid microphysics, through the stress-rate of strain relation in the momentum stress tensor. So we see that continuum reasoning alone leads to nonlinear partial differential equations for large-scale physical observables that are a phenomenological description of fluid flow. This description is experimentally quite accurate but theoretically incomplete. The coupling constants that determine the strength of the nonlinear terms—that is, the transport coefficients such as viscosity—have a direct physical interpretation in a microworld picture. In the continuum approach however, these must be measured and put in as data from the outside world. If we do not use some microscopic model for the fluid, the transport coefficients cannot be derived from first principles.

Solution Techniques—The Creation of a Microworld. The Navier-Stokes equations are highly nonlinear; this is prototypical of field-theoretical descriptions of large physical systems. The nonlinearity allows analytic solutions only for special cases and, in general, forces one to solve the system by approximation techniques. Invariably these are some form of perturbation methods in whatever small parameters can be devised. Since there is no systematic way of applying perturbation theory to highly nonlinear partial differential systems, the analysis of the Navier-Stokes equations has been, and still remains, a patchwork of ingenious techniques that are designed to cover special parameter regimes and limited geometries.

After an approximation method is chosen, the next step toward a solution is to discretize the approximate equations in a form suitable for use on a digital computer. This discretization is equivalent to introducing an artificial microworld. Its particular form is fixed by mathematical considerations of elegance and efficiency applied to simple arithmetic operations and the particular architecture of available machines. So, even if we adopt the view that the molecular kinetics of a fluid is unimportant for describing the general features of many fluid phenomena, we are nevertheless forced to describe the system by a microworld with a particular microkinetics. The idea of a partial differential equation as a physical model is tied directly to finding an analytic solution and is not particularly suited to machine computation. In a sense, the geometrically motivated continuum picture is only a clever and convenient way of encoding conservation laws into spaces with which we are comfortable.

Coming up from Below—
The Kinetic Theory Description

Kinetic theory models a fluid by using an atomic picture and imposing Newtonian mechanics on the motions of the atoms. Atomic interactions are controlled by potentials, and the number of atomic elements is assumed to be very large. This attempt at fluid realism has an immediate difficulty. We are unable to specify completely the initial state of the system or to follow its microdynamics. It follows that we cannot use a microdynamics that is this detailed. The obvious strategy is to make a smoothened model that reduces the number of degrees of freedom in the system to just a few. This reduction assumes maximum ignorance of the details of the system below some time and distance scale and replaces exact data on events by probabilistic outcomes. Measurements are assumed to be average values of quantities over large ensembles of representative systems. The assumption is that after a sufficiently long time these average observables are a close description of the fluid.

This approach seems very familiar and obvious from elementary courses in statistical mechanics. But it is unclear how to go from a statistical-mechanical description of an atomic system to the prediction of the details of collective motions that come from the evolution of that system. Fidelity to the atomic picture brings with it considerable mathematical difficulties. As we will see below and in "The Hilbert Contraction," the success of the derivation of the Navier-Stokes equations from the kinetic theory picture—that one derives the Navier-Stokes equations with the correct coefficients and not some other macrodynamics—is justified after the fact.

Kinetic Theory and the Boltzmann Transport Equations. Complete information on the statistical description of a fluid or gas at, or near, thermal equilibrium is assumed to be contained in the one-particle phase-space distribution function $f(t, \mathbf{r}, \Gamma)$ for the atomic constituents of the system. The variables t and $\mathbf{r}$ are the time and space coordinates of the atoms and Γ stands for all other phase-space coordinates (for example, momenta). In this rapid overview of kinetic transport theory, we will not dwell on the many and difficult questions raised by this description but keep to a level of precision consistent with a general understanding of the basic ideas.

The distribution function f is basically a weighting function that is used to define the mean values of physical observables. The relation

$$N(t, \mathbf{r}) \equiv \int f(t, \mathbf{r}, \Gamma) \, d\Gamma \tag{1}$$

defines the density function $N(t, \mathbf{r})$ for the particles in the system over all space. Therefore $N \, dV$ is the mean number of particles in the volume dV. Here dV is a physical volume $\propto L^3$ whose characteristic length L is much larger than l_m, the mean free path of a particle, and much smaller than L_g, some global length, such as the edge of a container for the whole gas. Thus $l_m \ll L \ll L_g$. The basic equation of kinetic theory is the evolution equation for $f(t, \mathbf{r}, \Gamma)$ in the presence of gas collisions. Imagine first that the system has no collisions. Conservation of phase-space volumes, or Liouville's theorem, tells us that

$$\frac{df}{dt} = 0, \tag{2}$$

where d/dt is a total derivative. In an isolated system with no external fields, we can expand the total derivative as

$$\frac{df}{dt} = \partial_t f + \mathbf{v} \cdot \nabla f \equiv \partial_t f + v_i \partial_i f. \tag{3}$$

(We use the convention that repeated indices are summed over.) Equation 3 defines the free-streaming operator, which represents the local change in f per unit time caused by the independent motion of particles alone.

The Boltzmann Form of the Collision Term. Let the particles in a two-body collision process have incoming distribution functions g_1 and g_2 and outgoing distribution functions $\bar{g}_1$ and $\bar{g}_2$. Fixing attention on particle 1, assume that before colliding it occupies a phase-space region $d\Gamma_1$, and after collision it occupies $d\bar{\Gamma}_1$; similarly, particle 2 occupies $d\Gamma_2$ before colliding and $d\bar{\Gamma}_2$ afterwards. If particle 1 undergoes a collision, $d\bar{\Gamma}_1$ will not in general be in $d\Gamma_1$, and particle 1 is said to be lost from $d\Gamma_1$. From these considerations we can compute the functional structure of the general loss term for a binary collision.

The probability of loss will be proportional to the product of four terms: (1) the number of particles of type 1 already in the volume, namely g_1; (2) the number of type-2 particles that enter the volume from some phase-space range $d\Gamma_2$, namely, $g_2 d\Gamma_2$; (3) the total volume of allowed outgoing phase space, $d\bar{\Gamma}_1 d\bar{\Gamma}_2$; and finally (4) a probability for the collision process $P_g\{\Gamma\}$. Now we sum over all possible allowed volumes of phase space. So the total number of losses $\mathcal{L}$ in the volume dV and from $d\Gamma$ due to binary collision processes is

$$\mathcal{L} = dV\, d\Gamma \int P_g\{\Gamma\} g_1 g_2\, d\Gamma_2 d\bar{\Gamma}_1 d\bar{\Gamma}_2.$$

Similarly, particle gain into the phase space volume $d\Gamma$ can only come from reversed channel processes $\bar{g}_1, \bar{g}_2 \rightarrow g_1, g_2$, with fixed Γ_1, and summed over all of $\bar{\Gamma}_1$, $\bar{\Gamma}_2$, and Γ_2, so

$$\mathcal{G} = dV\, d\Gamma \int P_g\{\Gamma\} \bar{g}_1 \bar{g}_2\, d\Gamma_2 d\bar{\Gamma}_1 d\bar{\Gamma}_2.$$

The Boltzmann form for $C(f)$ is the net flow into the region, which is $\mathcal{G} - \mathcal{L}$. Using this form, we get the Boltzmann transport equation, a highly nonlinear integro-differential equation:

$$\frac{df}{dt} = \partial_t f + v_i \partial_i f = \mathcal{G} - \mathcal{L}. \tag{5}$$

In Part II we will use the same reasoning to construct the Boltzmann equation for the discrete lattice gas. The explicit form of the lattice gas collision operator is much simpler than in standard kinetic models.

Note that the Boltzmann form for the $(\mathcal{G} - \mathcal{L})$ collision term implicitly assumes only two-body collisions. It also assumes the collisions are pairwise statistically independent events occurring at a single point with detailed, or at most semi-detailed, balance symmetry for collision probabilities.

Solutions to the Boltzmann Transport Equation. Even though the Boltzmann equation is intractable in general, by using entropy arguments (Boltzmann's H theorem), the following can be stated about possible functional forms for f, the one-particle distribution function. If the system is uniform in space, any form for f will relax monotonically to the global Maxwell-Boltzmann form:

$$f_{\text{global}} \sim \rho e^{-E(\rho, v)/T},$$

in which the macroscopic variables ρ, v, and T (density, macrovelocity, and temperature) are independent of position, or global. In the non-equilibrium case, with a soft space dependence, any distribution function will relax monotonically in velocity space to a *local* Maxwell-Boltzmann form. This means that ρ, v, and T will depend on space as well as time. These local distribution functions are solutions to the Boltzmann transport equation. For the non-uniform case, one gets a picture of the full solution as an ensemble of local Maxwell-Boltzmann distributions covering the description space of the fluid, with some gluing conditions providing the consistency of the patching.

Recovering Macrodynamics–The Euler Equations. If we assume a simple fluid and neglect all dissipative processes (viscosity, heat transfer, etc.), we can quickly derive the Euler equations (presented in "The Continuum Argument") from the Boltzmann transport equation. But first we need the notion of average quantities and some observations about collisions in a dissipation-free system.

As before, let $\rho(t, r) = \int f(t, \mathbf{r}, \Gamma)\, d\Gamma$ be the density field of the gas. Then a mean gas velocity v = $\frac{1}{\rho} \int \mathbf{v}' f(t, \mathbf{r}, \Gamma)\, d\Gamma$, where v' is a microvelocity. We will use v as a macroscopic variable that characterizes cells whose length L in any direction is much, much greater than the mean free path in the gas, l_m; that is, $L \gg l_m$.

Since, by assumption, collisions preserve conservation laws exactly, the moments of $C(f)$, in particular the integrals $\int C(f) d\Gamma$ and $\int \mathbf{v} \, C(f) d\Gamma$, are equal to zero (similarly for any conserved quantity). We use this fact by integrating the Boltzmann equation in two ways: $\int (B.E.) d\Gamma$ and $\int \mathbf{v}(B.E.) d\Gamma$ (where B.E. stands for the Boltzmann equation). The first integral gives the continuity equation:

$$\partial_t \rho + \partial_i (\rho \mathbf{v}_i) = 0. \tag{6}$$

The second integral gives the momentum tensor equation:

$$\partial_t (\rho \mathbf{v}_i) + \partial_k \Pi_{ik} = 0, \tag{7}$$

where the momentum flux tensor Π_{ik} is given by

$$\Pi_{ik} \equiv \int \mathbf{v}_i \mathbf{v}_k f \, d\Gamma.$$

In order to derive the Euler equation for ideal gases with the usual form for the momentum flux tensor, we need to assume that each region in the gas has a local Maxwell-Boltzmann distribution. With this assumption one can show that the momentum flux tensor in Eq. 7 has the following form:

$$\Pi_{ik} = \rho \mathbf{v}_i \mathbf{v}_k + \delta_{ik} p,$$

where p is the pressure. This form of Π_{ik} gives the same Euler equation that we found by general continuum arguments. (We will see in Part II that the form of Π_{ik} for the totally discrete fluid is not so simple but depends upon the geometry of the underlying lattice. Again by assuming a form for the local distribution function (the appropriate form will turn out to be Fermi-Dirac rather than Boltzmann), Π_{ik} will reduce to a form that gives the lattice Euler equation.)

Recovering the Navier-Stokes Equation. The derivation of the Navier-Stokes equation from the kinetic theory picture is more involved and requires us to face the full Boltzmann equation. Hilbert accomplished this through a beautiful argument that relies on a spatial-gradient perturbation expansion around some single-particle distribution function f_L assumed to be given at t_0. In "The Hilbert Contraction" we discuss the main outline of his argument emphasizing the assumptions involved and their limitations. Here we will summarize his argument. Hilbert was able to show that the evolution of f for times $t > t_0$ is given in terms of its initial data at t_0 by the first three moments of f, namely the familiar macroscopic variables ρ (density), $\mathbf{v}$ (mean velocity), and T (temperature). In other words, he was able to contract this many-degree-of-freedom system down to a low-dimensional descriptive space whose variables are the same as those used in the usual hydrodynamical description. The beauty of Hilbert's proof is that it is constructive. It explicitly displays a recursive closed tower of constraint relations on the moments of f that come directly from the Boltzmann equation. The zero-order relation gives the Euler equations and the second-order relation gives the Navier-Stokes equations. However, Hilbert's method is an asymptotic functional expansion, so that the higher order terms take one away from ordinary fluids rather than closer to them. Nevertheless, solving explicitly for the terms in the functional expansion provides a way of evaluating transport coefficients such as viscosity. (See the "Hilbert Contraction" for more discussion.)

Summary of the Kinetic Theory Picture. Our review of the kinetic theory description of fluids introduced a number of important concepts: the idea of local thermal equilibrium; the characterization of an equilibrium state by a few macroscopic observables; the Boltzmann transport equation for systems of many identical objects (with ordinary statistics) in collision; and the fact that a solution to the Boltzmann transport equation is an ensemble of equilibrium states. In "The Hilbert Contraction" we introduced the linear approximation to the Boltzmann equation with which one can derive the Navier-Stokes equations for systems not too far (in an appropriate sense) from equilibrium in terms of these same macroscopic observables (density, pressure, temperature, etc.). We then outlined a method for calculating the coupling constants in the Navier-Stokes system—that is, the strengths of the nonlinear terms—as a function of any particular microdynamics.

This review was intended to give a flavor for the chain of reasoning involved. We will use this chain again in the totally discrete lattice world. However, just as important as understanding the kinetic theory viewpoint is keeping in mind its limitations. In particular, notice that perturbation theory was the main tool used for going from the exact Boltzmann transport equation to the Navier-Stokes equations. We did not discover more powerful techniques for finding solutions to the Navier-Stokes equations than we had before. To go from the Boltzmann to the Navier-Stokes description, we made many smoothness assumptions in various probabilistic disguises; in other words, we recreated an approximation to the continuum. It is true one could compute (at least for relatively simple systems) the transport coefficients, but in a sense these coefficients are a property of microkinetics, not macrodynamics.

We are at a point where we can ask some questions about the emergence of macrodynamics from microscopic physics. It is clear by now that microscopic conservation laws, those of mass, momentum, and energy are crucial in fixing the form of large-scale dynamics. These are in a sense sacred. But one can question the importance of the description of individual collisions. How detailed must micromechanics be to generate the qualitative behavior predicted by the Navier-Stokes equations? Can it be done with simple collisions and very few classes of them? There exists a whole collection of equations whose functional form is very nearly that of the Navier-Stokes equations. What microworlds generate these? Do we have to be exactly at the Navier-Stokes equations to generate the qualitative behavior and numerical values that we derive from the Navier-Stokes equations or from real fluid experiments? Is it possible to design a collection of synthetic microworlds that could be considered local-interaction board games, all having Navier-Stokes macrodynamics? In other words, does the detailed microphysics of fluids get washed out of the macrodynamical picture under very rapid iteration of the deterministic system? If the microgame is simple enough to update it deterministically on a parallel machine, is the density of states required to see everything we see in ordinary Navier-Stokes simulations much smaller than the density of atoms in real physical fluids? If so, these synthetic microworlds become a potentially powerful analytic tool.

Our approach in building a cellular space is to move away from the idea of a fluid state and focus instead on the idea of the macrodynamics of a many-element system. In abstract terms, we want to devise the simplest deterministic local game made of a collection of few-bit, finite-state machines that has the Navier-Stokes equations as its macrodynamical description. From our brief look at kinetic transport theory, we can abstract the essential features of such a game (Fig. 3). The many-element system must be capable of supporting a notion of local thermodynamic equilibrium and must also include local microscopic conservation laws. The state of a real fluid can be imagined as a collection of equilibrium distribution functions whose macroscopic parameters are unconstrained. These distribution functions have a Maxwell-Boltzmann form, $e^{-E(\rho, v)/T}$. If these distribution functions are made to deviate slightly from equilibrium, then local conservation laws impose consistency conditions among their parameters, which become constrained variables. These consistency conditions are the macrodynamical equations necessary to put a consistent equilibrium function description onto the many-element system. In physical fluids they are the Navier-Stokes equations. This is the general setup that will guide us in creating a lattice model.

Evolution of Discrete Fluid Models

Continuous Network Models. The Navier-Stokes equations, however derived, are analytically intractable, except in a few special cases for especially clean geometries. Fortunately, one can avoid them altogether for many problems, such as shocks in certain geometries. The strategy is to rephrase the problem in a very simple phase space and solve the Boltzmann transport equation directly. If a single type of particle is constrained to move continuously only along a regular grid, the Boltzmann equation is so tightly constrained that it has simple analytic solutions. In the early 1960s Broadwell and others applied this simplified method of analysis to the dynamics of shock problems. Their numerical results agreed closely with much more elaborate computer modeling from the Navier-Stokes equations. However, there was no real insight into why such a calculation in such a simplified microworld should give such accurate answers. The accuracy of the limited phase-space approach was considered an anomaly.

Discrete Skeletal Models. The next development in discrete fluid theory was a discrete modification of the continuous-speed network models of the Broadwell class. By forming a loose analogy to the structure of the Ising model (spins on a lattice), Hardy, de Pazzis, and Pomeau created the first minimalist fluid model on a two-dimensional square lattice. It was a simple, binary-valued, nearest-neighbor gas with a single species of molecule, limited to binary collisions. The new feature was a totally discrete velocity and state space for the gas. Particles hopped from one site to the next without a notion of continuous movement between sites. Particles were confined to the vertices of the network, and the velocity vector of each particle could point in only one of four directions. Since there was no natural way to deal with bound states, these authors imposed the arbitrary rule that the maximum number of particles occupying any vertex be four.

This simple model possessed remarkable properties including local thermodynamic equilibrium and the emergence of a scale separation; that is, the typical collective motion scale L is much greater than the microscopic mean free path l_m; $L \gg l_m$. However, the macrodynamics that emerged was not that of the Navier-Stokes equations but a more complex one with unphysical features. The square model was the first example of rich dynamics emerging wholly on a cellular space. It had all the right ingredients except one: isotropy under the rotation group of the lattice. The momentum flux tensor must reduce to a scalar for isotropy, but this is impossible with a square lattice. In two dimensions the neighborhood that has the minimal required symmetry and tiles the plane is a hexagonal neighborhood. In Part II we will present the simple hexagonal model, analyze it mathematically, and describe the simulations of fluid phenomena that have been done so far.

The Continuum Argument

L et $v(x, t)$ be a vector-valued field referred to a fixed origin in space, which we identify with the velocity of a "macroscopic" fluid cell. The cell is not small enough to notice a particle structure for the fluid, but it is small enough to be treated as a mathematical point and still agree with physics.

To derive the properties of a flow defined by the vector field, one now invokes the generalized Stokes theorem:

$$\oint_{\partial\Sigma} A = \oint_{\Sigma} dA,$$

where Σ is a generalized surface or volume, $\partial\Sigma$ the boundary of Σ, A an n-differential form and dA an $(n + 1)$-differential form. This very general theorem has two familiar forms: one is the classical Stokes theorem from one to two dimensions,

$$\oint_{\partial\Sigma} A \cdot d\ell = \oint_{\Sigma} \nabla \times A \cdot dS,$$

and the other is the Gauss law from two to three dimensions,

$$\oint_{\partial\Sigma} A \cdot dS = \int_{\Sigma} (\nabla \cdot A) dV,$$

where ℓ is a curve, S is a surface, and V is a volume in three-dimensional Euclidean space $\mathbf{R}^3$.

Conservation Laws and Euler's Equation. First, we deal with the idea of continuity, or conservation of flow. If ρ is the density, or mass per unit volume, then the mass of the fluid in volume V (that is, Σ), is equal to $\int_{\Sigma} \rho dV$. A two-dimensional surface in $\mathbf{R}^3$ has an outward normal vector $\mathbf{n}$ which is defined to be positive. The total mass of fluid flowing out of a volume Σ can be written as

$$\oint_{\partial\Sigma} \rho v \cdot dS = \int_{\partial\Sigma} \rho v \cdot \mathbf{n} dS.$$

Continuity of the flow implies a balance between the flow through the surface and the loss of fluid from the volume. That is, the decrease in mass in the volume must equal the outflow of fluid mass through the surface of the volume, which implies by the Gauss law that

$$\oint_{\partial\Sigma} \rho v \cdot dS = -\partial_t \int_{\Sigma} \rho dV = \int_{\Sigma} \nabla \cdot (\rho v) dV.$$

This gives the first evolution equation for a fluid, the continuity, or mass-conservation, equation:

$$\partial_t \rho + \nabla \cdot (\rho v) = 0. \tag{1}$$

Now we introduce the idea of pressure p as the force exerted by the fluid on a unit surface area of an enclosed volume and use Newton's second law, $\mathbf{F} = m\mathbf{a}$. The total force acting on a volume of fluid due to the remainder of the fluid is given by $-\oint_{\partial\Sigma} p dS$. Using Stokes theorem we can write

$$-\oint_{\partial\Sigma} p dS = -\oint_{\Sigma} \nabla p dV.$$

The translation of $\mathbf{F} = m\mathbf{a}$ to a continuous medium is

$$-\nabla p = \rho dv/dt,$$

where dv/dt is a total derivative. The chain rule on $dv(\mathbf{x}, t)/dt$ gives

$$\rho dv/dt = \rho\{\partial_t v + v \cdot \nabla v\}$$

Substituting this result into the equation for $-\nabla p$ yields Euler's equation for an ideal, dissipation-free fluid:

$$\partial_t \mathbf{v} = -(\mathbf{v} \cdot \nabla)\mathbf{v} - \frac{1}{\rho}\nabla p. \tag{2}$$

One can generalize Euler's equation to a form more useful for a dissipative fluid. For this we look at the flux of momentum through a fluid volume. The momentum of fluid passing through an element dV is $\rho\mathbf{v}$, and its time rate of change expressed in components is

$$\partial_t(\rho v_i) = (\partial_t\rho)v_i + \rho(\partial_t v_i).$$

We can rewrite $\partial_t\rho$ and $\partial_t v_i$ as spatial derivatives by using Eqs. 1 and 2. Then

$$\partial_t(\rho v_i) = -\partial_k\Pi_{ik}, \tag{3}$$

where the momentum flux tensor $\Pi_{ik} \equiv p\delta_{ik} + \rho v_i v_k$.

The meaning of the momentum flux tensor can be seen immediately by integrating Eq. 3 and applying Stokes theorem.

$$\partial_t \int_\Sigma \rho v_i d\Sigma = -\int_\Sigma \partial_k\Pi_{ik}d\Sigma = -\oint_{\partial\Sigma} \Pi_{ik}n_k dS.$$

So

$$\partial_t \int_\Sigma \rho v_i dV = -\oint_{\partial\Sigma} \Pi_{ik}n_k dS,.$$

where the left-hand side is the rate of change of the ith component of momentum ρv_i in the volume and $\Pi_{ik}n_k d\Sigma$ is the ith component of momentum flowing through dS. Therefore, Π_{ik} is the ith component of momentum flowing in the kth direction. This is more easily seen by writing

$$\Pi_{ik}n_k = p\delta_{ik}n_k + \rho v_i v_k n_k = p\mathbf{n} + \rho\mathbf{v}(\mathbf{v}\cdot\mathbf{n}).$$

Equations 1, 2, and 3 are the basic formalism for classical Newtonian ideal fluids (fluids with no dissipation) and are also true for flows in general.

Classical Dissipative Fluids—The Navier-Stokes Equations. The general Euler's equation is $\partial_t(\rho v_i) = -\partial_k\Pi_{ik}$, where Π_{ik} is now the momentum stress tensor. The form of this tensor changes if the fluid is dissipative, for example, if viscous forces convert the energy in the flow into heat. Traditionally, Π_{ik} is modified in the following way. Take $\Pi_{ik} = p\delta_{ik} + \rho v_i v_k$ and introduce an unknown tensor σ'_{ik} that describes the effects of viscous stress. Then rewrite the momentum stress tensor as

$$\Pi_{ik} = p\delta_{ik} + \rho v_i v_k - \rho\sigma'_{ik} \equiv \sigma_{ik} + \rho v_i v_k,$$

where $\sigma_{ik} = p\delta_{ik} - \rho\sigma'_{ik}$ is called the stress tensor and σ'_{ik} the viscosity stress tensor.

The form of σ'_{ik} can be deduced on general grounds. First we assume that the gradient of the velocity changes slowly so σ_{ik} is linear in $\partial_k v_i$. Moreover, σ'_{ik} is zero for $\mathbf{v} = 0$, and under rotation it must vanish since uniform rotation produces no overall transport of momentum. The unique form that has these properties is

$$\sigma'_{ik} = a(\partial_k v_i + \partial_i v_k) + b\delta_{ik}\partial_j v_j,$$

where a and b are unknown coefficients. It is usually written in the form

$$\sigma'_{ik} = \nu(\partial_k v_i - \partial_i v_k - 2/3\delta_{ik}\partial_j v_j) + \varsigma\delta_{ik}\partial_j v_j,$$

where ν is the kinematic shear viscosity and ς is the kinematic bulk viscosity.

For an incompressible fluid (the density is constant so $\rho = \rho_0$) this tensor simplifies, and Euler's equation goes over to the incompressible Navier-Stokes equations:

$$\partial_t\mathbf{v} + (\mathbf{v}\cdot\nabla)\mathbf{v} = -\frac{1}{\rho}\nabla p + \nu\nabla^2\mathbf{v} \quad \text{and} \quad \nabla\cdot\mathbf{v} = 0.$$

The Hilbert Contraction

The Boltzmann equation is a microscopic equation for colliding-gas evolution valid in a very tight regime. It is first order in time and so requires a complete description of the one-particle distribution function at one time, say $t = 0$, after which its functional form is completely fixed by the Boltzmann transport equation.

Describing the one-particle distribution function completely is a hopeless procedure, since the amount of information is too large. However, one wants to recover hydrodynamics, which is essentially a partial differential equation for a macroscopic description of the fluid at long times and distances compared to molecular scales. So there must exist a contraction mechanism that reduces the number of degrees of freedom required to describe the solution to the Boltzmann transport equation at such long times and distances. It is not obvious how that can happen, but Hilbert gave a proof that is central to understanding that it must happen and in a rather surprising way. We will call this process the Hilbert contraction. All analyses of the Boltzmann equation are based on this contraction. We would like to give it in detail because it is a beautiful argument, but space forbids this, so we outline how Hilbert reasoned.

Since we don't know what else to do when faced with such a highly nonlinear system, we construct a perturbation expansion in a small variable around some distribution function f, assumed to be given to us at t_0. Under some very mild assumptions, and assuming the existence of such a general perturbation expansion in some parameter δ, Hilbert was able to show that the evolution of f for $t > t_0$ is given in terms of its initial data at t_0 by the first three moments of f, namely ρ, v, and T. The system has contracted down to a low-dimensional descriptive manifold whose coordinates are the same variables used by the hydrodynamic description. The beauty of Hilbert's proof is that it is constructive. It explicitly displays a recursive closed tower of constraint relations on the moments of f that come directly from the Boltzmann equation. The proof also shows that such a contracted description is unique—a very powerful result.

It must be pointed out that Hilbert's construction is on the time-evolved solution to the Boltzmann transport equation, not on the equation itself, which still requires a complete specification of f. It amounts to a hard mathematical statement on an effective field-theory description for times much greater than elementary collision times, but with space gradients still smooth enough to entertain a serious gradient perturbation expansion. As such, it says nothing about the turbulent regime, for example, where all these assumptions fail.

In standard physics texts one can read all kinds of plausibility arguments as to why this contraction process should exist, but they lack force, for, by arguing tightly, one can make the conclusion go the other way. This is why the Hilbert contraction is important. It is really a powerful and mathematically unexpected result about a highly nonlinear integro-differential equation of very special form. Beyond Hilbert's theorem and within the Boltzmann transport picture, we can say nothing more about the contraction of descriptions.

The construction of towers of moment constraints, coupled to a perturbation expansion that Hilbert developed for his proof of contraction, was used in a somewhat different form by Chapman and Enskog. Their main purpose was to devise a perturbation expansion with side constraints in such a way as to pick off the values of the coupling constants—which are called transport coefficients in standard terminology—for increasingly more sophisticated forms of macrodynamical equations.

One makes the usual kinetic assumptions: The gas reaches local equilibrium in a collision time or so; the one-particle distribution function has a local Maxwell-Boltzmann form (or whatever form is appropriate), call it f_L; a second time scale is assumed where space gradients are still small, but collective modes develop at large distances and at times much greater than molecular collision times. Then one assumes a general functional perturbation expansion exists of the form

$$f = f_L(1 + \xi^{(1)} + \xi^{(2)} + \cdots),$$

which turns out to be explicitly a spatial gradient expansion:

$$f = f_L(1 + c_1(v)(\lambda \nabla) + c_2(v)(\lambda \nabla)^2 + \cdots)$$

where λ is the mean free path in the system and v is the macrovelocity.

The perturbation expansion is set up so that at nth order, the correction to f_L obeys an integral equation of the form $f_L C(\xi^{(n)}) = L_n$, where C is the Boltzmann collision operator and L_n is an operator that depends only on lower order spatial derivatives. This generates a recursive tower of relations $\xi^{(n)}$ whose solubility conditions at order n are the $(n-1)$th-order hydrodynamical equations.

For example, assume

$$f = f_L \left(1 + \xi^{(1)}\right);$$

that is, we keep only 1st order in ξ. Then in the Boltzmann collision term keep consistently only order $\xi^{(1)}$ and in the streaming operator put $\xi^{(0)} \equiv f_L$. So we get

$$(\partial_t + v_\alpha \partial_\alpha + a_\alpha \frac{\partial}{\partial v_\alpha}) f_L = f_L C(\xi^{(1)}),$$

which is of the form

$$f_L C(\xi^{(1)}) = L_1.$$

The solubility conditions for this are that L_1 must be orthogonal to the five zero eigenmodes of $C(\xi^{(1)}) = 0$ (the solutions are $1, \mathbf{v}$, and $\mathbf{v}^2$). These solubility conditions are the Euler equation for ρ, v, and T and the ideal gas equation of state. In this way one derives a sequence of hydrodynamical equations with explicit forms for the transport coefficients. Order 0 gives the Euler equation, order 1 gives the Navier-Stokes equations, order 2 and greater give the generalized hydrodynamical equations, which have some validity only in special situations. The expansion is an asymptotic functional expansion, so going beyond Navier-Stokes takes one away from ordinary fluids rather than closer to them. Solving explicitly for the various $\xi^{(n)}$ gives a way to evaluate the transport quantities (viscosity, etc.).

There are many other ways to do the same thing—multiple time expansions, dispersion methods, etc. We have developed everything so far within the conceptual frame of the Boltzmann transport equation. Within that framework the problem of deriving macrodynamical equations and associated transport coefficients reduces to tedious but straightforward linear algebra that has absorbed the best efforts of excellent technical people since the turn of the century. It is a problem best suited to a computer but only recently have algebraic processors of sufficient power been available.

This asymptotic perturbation expansion is a way to compute measurable quantities from microdynamical properties, but the physical insight one gains from doing it is small. The other methods mentioned, especially correlation-function techniques, are much more revealing. All of these comments and approaches carry over directly to the discrete case of the lattice gas. Nothing conceptually new arises in the totally discrete case, except that explicit calculations are a great deal easier. ■

PART II
THE SIMPLE HEXAGONAL MODEL: THEORY AND SIMULATIONS

The Minimal Totally Discrete Model of Navier-Stokes in Two Dimensions

We can now list the ingredients we need to build the simplest cellular-space world with a dynamics that reproduces the collective behavior predicted by the compressible and incompressible Navier-Stokes equations:

1. A population of identical particles, each with unit mass and moving with the same average speed c.
2. A totally discrete phase space (discrete values of x, y and discrete particle-velocity directions) and discrete time t. Discrete time means that the particles hop from site to site.
3. A lattice on which the particles reside only at the vertices. In the simplest case the lattice is regular and has a hexagonal neighborhood to guarantee an isotropic momentum flux tensor. We use a triangular lattice for convenience.
4. A minimum set of collision rules that define symmetric binary and triple collisions such that momentum and particle number are conserved (Fig. 4).
5. An exclusion principle so that at each vertex no two particles can have identical velocities. This limits the maximum number of particles at a vertex to six, each one having a velocity that points in one of the six directions defined by the hexagonal neighborhood.

The only way to make this hexagonal lattice gas simpler is to lower the rotation symmetry of the lattice, remove collision rules, or break a conservation law. In a two-dimensional universe with boundaries, any such modification will not give Navier-Stokes dynamics. Left as it is, the model will. Adding attributes to the model, such as different types of particles, different speeds, enlarged neighborhoods, or weighted collision rules, will give Navier-Stokes behavior with different equations of state and different adjustable parameters such as the Reynolds number (see the discussion in Part III). The hexagonal model defined by the five ingredients listed above is the simplest model that gives Navier-Stokes behavior in a sharply defined parameter regime.

At this point it is instructive to look at the complete table of allowed states for the model (Fig. 5). The states and collision rules can be expressed by Boolean logic operations with the two allowed values taken as 0 and 1. From this organization scheme we see that the hexagonal lattice gas can be seen as a Boolean parallel computer. In fact, a large parallel machine can be constructed to implement part or all of the state table locally with Boolean operations alone. Our simulations were done this way and provide the first example of the programming of a cellular-automaton, or cellular-space, machine that evolves the dynamics of a many-degrees-of-freedom, nonlinear physical system.

Theoretical Analysis of the Discrete Lattice Gas

Before presenting the results of simulations with the lattice-gas automaton, we will analyze its behavior theoretically. The setup we work on is a regular triangular grid with hexagonal neighborhood. The natural explicit coordinate system for a single-speed, six-directional world (Fig. 6) is the set of unit vectors:

$$\hat{\mathbf{i}}_\beta = \left\{ \cos\left(\frac{2\pi\beta}{6}\right), \ \sin\left(\frac{2\pi\beta}{6}\right) \right\}, \quad \beta = 1, \cdots, 6. \tag{8}$$

One never requires this much detail except to work out explicit tensor structures and scalar products particular to the hexagonal model case, but the index conventions are important to avoid disorientation. From now on the

Greek indices $\alpha, \beta, \cdots$ label lattice direction indices; $\hat{\mathbf{i}}, \hat{\mathbf{j}}, \hat{\mathbf{k}}, \ldots$ are lattice unit vectors and i, j, k label space indices $(x_1, x_2, \ldots)$; on a square lattice we have $\mathbf{r} = (x_1, x_2) = (x, y)$.

The first thing we will look at is pure transport on the lattice with no collisions. Because the basic space is a discrete lattice with a fundamental lattice spacing, rather than a continuum, a shadow of the lattice is induced into the coupling constant of the theory, namely the viscosity. This lattice effect is not obvious, but we will make it so by looking at transport on the lattice in detail. As a corollary we will derive the usual Euler equations for the "macroscopic" flow of the lattice gas.

To do a quick analysis on lattice models we lift the restriction of a deterministic gas and pass to a probabilistic description familiar from kinetic theory; then we can use familiar stochastic and kinetic theory tools outlined in Part I of this article. In going from a continuous to a discrete probabilistic formalism we introduce the lattice form of the single-particle distribution function by making the identifications

$$f(\mathbf{r}, \Gamma, t) \to f_\beta(\mathbf{r}, t)$$

$$\int f(\mathbf{r}, \Gamma, t) d\Gamma \to \Sigma_\beta f_\beta(\mathbf{r}, t) \equiv \rho,.$$

and

$$\rho \mathbf{v} \to \Sigma_\beta \hat{\mathbf{i}}_\beta f_\beta$$

To begin we write the master equation for f_β in the absence of collisions. The master equation expresses conservation of probability. For simplicity we write it for a square lattice with the following conventions: $n_\beta(\mathbf{r} + \hat{\mathbf{i}}_\beta, t)$ = number of particles in the direction β at the node $\mathbf{r} + \hat{\mathbf{i}}_\beta$ at time t. The master equation for the system, neglecting collisions and written in a continuum notation for convenience, is

$$f_\beta(\mathbf{r} + h, t + k) - f_\beta(\mathbf{r}, t) = 0, \quad \text{with} \quad h = \hat{\mathbf{i}}_\beta d_x, k = d_t,$$

where $d_t, d_x \ll 1$.

If we expand the first term in the master equation out to $O^2(h, k)$ using the Taylor series expansion $f(x_0 + h, y_0 + k) = \sum_{\lambda=0}^{m-1} \frac{1}{\lambda!} (h\partial_x + k\partial_y)^\lambda f(x_0, y_0) + R_m$, we obtain

$$0 = d_t \partial_t f_\beta + d_x \hat{\mathbf{i}}_\beta \cdot \nabla f_\beta + \frac{1}{2} d_t^2 \partial_t^2 f_\beta + \frac{1}{2} d_x^2 (\hat{\mathbf{i}}_\beta \cdot \nabla)^2 f_\beta + d_x d_t (\hat{\mathbf{i}}_\beta \cdot \nabla) \partial_t f_\beta.$$

To lowest order in h and k, we have

$$\partial_t f_\beta + \hat{\mathbf{i}}_\beta \cdot \nabla f_\beta = 0, \tag{9}$$

which has the standard form of the kinetic theory transport equation in the absence of collisions. If we include collisions, the full Boltzmann transport equations schematically become

$$\partial_t f_\beta + \hat{\mathbf{i}}_\beta \cdot \nabla f_\beta = C_\beta(f) \tag{10}$$

where $C_\beta(f)$ is the collision operator on the lattice. The form of the lattice collision operator will tell us a great deal about how the model works, but for the moment we just look at the general structure of the "macroscopic" equations for the lattice gas to the lowest order in the lattice expansion parameters.

As in standard kinetic theory, the usual zero integrals of the motion hold, since the lattice model is assumed to have some kind of detailed balance (that is, microscopic reversibility of reaction pathways). Accordingly $\sum_\beta C_\beta(f) = 0$ and $\sum_\beta \hat{\mathbf{i}}_\beta C_\beta(f) = 0$ for a skeletal gas. Following the kinetic theory procedure, we write the continuity and momentum equations that follow from these conditions as:

$$\partial_t \rho + \partial_i (\rho v_i) = 0 \tag{11}$$

and

$$\partial_t(\rho v_i) + \partial_j \Pi_{ij} = 0, \tag{12}$$

where the tensor Π_{ij} is defined as

$$\Pi_{ij} = \sum_\beta (\hat{i}_\beta)_i (\hat{i}_\beta)_j f_\beta. \tag{13}$$

So far we have kept only the leading terms of the Taylor series expansion in the scaling factors that relate to the discreteness of the lattice. It's easy to show that keeping quadratic terms in this lattice-size expansion leaves the continuity equation invariant but alters the momentum equation by introducing a free-streaming correction to the measured viscosity. This rather elegant way of viewing this correction was first developed by D. Levermore. The correction comes from breaking the form of a Galilean covariant derivative and is a geometrical effect. Specifically, to second order in the lattice size expansion, the momentum equation does not decompose simply into factors of these covariant derivatives but instead the expansion introduces a nonvanishing covariant-breaking term:

$$\text{Noncovariant term} = \sum_\beta \{(\hat{i}_\beta)_i \partial_i \partial_t + ((\hat{i}_\beta)_i \partial_i)^2\} \hat{i}_\beta f_\beta. \tag{14}$$

This term is of the same order as those terms that contribute to the viscosity. Later we will show how to use the Chapman-Enskog expansion to compute an explicit form for the lattice-gas viscosity.

The Chapman-Enskog Expansion and the Direct Expansion. The form of Π_{ij} depends on the form of f, the solution to the full lattice Boltzmann transport equation. By Hilbert's construction we know that an efficient expansion can be developed in terms of the collision invariants of the model up to powers of terms linear in the gradient of the macroscopic velocity. In whatever perturbation expansion of f we choose, the coefficients in the expansion are fixed by solving for them under the Lagrange multiplier constraints of mass and momentum conservation: $\rho = \sum_\beta f_\beta$ and $(\rho v) = \sum_\beta \hat{i}_\beta f_\beta$. In the simple hexagonal model there is no explicit mechanism provided for storing energy in internal state space, so there is no independent energy equation.

For the lattice case, the Chapman-Enskog version of Hilbert's expansion reduces to an expansion in all available scalar products using the vectors $\hat{i}_\beta, v$ and the vector operator $\overrightarrow{\partial}$. The expansion is made around the global equilibrium solution for $v = 0$, which we will call $N_{eq}^{v=0}$ and terms are kept up to those linear in $\overrightarrow{\partial}$. The relevant scalar products are

$$(\hat{i}_\beta \cdot \hat{i}_\beta), \ (\hat{i}_\beta \cdot v), \ (\hat{i}_\beta \cdot \overrightarrow{\partial})(\hat{i}_\beta \cdot v), \ (\overrightarrow{\partial} \cdot v), \ (\hat{i}_\beta \cdot v)^2, \ (v \cdot v) + O(v^3).$$

The systematic expansion becomes

$$f_\beta = N_{eq}^{v=0} \left\{ 1 + \alpha \hat{i}_\beta \cdot v + \beta \left[(\hat{i}_\beta \cdot v)^2 - \frac{1}{2}|v|^2 \right] + \beta_1 \left[(\hat{i}_\beta \overrightarrow{\partial})(\hat{i}_\beta v) - \frac{1}{2} \overrightarrow{\partial} \cdot v \right] + O(v^3) \dots \right\}. \tag{15}$$

In the usual kinetic theory approach the coefficients α and β can be found by neglecting collisions and β_1, the gradient term, can be determined only by an explicit solution to the full Boltzmann equation including collision terms. In this way one obtains the viscosity in terms of β_1. For the discrete lattice, however, both β and β_1 depend on the explicit form of the solution to the full Boltzmann equation with collisions. We also need that form to recover the correction to the raw viscosity that, as mentioned in the last section, comes from pure translation effects on the lattice.

Given that we have to use the full solution to the Boltzmann transport equation almost immediately, we now derive its structure, find the general and equilibrium solution, and then use a direct expansion to fix both β and β_1. In the process we will recover the Euler equations for inviscid flow and the Navier-Stokes equations for the flow with dissipation.

The Lattice Collision Operator and the Solution to the Lattice Boltzmann Transport Equation. We will write down the discrete form of the Boltzmann equation, especially noting the collision operator, for a number of reasons. First, writing the explicit form of the collision kernel builds up an intuition of how the heart of the model works; second, we can show in a few lines that the Fermi-Dirac distribution satisfies the lattice gas Boltzmann equation; third, knowing this, we can quickly compute the lattice form of the Euler equations; fourth, we can see that many properties of the lattice-gas model are independent of the types of collisions involved and come only from the form of the Fermi-Dirac distribution.

Collision operators for lattice gases with continuous speeds were derived by Broadwell, Harris, and other early workers on continuum lattice-gas systems. For totally discrete lattice gases with an exclusion principle, we must be careful to apply this principle correctly. It is similar to the case of quasi-particles in quantum Fermi liquids. The construction reduces to following definitions of collision operators introduced in the section on classical kinetic theory and counting properly.

Taking any hexagonal neighborhood, let i be one of the six directions and use the convention $i, i+1, i+2, \ldots \equiv \hat{1}_{\beta=i}, \hat{1}_{\beta=i+1}, \hat{1}_{\beta=i+2}, \ldots$ for convenience. (Later we will return to our original notation.) First consider binary collisions alone, and assume detailed balance, which implies microscopic reversibility of a collision at each vertex. One need not use detailed balance, but other balancing schemes are algebraically tedious and conceptually similar extensions of this basic case. Given a vertex at $(\mathbf{r}, t)$ we compute the gain and loss of particles into a neighborhood along a fixed direction, say i. This is, by definition, the collision kernel for binary processes. First compute the number of particles thrown in a collision into a phase-space region along the direction i. Let $n_i(\mathbf{r}, t)$ be the probability that a particle is at the node $(\mathbf{r}, t)$ and has a velocity in the ith direction.

If a particle scatters into a vector direction i, it must have come from binary processes along directions $(i + 1$ and $i + 4)$ or $(i + 2$ and $i + 5)$ (see the two-body scattering rules in Fig. 4). Interpreted as probabilities for the two events to happen, the probability for gain in the i direction due to binary processes alone is

$$P_i^{\text{binary}} = n_{i+1} n_{i+4} \bar{n}_i \bar{n}_{i+2} \bar{n}_{i+3} \bar{n}_{i+5} + n_{i+2} n_{i+5} \bar{n}_i \bar{n}_{i+1} \bar{n}_{i+3} \bar{n}_{i+4},$$

where $\bar{n}_k \equiv (1 - n_k)$. The $\bar{n}_k$'s impose the exclusion rule in the output channel, namely, that a particle cannot scatter there if one is already present.

Loss of a particle from direction i can occur only by the binary collision $(i + 3, i)$, and this can happen for each of the two choices of gain collisions separately. So we have $(-2n_i n_{i+3} \bar{n}_{i+1} \bar{n}_{i+2} \bar{n}_{i+4} \bar{n}_{i+5})$ as the probability for loss in the i direction due to binary collisions alone. Note that these products can be compactly expressed as $\hat{n}_i \hat{n}_{i+3} \Pi_{i=0}^5 (1 - n_i)$ where $\hat{n}_i \equiv \frac{n_i}{1-n_i}$.

The three-body gain-loss term can be written down by inspection in the same way as the binary term. The complete two- and three-body collision term for the ith direction, in compact notation, is

$$C_i^{2+3} = [(\hat{n}_{i+1} \hat{n}_{i+4} + \hat{n}_{i+2} \hat{n}_{i+5} - 2\hat{n}_i \hat{n}_{i+3}) + (\hat{n}_{i+1} \hat{n}_{i+3} \hat{n}_{i+5} - \hat{n}_i \hat{n}_{i+2} \hat{n}_{i+4})] \, \Pi_{i=0}^5 (1 - n_i).$$

For extensive calculations more compact notations are easily devised, but this one clearly brings out the essential idea in constructing arbitrary collision schema. With some minor modifications this form for the collision operator can be reinterpreted as a master equation for a transition process, which is useful as a starting point for a detailed microkinetic analysis by stochastic methods.

Given the $C(f)$ for two- and three-body collisions in the above compact form, and given detailed balance, we show that $C(f_\beta) = 0$ for the Fermi-Dirac distribution. The proof is simple and well known from quantum Fermi-liquid theory where the same functional form for the collision operator appears but with a different interpretation.

If n is a Fermi-Dirac distribution, it has the form $(1 + e^E)^{-1} = n(E)$ where E is expanded in collision invariants, in this case particle number and momentum. Then note that $\frac{n}{1-n} \equiv \hat{n} = e^{-E}$, the form of the Maxwell-Boltzmann distribution. This is also the form of the collision kernel, and the exponential terms just contain the sum of momenta in the collision. Since this sum is conserved, each collision term (binary, triple, etc.) vanishes separately, because of the exclusion principle. So the solution is a Fermi-Dirac distribution. This proof also shows that as long as conservation laws of any kind are embodied in the collision term, each type of collision is separately zero under the

Fermi-Dirac distribution. Accordingly, the Fermi-Dirac solution is universal across collision types. This implies that one cannot alter the character of the Fermi-Dirac distribution in the lattice gas by adding collision types that respect collision invariants.

Since f_β is now assumed to be a Fermi-Dirac distribution, we take it as

$$f_\beta = (1 + e^E)^{-1},$$

with
$$E = \alpha(\rho, v) + \overrightarrow{\beta}(\rho, v) \cdot \hat{i}_\beta.$$

(Here we have returned to our original conventions for $\hat{i}_\beta$.) The equilibrium value for f_β at $v = 0$, namely $N_{eq}^{v=0}$, is $\frac{\rho}{6}$ where ρ is the density. Expanding the Fermi-Dirac form for f_β about this equilibrium value gives us

$$f_\beta = \frac{\rho}{6} \left\{ 1 + \alpha(\rho)(\hat{i}_\beta \cdot v) + \beta(\rho) \left[(\hat{i}_\beta \cdot v)^2 - \frac{1}{2}|v|^2 \right] + \cdots \right\}, \tag{16}$$

the same form as the Chapman-Enskog expansion (Eq. 15). To fix α and β we use number and momentum conservation as constraints, so that f_β becomes

$$f_\beta = \frac{\rho}{6} \left\{ 1 + 2(\hat{i}_\beta \cdot v) + 4g(\rho) \left[(\hat{i}_\beta \cdot v)^2 - \frac{1}{2}|v|^2 \right] + \cdots \right\},$$

where we have taken the particle speed as 1 ($c = 1$). The coefficient $g(\rho)$ is

$$g(\rho) = \frac{3 - \rho}{6 - \rho}.$$

If we substitute this result for f_β in the momentum tensor (Eq. 13) and do the sum over β, the particle directions, we have

$$\Pi_{ik} = \frac{\rho}{2} \left(1 - g(\rho)v^2 \right) \delta_{ik} + \rho g(\rho)v_i v_k.$$

The lattice Euler equation (Eq. 12) thus becomes

$$\partial_t(\rho v_i) + \partial_j \left[\rho g(\rho)v_i v_j + \cdots \right] = -\partial_i p. \tag{17}$$

In the usual Euler equation $g(\rho) = 1$. Here $g(\rho)$ is the lattice correction to the convective term due to the explicit lattice breaking of Galilean invariance. The equation of state for Eq. 17 is

$$p = \frac{\rho}{2} \left(1 - g(\rho)\frac{v^2}{2} \right).$$

For general single-speed models with particle speed c and b velocity vectors in D dimensions, the result above generalizes to

$$f_\beta = \frac{\rho}{b} \text{ when } v = 0$$

and
$$g(\rho) = \frac{D}{D+2} \frac{b - 2\rho}{b - \rho}.$$

These forms depend only on the structure of tensor products of $\hat{i}_\alpha$ in D dimensions.

When we discuss the full Navier-Stokes equations, we will show how to absorb the $g(\rho)$ Galilean-invariance-breaking term in Eq. 17 into a rescaling of variables.

Isotropy and The Momentum Tensor. We will go on to discuss viscosity and the lattice form of the Navier-Stokes equation, but first we comment briefly on how the structure of the momentum tensor depends on the geometry of the lattice. Those interested in all the details can find them discussed from several viewpoints in Frisch, d'Humières, Hasslacher, Lallemand, and Pomeau ~~1987.~~ and Rivet, 1987.

By definition $\Pi_{ij} \equiv \sum_\beta (\hat{i}_\beta)_i (\hat{i}_\beta)_j f_\beta$, where f_β is determined by the Chapman-Enskog, or direct, expansion (Eq. 15). Isotropy implies invariance under rotations and reflections; tensors that are isotropic are proportional to a scalar. Define the tensors $E^{(n)} = \sum_\beta (\hat{i}_\beta)_{i_1} \dots (\hat{i}_\beta)_{i_n}$. For $E^{(n)}$ with regular b-sided polygons, we can derive conditions on b for $E^{(n)}$ to be isotropic. These conditions are

$$\left(E^{(2)}|b>2\right), \ \left(E^{(3)}|b\geq 2, b\neq 3\right), \ \left(E^{(4)}|b>2, b\neq 4\right), \ \left(E^{(5)}|b\geq 2, b\neq 3,5\right), \dots .$$

For $b = 4$, the case of the HPP (Hardy, de Pazzis, and Pomeau) square lattice, $E^{(4)}$ is not isotropic. For $b = 6$, the hexagonal neighborhood case, all tensors up to $n = 5$ are isotropic.

Using the Chapman-Enskog expansion for f_β and the notation above for tensors, Π_{ij} has the following tensor structure.

$$\Pi_{ij} \approx N_{eq}^{v=0}\left(E_{ij}^{(2)} + \alpha E_{ijk}^{(3)}v_k + \beta\left[E_{ijkl}^{(4)}v_kv_l, E_{ij}^{(2)}v_kv_k\right] + \beta_1\left[E_{ijkl}^{(4)}\partial_kv_l, E_{ij}^{(2)}\partial_kv_k\right]\right)$$

where we are following the discussion of Wolfram. The momentum stress tensor must be isotropic up to $E^{(4)}$ in order that the leading terms in the momentum equation (corresponding to the convective and viscous terms in the Navier-Stokes equation) be isotropic. For the square model, the original discrete-lattice model, we have nonisotropy manifested in two places through the momentum flux tensor.

$$\Pi_{11} = \rho g(\rho)(v_1^2 - v_2^2) + \frac{\rho}{2} + O(v^4),$$

$$\Pi_{22} = \rho g(\rho)(v_2^2 - v_1^2) + \frac{\rho}{2} + O(v^4),$$

$$\Pi_{12} = \Pi_{21} = 0,$$

where

$$g(\rho) = \frac{2-\rho}{4-\rho}.$$

See Frisch et al. for further discussion.

The nonisotropy implies that we do not get a Navier-Stokes type equation for the square lattice. For the hexagonal model, $\beta = 6$, isotropy is maintained through order $E^{(4)}$. By using general considerations on tensor structures for polygons and polyhedra in D-dimensional space, one can quickly arrive at probable models for Navier-Stokes dynamics in any dimension. The starting point is that isotropy, or the lack of it, in both convective and viscous terms (the Euler and the Navier-Stokes equations), is controlled completely by the geometry of the underlying lattice. This crucial point was missed by all earlier workers on lattice models who thought that the geometry of the underlying lattice was irrelevant.

Viscosity for Lattice-Gas Models. In "The Continuum Argument" we saw that the general form of the compressible Navier-Stokes equation with bulk viscosity $\varsigma = 0$ is

$$\partial_t(\rho v_i) + \partial_j(\rho(v_i v_j)) = -\partial_i p + \partial_j\left(\nu\rho(\partial_j v_i + \partial_i v_j - \frac{2}{3}\delta_{ij}\partial_k v_k)\right),$$

where ν is the kniematic shear viscosity. To derive this form for the discrete model, one must solve for Π_{ij} using both the Chapman-Enskog approximation for f_β and the momentum-conservation equation. We noted earlier that the momentum equation contained corrections as powers of the lattice spacing but chose to ignore these at first pass. However, if we use the full Taylor expansion developed in the lattice-size scaling, we find that the contribution to the viscous term of the momentum equation is $-\frac{1}{8}\rho\nabla^2 v$. Note that the correction to the viscosity is a constant (see

Eq. 19) that depends only on the lattice and dimension and is independent of the scattering-rule set. This extra noncovariant-derivative contribution to the viscosity must be subtracted from the bare viscosity calculated from the normal perturbation expansion to get the renormalized viscosity, which is the one actually measured in the lattice gas. In other words, the bare coupling constant of the lattice gas model gets "dressed" by this constant amount, owing to the discrete vacuum that the particle must pass through, to become the physical lattice-gas viscosity.

Viscosity is a coupling constant and can be found by any method that can isolate the β_1 term in the Chapman-Enskog expansion. The simplest methods involve solving for the eigenvalues and right eigenvectors of the linearized collision operator, which is a tedious exercise in linear algebra. Using the results of such a calculation, we can write the Navier-Stokes form of the momentum equation in which the viscosity $\nu(\rho)$ appears explicitly:

$$\partial_t(\rho v_i) + \partial_j \Pi_{ij} = \partial_j S_{ij}, \tag{18}$$

where the momentum tensor Π_{ij} and the viscosity stress tensor are

$$\Pi_{ij} = c_s^2 \rho (1 - g(\rho)\frac{v^2}{c^2})\delta_{ij} + \rho g(\rho) v_i v_j$$

and

$$S_{ij} = \nu(\rho)\left(\partial_i(\rho v_j) + \partial_j(\rho v_i) - \frac{2}{D}\delta_{ij}\partial_k(\rho v_k)\right).$$

The coefficient c_s^2 is given by

$$c_s^2 = \frac{c^2}{D},$$

and c_s can be identified as the speed of sound. For the simple hexagonal model $c_s = 1/\sqrt{2}$, and the viscosity is given by

$$\nu = \frac{1}{12}\frac{1}{d(1-d)^3} - \frac{1}{8}, \tag{19}$$

where $d = \frac{\rho_0}{6}$, that is, the mass density per cell. (The $-\frac{1}{8}$ in the viscosity was mentioned above as the noncovariant correction due to the finite lattice size.)

The Incompressible Limit. Many features of low Mach number ($M = v/c_s \ll 1$) flows in an ordinary gas can be described by the incompressible Navier-Stokes equations:

$$\partial_t v + v \cdot \nabla v = -\nabla p + \nu \nabla^2 v \tag{20}$$

and

$$\nabla \cdot v = 0.$$

We end this theoretical analysis by showing under what conditions we recover these equations for lattice gases. One way is to freeze the density everywhere except in the pressure term of the momentum equation (Eq. 18). Then, in the low-velocity limit, we can write the lattice Navier-Stokes equations as

$$\rho_0 \partial_t v + \rho_0 g(\rho_0) v \cdot \nabla v = -c_s^2 \nabla \rho' + \rho_0 \nu(\rho_0)\nabla^2 v, \tag{21}$$

and

$$\nabla \cdot v = 0$$

where $\rho = \rho_0 + \rho'$ and we allow density fluctuations in the pressure term only. As it stands, Eq. 21 is not Galilean invariant. To make it so, we must scale away the $g(\rho_0)$ term in a consistent way. We rescale time and viscosity as follows:

$$t \to \frac{t}{g(\rho_0)} \quad \text{and} \quad \nu \to g(\rho_0)\nu.$$

To be more precise, we do an ϵ expansion of the momentum equation, where ϵ^{-1} is the same order as the global lattice size L_g (see Frisch et al. for details), and rescale the variables as follows:

$$\mathbf{r} = \epsilon^{-1}\mathbf{r}_1, \qquad t = \frac{1}{g(\rho_0)}\epsilon^{-2}T,$$

$$\mathbf{v} = \epsilon\mathbf{V}, \qquad \rho' = \frac{\rho_0 g(\rho_0)}{c_s^2}\epsilon^2 P',$$

and

$$\nu = g(\rho_0)\nu'.$$

where ϵ^{-1} is on the order of the global lattice size L_g. (Note that this rescaling of variables keeps the Reynolds number fixed.) Now all the relevant terms in the momentum equation are of $O(\epsilon^3)$ and higher order terms are $O(\epsilon^4)$ or smaller. So to leading order (where ∇_1 means $\frac{\partial}{\partial r_1}$) we get

$$\partial_T\mathbf{V} + \mathbf{V} \cdot \nabla_1\mathbf{V} = -\nabla_1 P' + \nu'\nabla_1^2\mathbf{V} \text{ and } \nabla_1 \cdot \mathbf{V} = 0.$$

Thus we recover the incompressible Navier-Stokes equations. To obtain this result, we have done a fixed-Reynolds-number, large-scale, low-Mach-number expansion and Galilean invariance has been restored, at least formally, by a time rescaling.

Simulations of Fluid Dynamics
with the Hexagonal Lattice-Gas Automaton

In the last two years several groups in the United States and France have done simulations of fluid-dynamical phenomena using the hexagonal lattice-gas automaton. The purpose of these simulations was twofold: first, to check the internal consistency of the automaton, and second, to determine, by both qualitative and quantitative measures, whether the model behaves the same or nearly the same as the known analytic and numerical solutions of the Navier-Stokes equations.

The classes of experiments done can be grouped roughly as free flows, flow instabilities, flows past objects, and flows in channels or pipes. These simulations were run in a range of Reynolds numbers between 100 and 700 (and for relatively low mean flow velocities, so that the fluid is nearly incompressible). We first checked to see whether the automaton developed various classic instabilities when triggered by two types of mechanisms, external perturbations and internal noise. The two classic instabilities studied were the Kelvin-Helmholtz instability of two opposing shear flows and the Rayleigh-Taylor instability. We describe the Kelvin-Helmholtz instability in some detail.

In the Kelvin-Helmholtz instability one is looking for the development of a final-state vortex structure of appropriate vortex polarity. From an initial state of two opposing flows undergoing shear, the detailed development of the instability depends on the initial perturbation of the flows. Left unperturbed, except by internal noise in the automaton, at first the two opposing flows develop velocity fields that signal the development of a boundary layer, then sets of vortices develop in these boundary layers, and finally vortex interactions occur that trigger a large-scale instability and the development of large-vortex final states. The same pattern appears in standard two-dimensional numerical simulations of the Navier-Stokes equations near the incompressible regime. No pathological non-Navier-Stokes behavior was observed. These results extend over the entire range of Reynolds numbers (100–700) run with the simple hexagonal model. It is notable that the Kelvin-Helmholtz instability is self-starting due to the automaton internal noise, and the instability proceeds rapidly.

The Rayleigh-Taylor instability was simulated by a French group in a slightly compressible fluid range, where it behaves like a Navier-Stokes fluid with no anomalies.

These global topological tests check whether automaton dynamics captures the correct overall structure of fluids. In general, whenever the automaton is run in the Navier-Stokes range, it produces the expected global topological behavior and correct functional forms for various fluid dynamical laws. The question of quantitative accuracy of various known constants is harder to answer, and we will take it up in detail later.

The next broad class of flows studied are flows past objects. Here, we look for distinctive qualitative behavior characteristic of a fluid or gas obeying Navier-Stokes dynamics. The geometries studied, through a wide range of Reynolds numbers, were flows past flat plates placed normal to the flow, flows past plates inclined at various angles to the flow, and flows past cylinders, 60-degree wedges, and typical airfoils. The expected scenario changes as a function of increasing Reynolds number: recirculating flow behind obstacles should develop into vortices, growing couples of vortices should eventually break off to form von Karman streets with periodic oscillation of the von Karman tails; finally, and as the Reynolds number increases, the periodic oscillations should become aperiodic, and the complex phenomena characteristic of turbulent flow should appear. The lattice gas exhibits all these phenomena with no non-Navier-Stokes anomalies in the range of lattice-gas parameters that characterize near incompressibility.

The next topic is quantitative self-consistency. We used the Boltzmann transport approximation for the discrete model to calculate viscosities for the simple hexagonal automaton as well as models with additional scattering rules and rest particles. We then checked these analytic predictions against the viscosities deduced from two kinds of simulations. We ran plane-parallel Poiseuille flow in a channel, saw that it developed the expected parabolic velocity profile (Fig. 6) and then deduced the viscosity characteristic of this type of flow. We also ran an initially flat velocity distribution and deduced a viscosity from the observed velocity decay. These two simulations agree with each other to within a few percent and agree with the analytic predictions from the Boltzmann transport calculation to within 10 percent. Viscosity was also measured by observing the decay of sound waves of various frequencies (Fig. 7). The level of agreement between simulation and the computed Boltzmann viscosity is generic: we see a systematic error of approximately 10 percent. Monte Carlo calculations of viscosities computed from microscopic correlation functions improve agreement with simulations to at least 3 percent and indicate that the Boltzmann description is not as accurate an analytic tool for the automaton as are microscopic correlation techniques. One would call this type of viscosity disagreement a Boltzmann-induced error. Other consistency checks between the automaton simulation and analytic predictions display the same level of agreement.

Detailed quantitative comparisons between conventional discretizations of the Navier-Stokes equations and lattice-gas simulations have yet to be done for several reasons. The simple lattice-gas automaton has a Fermi-Dirac distribution rather than the standard Maxwell-Boltzmann distribution. This difference alone causes deviations of $O(v^2)$ in the macrovelocity from standard results. For the same reason and unlike standard numerical spectral codes for fluid dynamics, the simple lattice-gas automaton has a velocity-dependent equation of state. A meaningful comparison between the two approaches requires adjusting the usual spectral codes to compute with a velocity-dependent equation of state. This rather considerable task has yet to be done. So far our simulations can be compared only to traditional two-dimensional computer simulations and analytic results derived from simple equations of state.

Some simple quantities such as the speed of sound and velocity profiles have been measured in the automaton model. The speed of sound agrees with predicted values and functional forms for channel velocity profiles and D'Arcy's law agree with calculations by standard methods. The automaton reaches local equilibrium in a few time steps and reaches global equilibrium at the maximum information-transmission speed, namely, at the speed of sound.

Simulations with the two-dimensional lattice-gas model hang together rather well as a simulator of Navier-Stokes dynamics. The method is accurate enough to test theoretical turbulent mechanisms at high Reynolds number and as a simulation tool for complex geometries, provided that velocity-dependent effects due to the Fermi nature of the automaton are correctly included. Automaton models can be designed to fit specific phenomena, and work along these lines is in progress.

Three-dimensional hydrodynamics is being simulated, both on serial and parallel machines, and early results show that we can easily simulate flows with Reynolds numbers of a few thousand. How accurately this model reproduces known instabilities and flows remains to be seen, but there is every reason to believe agreement will be good since the ingredients to evolve to Navier-Stokes dynamics are all present.

Reynolds Numbers and Lattice Gas Computations

The only model-dependent coupling constant in the Navier-Stokes equation is the viscosity. Its main role in lattice gas computations is its influence on the Reynolds number, an important scaling concept for flows. Given a system with a fixed intrinsic global length scale, such as the size of a pipe or box, and given a flow, then the Reynolds number can be thought of as the ratio of a typical macrodynamic time scale to a time scale set by elementary molecular processes in the kinetic model.

Reynolds numbers characterize the behavior of flows in general, irrespective of whether the system is a fluid or a gas. At high enough Reynolds numbers turbulence begins, and turbulence quickly loses all memory of molecular structure, becoming universal across liquids and gases. For this reason and because many interesting physical and mathematical phenomena happen in turbulent regimes, it is important to be able to reach these Reynolds numbers in realistic simulations without incurring a large amount of computational work or storage.

Some simple arguments based on dimensional analysis and phenomenological theories of turbulence indicate, at first glance, that any cellular automaton model has a high cost in computer resources when simulating high-Reynolds-number flows. These arguments appeared in the first paper on the subject (Frisch, Hasslacher, and Pomeau 1987) and were later elaborated on by other authors. We will go through the derivation of some of the more severe constraints on simulating high-Reynolds-number flows with cellular automata, then discuss some possible ways out, and finally estimate the seriousness of the situation for a realistic large-scale simulation.

The turbulent regime has many length scales, bounded above by the length of the simulation box and below by the scale at which turbulent dynamics degenerates into pure dissipation, the so-called dissipation scale. We focus on these extreme scales and, with a few definitions, derive a bound on the computational storage and work needed for simulating high-Reynolds-number flows with cellular automata.

The Reynolds number R is usually defined not in terms of time but simply as $R = vL/\nu$ where L is a characteristic length, v is characteristic speed, and ν is the kinematic shear viscosity. One sees immediately why calculating viscosity functions for particular models is important. It is the only variable one can adjust in a flow problem, given a fixed flow in a fixed geometry. First, we calculate a rough upper bound on Reynolds numbers attainable with lattice models. If the speed of sound in the lattice gas is c_s and the spacing between lattice nodes is ℓ, then by definition the kinematic viscosity $\nu \geq c_s \ell$. Now viscosity estimated this way must agree with that fixed by the scale of hydrodynamic modes. Given a global length L and a global velocity V associated with these modes, $R = VL/\nu$ at best. In terms of the Mach number ($M = V/c_s$), the Reynolds number is equal to ML/ℓ. But M also characterizes fluid flow, and L and ℓ are model-dependent. In a lattice gas we can relate the ratio L/ℓ to the number of nodes in the gas simulator, namely $n = (L/\ell)^d$, where d is the space dimension of the model. Therefore, the number of nodes in a lattice model must grow at least as $n \sim (R/M)^d$. Computational work is the number of lattice nodes per time step multiplied by the number of time steps required to resolve hydrodynamical features. This is $L/\ell M$ steps (to cross the hydrodynamical feature at the given Mach number), and so we find the computational work is of order R^{d+1}/M^{d+2}. For a so-called normal simulation based on the usual ways of discretizing the Navier-Stokes equation, the growth in storage is roughly proportional to one power lower in the Reynolds number than the growth in storage for the lattice gas. So at first it seems that simulating high-Reynolds-number flows by lattice gas techniques is costly compared to ordinary methods.

This argument is not only approximate; it is also tricky and must be applied with great care. The normal way of simulating flows escapes power-law penalties by cutting off degrees of freedom at the turbulence-dissipation scale, which the lattice gas does not. The gas computes within these scales and so wastes computational resources for some problems. Actually computation of these very small scales is the source of the noisy character of the gas and is responsible for its power to avoid spurious mathematical singularities. One way around this is to find an effective gas with new collision rules for which the dissipation length scales are averaged out. A possible technique uses the renormalization group, but it is useful only if the effective gas is not too complex and has the attributes that made the original gas attractive, including locality. Work is going on at present to explore this possibility, and it seems likely that some such method will be developed.

The more serious consideration is what happens in a realistic large-scale simulation, and here we will find the lattice gas does very well indeed.

First, we note that a dissipation length ℓ_d with the behavior $\ell_d \to \infty$ as $R \to \infty$ is actually required to guarantee the scale separation between the lattice spacing and the hydrodynamic modes that is necessary to develop hydrodynamic behavior.

The actual Reynolds number in lattice gas models is much more complex than in normal fluid models. An accurate form is $R = Lvg(\rho_0)/\nu(\rho_0)$, where v is an averaged velocity and the fundamental unit of distance (the lattice spacing ℓ) and the fundamental unit of time (the speed required to traverse the lattice spacing ℓ) have been set to 1. To remain nearly incompressible, the velocities in the model should remain small compared to the speed of sound c_s, but c_s in lattice gases is model-dependent. So we factor the Reynolds number into model-dependent and invariant factors this way: we define $\hat{R}(\rho_0) = c_s \left(g(\rho_0)/\nu(\rho_0)\right)$ so that $R = ML\hat{R}(\rho_0)$. The value of $\hat{R}$ depends critically on the model used. In two dimensions it ranges from 0.39 to about 6 times that, depending on the amount of the state table we want to include. For the three-dimensional projection of the four-dimensional model, it is known that $\hat{R}$ is about 9.

By repeating essentially the same dimensional arguments, only more carefully, we find that the dissipation length $l_d = (M\hat{R})^{-1}R^{-1/2}$ for two dimensions and $l_d = (M\hat{R})^{-1}R^{-1/4}$ for three dimensions.

For a typical simulation in three dimensions, we take $M = 0.3$ for incompressibility, $\hat{R} \sim 9$, and $L = 10^3$, which is a large simulation, possible only on the largest Cray-class machines. Then l_d is about three lattice spacings, and the simulation wastes very little computational power. The subtle point is that the highly model-dependent factor $\hat{R}$ is not of order 1, as is usually estimated. It depends critically on the complexity of the collision set, going up a factor of 20 from the elementary hexagonal model in two dimensions to the projected four-dimensional case with an optimal collision table.

There is a great deal of work to be done on the high-Reynolds-number problem, but it is clear that the situation is complicated and rich in possibilities for evading simple dimensional arguments.

PART III
THE PROMISE OF LATTICE GAS METHODS

Adjusting the Model To Fit
the Phenomenon

There are several reasons for altering the geometry and rule set of the fundamental hexagonal model. To understand the mathematical physics of lattice gases, we need to know the class of functionally equivalent models, namely those models with different geometries and rules that produce the same dynamics in the same parameter range.

To explore turbulent mechanisms in fluids, the Reynolds number must be significantly higher than for smooth flow, so models must be developed that increase the Reynolds number in some way. The most straightforward method, other than increasing the size of the simulation universe, is to lower the effective mean free path in the gas. This lowers the viscosity and the Reynolds number rises in inverse proportion. Increasing the Reynolds number is also important for practical applications. In "Reynolds Number and Lattice Gas Computations" we discuss the computational storage and work needed to simulate high-Reynolds-number flows with cellular automata.

To apply lattice gas methods to systems such as plasmas, we need to develop models that can support widely separated time scales appropriate to, for example, both photon and hydrodynamical modes. The original hexagonal model on a single lattice cannot do so in any natural way but must be modified to include several lattices or the equivalent (see below).

Within the class of fluids, problems involving gravity on the gas, multi-component fluids, gases of varying density, and gases that undergo generalized chemical reactions require variations of the hexagonal model. Once into the subject of applications rather than fundamental statistical mechanics, there is an endless industry in devising clever gases that can simulate the dynamics of a problem effectively.

We outline some of the possible extensions to the hexagonal gas, but do so only to give an overview of this developing field. Nothing fundamental changes by making the gas more complex. This model is very much like a language. We can build compound sentences and paragraphs out of simple sentences, but it does not change the fundamental rules by which the language works.

The obvious alterations to the hexagonal model are listed below. They comprise almost a complete list of what can be done in two dimensions, since a lattice gas model contains only a few adjustable structural elements.

Indistinguishable particles can be colored to create distinguishable species in the gas, and the collision rules can be appropriately modified. Rules can be weighted to different outcomes; for example, one can create a chiral gas (left- or right-handed) by biasing collisions to make them asymmetric. In three dimensions there is an instability at any Reynolds number caused by lack of microscopic parity, so the chiral gas is an important model for simulating this instability.

At the next order of complexity, multi-speed particles can be introduced, either alone or with changes in geometry. The simplest example is a square neighborhood in two dimensions in which the collision domain is enlarged to include next-to-nearest neighbors, and a diagonal particle with speed $\sqrt{2}$ is introduced to force an isotropic lattice gas. In general, any lattice model with only two-body collisions and a single speed will contain spurious conservation laws. But if multiple speeds are allowed, models with binary collisions can maintain isotropy. In other words, models with multiple speeds are equivalent to single-speed models with a higher order rotation group and extended collision sets. Many variations are possible and each can be designed to a problem where it has a special advantage.

Finally, colored multiple-speed models are in general equivalent to single-species models operating on separate lattices. Colored collision rules couple the lattices so that information can be transferred between them at different time scales. Certain statistical-mechanical phenomena such as phase transitions can be done this way.

By altering the rule domain and adding gas species with distinct speeds, it is possible to add independent energy conservation. This allows one to tune gas models to different equations of state. Again, we gain no fundamental insight into the development of large collective models by doing so, but it is useful for applications.

In using these lattice gas variations to construct models of complex phenomena, we can proceed in two directions. The first direction is to study whether or not complex systems with several types of coupled dynamics are described by skeletal gases. Can complex chemical reactions in fluids and gases, for example, be simulated by adding collision rules operating on colored multi-speed lattice gases? Complex chemistry is set up in the gas in outline form, as a gross scheme of closed sets of interaction rules. The same idea might be used for plasmas. From a theoretical viewpoint one wants to study how much of the known dynamics of such systems is reproduced by a skeletal gas; consequently both qualitative and quantitative results are important.

Models of complex gas or fluid systems, like other lattice gas descriptions, may either be a minimalist description of microphysics or simply have no relation to microphysics other than a mechanism for carrying known conservation laws and reactions. We can always consider such gas models to be pure computers, where we fit the wiring, or architecture, to the problem, in the same fashion that ordinary discretization schemes have no relation to the microphysics of the problem. However for lattice gas models, or cellular-automaton models in general, there always seems to be a deep relation between the abstract computer embodying the gas algorithm for a physical problem and the mathematical physics of the system itself.

This duality property is an important one, and it is not well understood. One of the main aims of lattice gas theory is to make the underlying mathematics of dynamical evolution clearer by providing a new perspective on it. One would, for example, like to know the class of all lattice gas systems that evolve to a dynamics that is, in an appropriate sense, nearby the dynamics actually evolved by nature. Doing this will allow us to isolate what is common to such systems and identify universal mathematical mechanisms.

Engineering Design Applications. The second direction of study is highly applied. In most engineering-design situations with complicated systems, one would like to know first the general qualitative dynamical behavior taking place in some rather involved geometry and then some rough numerics. Given both, one can plot out the zoo of dynamical development within a design problem. Usually, one does not know what kinds of phenomena can occur as a parameter in the system varies. Analytic methods are either unavailable, hard to compute by traditional methods, or simply break down. Estimating phenomena by scaling or arguments depending on order-of-magnitude dimensional analysis is often inaccurate or yields insufficient information. As a result, a large amount of expensive and scarce supercomputer time is used just to scan the parameter space of a system.

Lattice gas models can perform such tasks efficiently, since they simulate at the same speed whether the geometry and system are simple or complex. Complicated geometries and boundary conditions for massively parallel lattice gas simulators involve only altering collision rules in a region. This is easily coded and can be done interactively with a little investment in expert systems. There is no question that for complex design problems, lattice gas methods can be made into powerful design tools.

Beyond Two Dimensions

In two dimensions there exists a single-speed skeletal model for fluid dynamics with a regular lattice geometry. It relies on the existence of a complete tiling of the plane by a domain of sufficiently high symmetry to guarantee the isotropy of macroscopic modes in the model. In three dimensions this is not the case, for the minimum appropriate domain symmetry is icosahedral and such polyhedra do not tile three-space. If we are willing to introduce multiple-speed models, there may exist a model with high enough rotational symmetry, as in the square model with nearest and next-to-nearest neighbor interaction in two dimensions, but it is not easy to find and may not be efficient for simulations.

A tactic for developing an enlarged-neighborhood, three-dimensional model, which still admits a regular lattice, is to notice that the number of regular polyhedra as a function of dimension has a maximum in four dimensions. Examination of the face-centered four-dimensional hypercube shows that a single-speed model connected to each of twenty-four nearest neighbors has exactly the right invariance group to guarantee isotropy in four dimensions. So four-dimensional single-speed models exist on a regular tiling. Three-dimensional, or regular, hydrodynamics can be recovered by taking a thin one-site slice of the four-dimensional case, where the edges of the slice are identified. Projecting such a scheme into three-dimensional space generates a two-speed model with nearest and next-to-nearest neighbor interactions of the sort guaranteed to produce three-dimensional Navier-Stokes dynamics.

Such models are straightforward extensions of all the ideas present in the two-dimensional case and are being simulated presently on large Cray-class machines and the Connection Machine 2. Preliminary results show good agreement with standard computations at least for low Reynolds numbers. In particular, simulation of Taylor-Green vortices at a Reynolds number of about 100 on a $(128)^3$ universe (a three-dimensional cube with 128 cells in each direction) agrees with spectral methods to within 1 percent, the error being limited by Monte Carlo noise. The ultimate comparison is against laboratory fluid-flow experiments. As displayed at the end of Part II, three-dimensional flows around flat plates have also been done.

A more intriguing strategy is to give up the idea of a regular lattice. Physical systems are much more like a lattice with nodes laid down at random. At present, we don't know how to analyze such lattices, but an approximation can be given that is intermediate between regular and random grids. Quasi-tilings are sets of objects that completely tile space but the grids they generate are not periodic. Locally, various types of rotation symmetry can be designed into such lattices, and in three dimensions there exists such a quasi-tiling that has icosahedral symmetry everywhere. The beauty of quasi-tilings is that they can all be obtained by simple slices through hypercubes in the appropriate dimension. For three dimensions the parent hypercube is six-dimensional.

The idea is to run an automaton model containing the conservation laws with as simple a rule set as possible on the six-dimensional cube and then take an appropriately oriented three-dimensional slice out of the cube so arranged as to generate the icosahedral quasi-tiling. Since we only examine averaged quantities, it is enough to do all the averaging in six dimensions along the quasi-slice and image the results. By such a method we guarantee exact isotropy everywhere in three dimensions and avoid computing directly on the extremely complex lattices that the quasi-tiling generates. Ultimately, one would like to compute on truly random lattices, but for now there is no simple way of doing that efficiently.

The simple four-dimensional model is a good example of the limits of present super-computer power. It is just barely tolerable to run a $(1000)^3$ universe at a Reynolds number of order a few thousand on the largest existing Cray's. It is far more efficient to compute in large parallel arrays with rather inexpensive custom machines, either embedded in an existing parallel architecture or on one designed especially for this class of problems.

Lattice Gases as Parallel Computers

Let us review the essential features of a lattice gas. The first property is the totally discrete nature of the description: The gas is updated in discrete time steps, lattice gas elements can only lie on discrete space points arranged in a space-filling network or grid, velocities can also have only discrete values and these are usually aligned with the grid directions, and the state of each lattice gas site is described by a small number of discrete bits instead of continuous values.

The second crucial property is the local deterministic rules for updating the array in space and time. The value of a site depends only on the values of a few local neighbors so there is no need for information to propagate across the lattice universe in a single step. Therefore, there is no requirement for a hardwired interconnection of distant sites on the grid.

The third element is the Boolean nature of the updating rules. The evolution of a lattice gas can be done by a series of purely Boolean operations, without ever computing with radix arithmetic.

To a computer architect, we have just described the properties of an ideal concurrent, or parallel, computer. The identical nature of particles and the locality of the rules for updating make it natural to update all sites in one operation—this is what one means by concurrent or parallel computation. Digital circuitry can perform Boolean operations naturally and quickly. Advancing the array in time is a sequence of purely local Boolean operations on a deterministic algorithm.

Most current parallel computer designs were built with non-local operations in mind. For this reason the basic architecture of present parallel machines is overlaid with a switching network that enables all sites to communicate in various degrees with all other sites. (The usual model of a switching network is a telephone exchange.) The complexity of machine architecture grows rapidly with the number of sites, usually as $n \log n$ at best with some time tradeoff and as $O(n^2)$ at worst. In a large machine, the complexity of the switching network quickly becomes greater than the complexity of the computing configuration.

In a purely local architecture switching networks are unnecessary, so two-dimensional systems can be built in a two-dimensional, or planar, configuration, which is the configuration of existing large-scale integrated circuits.

Such an architecture can be made physically compact by arranging the circuit boards in an accordion configuration similar to a piece of folded paper. Since the type of geometry chosen is vital to the collective behavior of the lattice gas model and no unique geometry fits all of parameter space, it would be a design mistake to hardwire a particular model into a machine architecture. Machines with programmable geometries could be designed in which the switching overhead to change geometries and rules would be minimal and the gain in flexibility large (Fig. 8).

In more than two dimensions a purely two-dimensional geometry is still efficient, using a combination of massive parallel updating in a two-dimensional plane and pipelining for the extra dimensions. As technology improves, it is easy to imagine fully three-dimensional machines, perhaps with optical pathways between planes, that have a long mean time to failure.

The basic hardware unit in conventional computers is the memory chip, since it has a large storage capacity (256 K bytes or 1 M bytes presently) and is inexpensive, reliable, and available commercially in large quantities. In fact, most modern computers have a memory-bound architecture, with a small number of controlling processors either doing local arithmetic and logical operations or using fast hashing algorithms on large look-up tables. An alternative is the local architecture described above for lattice gas simulators. In computer architecture terms it becomes very attractive to build compact, cheap, very fast simulators which are general over a large class of problems such as fluids. Such machines have a potential processing capacity much larger than the general-purpose architectures of present or foreseen vectorial and pipelined supercomputers. A number of such machines are in the process of being designed and built, and it will be quite interesting to see how these experiments in non-von Neumann architectures (more appropriately called super-von Neumann) turn out.

At present, the most interesting machine existing for lattice gas work is the Connection Machine with around 65,000 elementary processors and several gigabytes of main memory. This machine has a far more complex architecture than needed for pure lattice-gas work, but it was designed for quite a different purpose. Despite this, some excellent simulations have been done on it. The simulations at Los Alamos were done mainly on Crays with SUN workstations serving as code generators, controllers, and graphical units. The next generation of machines will see specialized lattice gas machines whether parallel, pipelined, or some combination, running either against Connection Machine style architectures or using them as analyzing engines for processing data generated in lattice gas "black boxes." This will be a learning experience for everyone involved in massive simulation and provide hardware engines that will have many interesting physics and engineering applications.

Unfortunately, fast hardware alone is not enough to provide a truly useful exploration and design tool. A large amount of data is produced in a typical many degree of freedom system simulation. In three dimensions the problems of accessing, processing, storing, and visualizing such quantities of data are unsolved and are really universal problems even for standard supercomputer technology. As the systems we study become more complex, all these problems will also. It will take innovative engineering and physics approaches to overcome them.

Conclusion

To any system naturally thought of as classes of simple elements interacting through local rules there should correspond a lattice-gas universe that can simulate it. From such skeletal gas models, one can gain a new perspective on the underlying mathematical physics of phenomena. So far we have used only the example of fluids and related systems that naturally support flows. The analysis of these systems used the principle of maximum ignorance: Even though we know the system is deterministic, we disregard that information and introduce artificial probabilistic methods. The reason is that the analytic tools for treating problems in this way are well developed, and although tedious to apply, they require no new mathematical or physical insight.

A deep problem in mathematical physics now comes up. The traditional methods of analyzing large probabilistic systems are asymptotic perturbation expansions in various disguises. These contain no information on how fast large-scale collective behavior should occur. We know from computer simulations that local equilibrium in lattice gases takes only a few time steps, global equilibrium occurs as fast as sound propagation will allow, and fully developed hydrodynamic phenomena, including complex instabilities, happen again as fast as a traverse of the geometry by a sound wave. One might say that the gas is precociously asymptotic and that this is basically due to the deterministic property that conveys information at the greatest possible speed.

Methods of analyzing the transient and invariant states of such complex multi-dimensional cellular spaces, using determinism as a central ingredient, are just beginning to be explored. They are non-perturbative. The problem seems as though some of the methods of dynamical systems theory should apply to it, and there is always the tempting shadow of renormalization-group ideas waiting to be applied with the right formalism. So far we have been just nibbling around the edges of the problem. It is an extraordinarily difficult one, but breaking it would provide new insight into the origin of irreversible processes in nature.

The second feature of lattice gas models, for phenomena reducible to natural skeletal worlds, is their efficiency compared to standard computational methods. Both styles of computing reduce to inventing effective microworlds, but the conventional one is dictated and constrained by a limited vocabulary of difference techniques, whereas the lattice gas method designs a virtual machine inside a real one, whose architectural structure is directly related to physics. It is not a priori clear that elegance equals efficiency. In many cases, lattice gas methods will be better at some kinds of problems, especially ones involving highly complex systems, and in others not. Its usefulness will depend on cleverness and the problem at hand. At worst the two ways of looking at the microphysics are complementary and can be used in various mixtures to create a beautiful and powerful computational tool.

We close this article with a series of conjectures. The image of the physical world as a skeletal lattice gas is essentially an abstract mathematical framework for creating algorithms whose dynamics spans the same solution spaces as many physically important nonlinear partial differential equations that have a microdynamical underpinning. There is no intrinsic reason why this point of view should not extend to those rich nonlinear systems which have no natural many-body picture. The classical Einstein-Hilbert action, phrased in the appropriate space, is no more complex than the Navier-Stokes equations. It should be possible to invent appropriate skeletal virtual computers for various gauge field theories, beginning with the Maxwell equations and proceeding to non-Abelian gauge models. Quantum mechanics can perhaps be implemented by using a variation on the stochastic quantization formulation of Nelson in an appropriate gas. When such models are invented, the physical meaning of the skeletal worlds is open to interpretation. It may be they are only a powerful mathematical device, a kind of virtual Turing machine for solving such problems. But it may also be that they will provide a new point of view on the physical origin and behavior of quantum mechanics and fundamental field-theoretic descriptions of the world.

Further Reading

G. E. Uhlenbeck and G. W. Ford. 1963. *Lectures in Statistical Mechanics*. Providence, Rhode Island: American Mathematical Society.

Paul C. Martin. 1968. *Measurements and Correlation Functions*. New York: Gordon and Breach Science Publishers.

Stewart Harris. 1971. *An Introduction to the Theory of the Boltzmann Equation*. New York: Holt, Rinehart, and Winston.

E. M. Lifshitz and L. P. Pitaevskii. 1981. *Physical Kinetics*. Oxford: Pergamon Press.

Stephen Wolfram, editor. 1986. *Theory and Applications of Cellular Automata*. Singapore: World Scientific Publishing Company.

L. D. Landau and E. M. Lifshitz. 1987. *Fluid Mechanics*, second edition. Oxford: Pergamon Press.

The entirety of volume 1, number 4 of the journal *Complex Systems*, which consists of papers presented at the Workshop on Modern Approaches to Large Nonlinear Systems, Santa Fe, New Mexico, October 27–29, 1986.

U. Frisch, B. Hasslacher, and Y. Pomeau. 1986. Lattice-gas automata for the Navier-Stokes equation. *Physical Review Letters* 56: 1505–1508.

Stephen Wolfram. 1986. Cellular automaton fluids 1: Basic theory. *Journal of Statistical Physics* 45: 471–526.

Victor Yakhot and Stephen A. Orszag. 1986. *Physical Review Letters* 57: 1722.

U. Frisch, D. d'Humières, B. Hasslacher, P. Lallemand, Y. Pomeau, and J. P. Rivet. 1987. *Complex Systems* 1: 649–707.

Leo P. Kadanoff, Guy R. McNamara, and Gianluigi Zanetti. 1987. A Poiseuille viscometer for lattice gas automata. *Complex Systems* 1: 791–803.

Dominique d'Humières and Pierre Lallemand. 1987. Numerical simulations of hydrodynamics with lattice gas automata in two dimensions. *Complex Systems* 1: 599–632.

PHYSICAL FLUID ⟶ ABSTRACT FLUID

Physical Fluids are atomically complex to maintain fluid state.

<u>Necessary ingredients for the emergence of a Macrodynamics</u>

1| Local Thermodynamic Equilibria

    (Density, Macrovelocity, Energy,-- The Hilbert Reduction)

2| Patching Conditions -- Microscopic conservation Laws

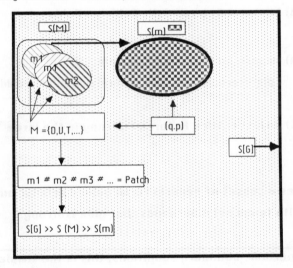

3| Dynamic Scale Separation   **FIG 1**

    S[G] >> S(M) >> S(m)

Properties of Simplest Lattice Automata

:IDEA: Reduce the Universe to Simplest Possible Setting

(1) All Descriptive Microscopic Variables are taken DISCRETE

Discrete time, phase space (q,p), state space⟶ LATTICE

(2) Impose Minimum Conservation Laws

In general, the first integrals of the motion.

(3) Minimum Deterministic Collision Rules

(4) Impose an Exclusion Principle

[physical non-linearities, natural rule cutoff, rapid asymptotics]

(5) Adjust Lattice to force isotropy in the large

Implies a subset of general Fermi valued cellular automata
with tuned neighborhoods that give isotropy.

> **With periodic boundary conditions, this is the simplest**
> **mapping of the torus (S1 X S1) onto itself**
> **that has complete fluid behavior.**

FIG   2

FHP HEXAGONAL MODEL STATE TABLE

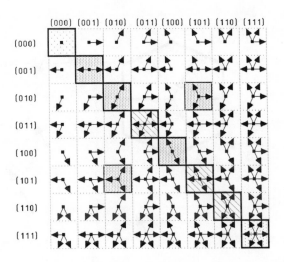

FHP HEXAGONAL MODEL

(001,001) ⟷ (100,100)
           ⟷ (010,010)   [2- BODY RULE]

(010,101) ⟷ (101,010)   [3- BODY RULE]

DUAL RULES

(110,110) ⟷ (011,011)
           ⟷ (101,101)

FIG 3

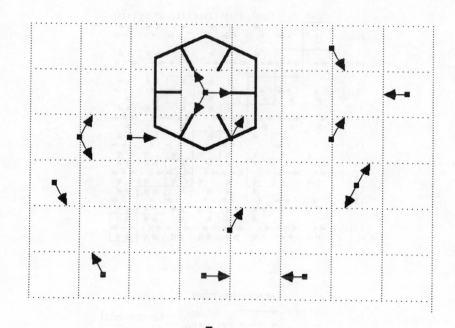

T

FIG 4

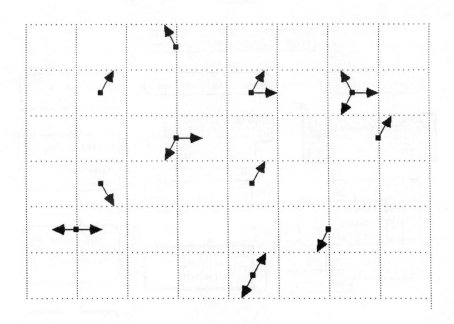

I + 1

FIG 5

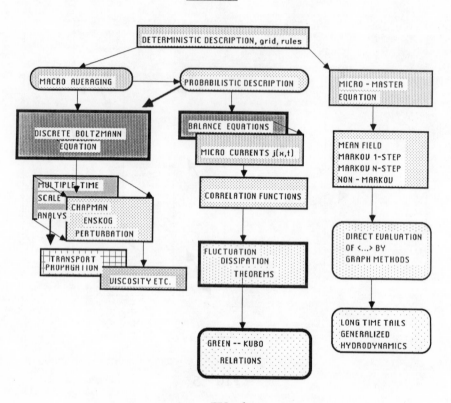

FIG   6

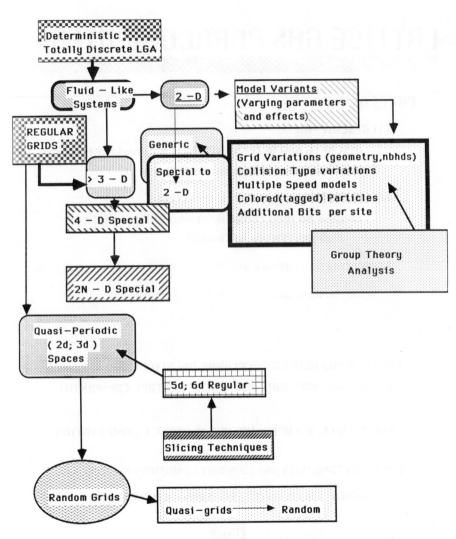

FIG 7

LATTICE GAS PARACOMPUTERS

PROPERTIES

(1) DESCRIPTIVE

DISCRETE and DETERMINISTIC [t,x,v] and state space,
Boolean valued, on a space filling grid.

(2) LOCALITY in x, t and all interactions

Implies no long range interconnections and so
NO SWITCHING NETWORKS

(3) All rule operations are Boolean and naturally parallel

Can be done quickly on parallel bit planes

**LEADS TO MEMORY BOUND PLANAR ARCHITECTURES
WITH THE MEMORY CHIP AS A CHEAP CENTRAL COMPONENT**

FAST, SIMPLE, INEXPENSIVE AND VERSATILE COMPUTATION

COMPLEX GEOMETRIES AND BOUNDARY CONDITIONS ARE JUST

     DIFFERENT COLLISION RULES--- CHANGED INTERACTIVELY

FIG   8

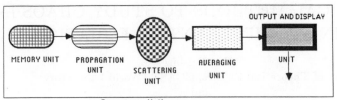

1. Processor Unit

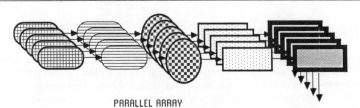

2. PARALLEL ARRAY

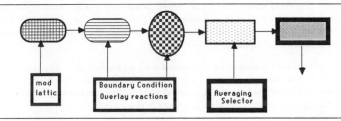

3.

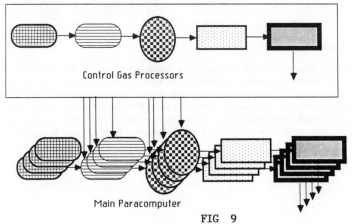

4. Main Paracomputer

FIG 9

NUMERICAL METHODS TO STUDY CHAOS IN ORDINARY DIFFERENTIAL EQUATIONS

Hao Bai-lin

Center of Theoretical Physics, CCAST (World Laboratory)

and

Institute of Theoretical Physics, Academia Sinica

1   INTRODUCTION

Many interesting models in physical sciences are described by systems of ordinary differential equations. While bifurcation and chaos in discrete mappings can be studied more or less thoroughly by both analytical and numerical means, similar task for differential equations may encounter great difficulty: analytical tools can be of some help only in very limited cases, e.g., when the criterion of Melnikov or Silnikov is applicable, and numerical study happens to be very time-consuming in most cases. Nevertheless, our present understanding of chaotic behaviour in differential equations relies heavily on numerical work. The goal of these lectures is to give an overview of the most frequently used numerical methods without going into programming details. We shall confine ourselves to the following types of differential equations.

1. Autonomous differential equations, i.e., equations with no explicit time dependence in their right-hand side. In this case one needs at least three independent variables in order to observe period-doubling or chaotic transitions. A classical and much-studied example of autonomous equations is the Lorenz model [13] obtained from a three-mode truncation of the thermal convection problem:

$$\dot{x} = \sigma(y - x) \qquad \dot{y} = rx - y - xz \qquad \dot{z} = xy - bz \qquad (1)$$

   where σ, b and r are control parameters.

2. Non-autonomous systems with at least two independent variables. It is well-known that by adding one or more independent variables a non-autonomous system can be transformed into an autonomous one. The most important class of non-autonomous systems are those driven by periodical external force. As we shall see a control frequency at one's disposal opens the possibility to reach very high frequency resolution in numerical study. In these lectures we shall occasionally refer to some results for so-called periodically forced Brusselator[6]:

$$\dot{x} = A - (B + 1)x + x^2 y + \alpha \cos(\omega t) \qquad \dot{y} = Bx - x^2 y \qquad (2)$$

The "free" part of Equations (2) describes the kinetics of a three-molecular chemical reaction and is capable to develop a limit cycle type oscillation when $B > A^2 + 1$.

3. **Time-delayed differential equations.** Formally it is enough to have a single independent variable to display complicated bifurcating and chaotic behaviour. However, time-delayed equations are in essence functional equations with an infinite number of degrees of freedom. This can be seen easily by rewriting the time delay as a sum over high order derivatives:

$$f(t - T) = \exp\left(-T\frac{d}{dt}\right) f(t) = \sum_{i=0}^{\infty} (-T)^i \frac{d^i}{dt^i} f(t) \tag{3}$$

or by considering the dependence of the solution on an initial function instead of an initial point. Time-delayed equations have been used in, e.g., ecological models and in describing optical bistable devices, one of the simplest cases being [12]

$$\tau \dot{x}(t) + x(t) = 1 - \mu x(t - T)^2 \tag{4}$$

Since we shall no longer treat time-delayed equations in these lectures I would like to make two remarks on the example of Equation (4).

First, in the long delay limit $T \gg \tau$ one can neglect the time derivative and, measuring time in units of T, transform Equation (4) into a mapping. However, no matter how detailed knowledge one possesses for this mapping, it can not be extrapolated into the $\tau \sim \mu$ parameter plane for small τ, because there are an infinite number of "linear" modes excitable near $\tau = 0$.

Second, in the opposite small delay limit, although one may use a truncated expansion (3) to transfer Equation (4) into a system of finite order ODE's, but the resulted system may resemble the original equation only for small enough t. Therefore, it is not of much help in the study of the asymptotic $t \to \infty$ behaviour.

The strategy of studying chaos in ODE's numerically consists in first identifying periodic solutions and their systematics and then characterizing the attractors. This is why we shall pay much attention to numerical calculation of periodic orbits. To anticipate a little, we point out that the succees of characterizing chaotic attractors in computer studies relies significantly on extending the relevant space to include the tangent space.

2 ON NUMERICAL INTEGRATION OF ODE'S

A prerequisite for applying any method that we are going to discuss consists in there being a good algorithm to integrate the equations. Gone has the time when a scientist must write his own mathematical routines, since so many tested programs have been accumulated that surpass any amateur's handiwork in stability, precision and efficiency. All one has to do is to choose an appropriate subroutine from one of the existing libraries (e.g., IMSL, NAG, CACM, etc.) which should be available on-line at any computing center. Therefore, I shall confine myself to a few comments.

To fix the notations we write a general autonomous system of nonlinear ODE's in the standard vector form:

$$\frac{dx}{dt} = \mathbf{F}(x) \tag{5}$$

It should be supplemented with the initial condition:

$$x(t = 0) = x_0$$

The solution of this initial value problem is an integral curve passing through x_0. All possible integral curves taken together constitute a flow Φ in the phase space and the solution of the above problem picks up a particular curve

$$x(t) = \Phi_t(x_0), \tag{6}$$

satisfying, of course, the condition

$$x_0 = \Phi_0(x_0)$$

The system (5) can be linearized at a point, say, x_1 by letting

$$x = x_1 + \mathbf{W}$$

where $\mathbf{W}$ is a small, hence tangential vector to the integral curve at x_1 and satisfies the linearized equation

$$\frac{d\mathbf{W}(t)}{dt} = \mathbf{J}(x_1)\mathbf{W}(t) \tag{7}$$

where

$$\mathbf{J}(x_1) = \frac{\partial \mathbf{F}(x)}{\partial x}\Big|_{x=x_1}$$

is a $n \times n$ matrix. We sometimes denote its elements as a_{ij}.

Being a linear system, the solutions of Eqs. (7) may be expressed by way of a linear evolution operator $\mathbf{U}(t) \equiv \mathbf{U}(t, 0)$

$$\mathbf{W}(t) = \mathbf{U}(t)\mathbf{W}(0)$$

It is easy to see that $\mathbf{U}(t)$ satisfies the same equation (7)

$$\frac{\mathbf{U}(t)}{dt} = \mathbf{J}(x_1)\mathbf{U}(t)$$

with the initial conditions

$$\mathbf{U}(t)\,|_{t=0} = \mathbf{I},$$

I being the unit matrix.

Numerical algorithms for integrating ODE's can be subdivided into two classes. The first class is based on Taylor's expansion and one goes step by step from the initial point. The second class makes use of some numerical quadrature on a small interval. Algorithms in the first class usually only require the result of the previous step and are called one-step or single-step methods. The second class usually leads to multi-step algorithms and, in particular, is useful for treating time-delayed problems when the initial function is known on an interval, e.g., we have used the 4th order Adam's method to study Eq. (4) [12]. As a rule, multistep methods require less arithmetic operations per step than single-step algorithms, but they need initial values at many points to get started. On the other hand, multi-step schemes are usually derived for equidistant partitions of the integration interval, whereas single-step methods are flexible enough to allow for varying step lengths.

A very frequently used single-step method is the Runge-Kutta algorithm. A fourth order Runge-Kutta scheme has a local truncation error of the order h^5, h being the integration step. It measures the difference between the discretized iteration equations and the original differential equations at one single step, hence the adjective local. However, small truncation error at each step tells us not very much about the degree of global approximation to our ODE system. In addition, being in the chaotic regime implies sensitive dependence on initial values as well as on local truncation errors. Nevertheless, as we shall see later, a number of global characteristics of the motion such as the Liapunov exponents, dimensions and entropies, remains quite insensitive to the algorithm or local truncation, provided small enough integration steps are used.

The last, but not the least point to be mentioned concerns so-called stiff equations, when the eigenvalues of the matrix $\mathbf{J}$ in Eq. (7) differ in orders of magnitude. One should consult the literature, e.g., [11], to avoid numerical puzzles.

3   CALCULATION OF POINCARÉ MAPS

In studying bifurcation and chaos in ODE's one does not deal with a single system, but has to treat family of equations, varying the control parameters. Furthermore, on a computer of finite word length and in finite run time it is impossible to distinguish, say, a very long periodic orbit from a quasiperiodic or chaotic one, if one keeps watching the trajectory in the phase (configuration) space only. However, by invoking the tangent space along with the phase space, one can calculate with confidence such quantitative characteristic of the motion as the Liapunov exponents which in turn enables us to make distinction between chaotic and non-chaotic behaviour. In addition, many systems show multistable solutions for one and the same set of parameter values, i.e., different attractors, trivial and strange ones, may coexist. There appears basin dependence: the destiny of the motion depends on the initial values chosen.

Put together what just said, we see that a full-scale numerical study of an ODE system would require a scanning of the product space $\Re^n \otimes \Re^n \otimes \Re^n \otimes \Re^m$ where n is the dimension of the phase (hence the initial value and tangent) space, and m is the dimension of the control parameter space. For the simplest autonomous system one should take at least $n = 3, m = 2$ (a codimension 2 study). This would be a job which exceeds the capability of many present-day supercomputers. One has to be contented with the knowledge on a few sections of this huge space. Fortunately, this happens to be sufficient in many cases and the Poincaré sections are the most important ones to study.

The Poincaré section is a low-dimensional (usually two-dimensional, but not necessarily so) intersection of the phase space chosen in such a way that the trajectories intersect it transversally, i.e., do not touch it tangentially. The choice of a Poincaé section must be preceded at least by a linear stability analysis of the system (5) to ensure that all qualitatively interesting trajectories do intersect it. With the section chosen we focus on consecutive intersecting points in the Poincaré section and view the motion as a point-to-point mapping in the section itself. Only in local sense this map sometimes may be written down approximately. In general, one has to resort to numerical calculations. Anyway, taking the Poincaré sections reduces the description of the dynamics signif- icantly and still reflects the essential feature of the motion. For example, a simple periodic orbit would become a single fixed point in the map; a periodic orbit with two commensurable frequency components would give rise to a finite number of points; a quasiperiodic trajectory would draw a closed curve in the Poincaré section, and chaotic motion would show off as erratically scattered points, etc.

3.1 Autonomous Systems

3.1.1 Hénon's method

A simple-minded way to calculate the Poincaré map would consists in integrating the equations step by step and testing the sign change of certain component, say, z when the $z = 0$ plane is used as the Poincaré section. However, to locate the intersection point with precision high enough to match that of the integration algorithm, one must use a high order interpolation scheme which requires saving and updating several consecutive points at each step of integration. M. Hénon [10] described a clever method to determine the intersection point at one shot. We explain the idea on the following example:

$$\dot{x} = f_1(x, y, z) \qquad\qquad \dot{y} = f_2(x, y, z) \qquad\qquad \dot{z} = f_3(x, y, z) \qquad (8)$$

with $z = 0$ being the Poincaré section. Suppose at the nth step we have got

$$t_n, \quad x_n, \quad y_n, \quad z_n < 0$$

and at the next step we find

$$t_n + \Delta t, \; x_{n+1}, \; y_{n+1}, \; z_{n+1} > 0$$

The intersection must take place in between these two steps. Now interchange the role of z and t, dividing the first two equations by the third and inverting the third one:

$$\frac{dx}{dz} = \frac{f_1(x, y, z)}{f_3(x, y, z)} \qquad\qquad \frac{dy}{dz} = \frac{f_2(x, y, z)}{f_3(x, y, z)} \qquad\qquad \frac{dt}{dz} = \frac{1}{f_3(x, y, z)} \qquad (9)$$

Integrating these equations in the new independent variable z backwards by one step $\Delta z = -z_{n+1}$ using

$$x_{n+1}, \quad y_{n+1}, \quad t_n + \Delta t$$

as initial values, or, integrating forwards by one step $\Delta z = -z_n$ from the $z_n < 0$ point (in practice the second way works a bit better) one settles down exactly in the $z = 0$ plane.

Hénon's method can be used to locate the intersecting point of an orbit with a general surface

$$S(x, y, z) = const \tag{10}$$

It is sufficient to introduce an additional function

$$u = S(x, y, z) - const$$

and to derive the differential equation for u using the original system (8):

$$\frac{du}{dt} = f_1 \frac{\partial S}{\partial x} + f_2 \frac{\partial S}{\partial y} + f_3 \frac{\partial S}{\partial z}$$

Now we are brought back to the old problem of looking for the intersection with the $u = 0$ plane for a system with one more equations.

In practice, due to the presence of transients, which may be very long when one gets close to a bifurcation point, it is a time-consuming task to calculate a Poincaré map with high precision even with the help of Hénon's procedure. However, in the case of periodic solutions there exists a very efficient method to locate the exact periodic orbit starting from an approximate one.

3.1.2 Locating a periodic orbit

Suppose in calculating a Poincaré map using Hénon's method we have come up to a suspected periodic orbit, i.e., starting from a point x_0 in the Poincaré section, we get back to its neighbourhood x_1 after integrating the system for some time T and we suspect that there might be an exact periodic point x^* with period T^*. How to find these unknown x^* and T^*, using our knowledge of x_0, x_1 and T?

The answer is based on an extension of Newton's method (method of tangents) to ODE's. This idea was first used by O. Lanford to the Lorenz equations (1) and then discussed by J. H. Curry [1]. We follow closely the description in Appendix in to the book of Sparrow [18].

We shall find it quite convenient to use the concise vector notations of § 2. Thus we have in terms of the flow $\Phi_\tau(x)$ an exact periodic solution

$$x^* = \Phi_{T^*}(x^*) \tag{11}$$

with both x^* and T^* unknown, and a near-by trajectory

$$x_1 = \Phi_T(x_0) \tag{12}$$

with x_0 known and x_1, T calculable by integrating the system. Being close to the true periodic solution, we can write

$$x^* = x_0 + dx \qquad\qquad T^* = T + dT \tag{13}$$

and insert them into the periodic flow (11):

$$x_0 + dx = \Phi_{T+dT}(x_0 + dx) \tag{14}$$

Now, expanding the right-hand side of the above equation and using the relation (12) as well as the original system (5), we get

$$\mathbf{x}_0 + d\mathbf{x} = \Phi_T(\mathbf{x}_0) + \frac{d}{d\tau}\Phi_T(\mathbf{x}_0)dT + D_{\mathbf{x}_0}\Phi_T(\mathbf{x}_0)d\mathbf{x} \quad = \mathbf{x}_1 + \frac{d\mathbf{x}_1}{dt}dT + D_{\mathbf{x}_0}\Phi_T(\mathbf{x}_0)d\mathbf{x} \quad (15)$$

that is

$$\mathbf{x}_0 - \mathbf{x}_1 = \mathbf{F}(\mathbf{x}_1)dT + (D_{\mathbf{x}_0}\Phi_T(\mathbf{x}_0) - \mathbf{I})\,d\mathbf{x} \tag{16}$$

where $\mathbf{F}(\mathbf{x}_1)$ is the right-hand side of the original system (5) calculated at $\mathbf{x}_1$.

Eq. (16) is actually a system of equations which we write down explicitly for the case $n = 3$, i.e.,

$$\mathbf{x} = (x, y, z) \qquad\qquad\qquad \mathbf{F} = (f_1, f_2, f_3)$$

$$\begin{pmatrix} x_0 - x_1 \\ y_0 - y_1 \\ z_0 - z_1 \end{pmatrix} = \begin{pmatrix} \frac{\partial x}{\partial x_0} - 1 & \frac{\partial x}{\partial y_0} & \frac{\partial x}{\partial z_0} & f_1(\mathbf{x}_1) \\ \frac{\partial y}{\partial x_0} & \frac{\partial y}{\partial y_0} - 1 & \frac{\partial y}{\partial z_0} & f_2(\mathbf{x}_1) \\ \frac{\partial z}{\partial x_0} & \frac{\partial z}{\partial y_0} & \frac{\partial z}{\partial z_0} - 1 & f_3(\mathbf{x}_1) \end{pmatrix} \begin{pmatrix} dx \\ dy \\ dz \\ dT \end{pmatrix} \tag{17}$$

The three unknown displacements dx, dy and dz must be confined to the surface of section, i.e., subject to the condition

$$dS = \frac{\partial S}{\partial x}dx + \frac{\partial S}{\partial y}dy + \frac{\partial S}{\partial z}dz = 0$$

where S is the left-hand side of the intersecting surface (10). In the simplest case $S = z$ we have $dz = 0$ and

$$\begin{pmatrix} dx \\ dy \\ dT \end{pmatrix} = \begin{pmatrix} \frac{\partial x}{\partial x_0} - 1 & \frac{\partial x}{\partial y_0} & f_1(\mathbf{x}_1) \\ \frac{\partial y}{\partial x_0} & \frac{\partial y}{\partial y_0} - 1 & f_2(\mathbf{x}_1) \\ \frac{\partial z}{\partial x_0} & \frac{\partial z}{\partial y_0} & f_3(\mathbf{x}_1) \end{pmatrix}^{-1} \begin{pmatrix} x_0 - x_1 \\ y_0 - y_1 \\ z_0 - z_1 \end{pmatrix} \tag{18}$$

Therefore, the exact periodic orbit $\mathbf{x}^*$ and period T^* can be determined from Eqs. (18) and (13). In practice, one has to iterate a few times to reach high enough precision. The convergence of this method is quadratic, just as the Newton's method usually does, provided one has good enough intial estimate for $\mathbf{x}_0$ and T.

Now we can summarize the orbit locating procedure as follows.

1. Get an initial estimate of $\mathbf{x}_0$ and T by using, e.g., the Hénon's method.

2. Integrate the original system to get the next return point $\mathbf{x}_1 = \Phi_T(\mathbf{x}_0)$, calculate T and $\mathbf{F}(\mathbf{x}_1)$.

3. Integrate the linearized system to calculate $(n - 1) \times n$ (in the case of $z = const$ section) or $n \times n$ (in the case of a general surface) components of $\frac{\partial x_i}{\partial x_{0j}}$:

4. Use these $\mathbf{x}_0, \mathbf{x}_1, \mathbf{F}(\mathbf{x}_1)$ and $\frac{\partial x_i}{\partial x_{0j}}$ to calculate dx and dT, and, consequently, $\mathbf{x}^*$ and T^*.

5. If necessary, use these as new estimates for $\mathbf{x}_0$ and T to start anew.

This method is equally well applicable to locate stable and unstable orbits. If one is interested only in stable orbits, then the stability must be checked by, e.g., calculating the largest Liapunov exponent (see § 5).

3.2 Non-autonomous Systems

We shall only consider the case of periodically driven systems, where the external frequency ω provides us with a reference period $T = 2\pi/\omega$. The extended phase space, i.e., the $x \sim y$ plane plus time t in the case of Eqs. (2) can be considered as closed in the t -direction to form a torus or a truncated cylinder of height T. The calculation of Poincaré maps now reduces to sampling x, y values at multiples of T (up to an unessential shift of the starting time). This procedure corresponds exactly to the stroboscopic sampling techniques used by experimentalists.

3.2.1 Subharmonic stroboscopic sampling

The subharmonic stroboscopic sampling (SSS) method is a very simple, yet quite effective extension [7] of the stroboscopic sampling idea mentioned above. When properly used this method may lead to very high frequency resolution at the cost of long computing time. Period-doubling cascade up to the 8192th subharmonics has been resolved in this way [7], and hierarchy of chaotic bands as well as the systematics of periodic orbits according to the U-sequence [15] have been observed [8] [9]. Similar resolution can usually be reached only for one-dimensional mappings, not for ODE's. When sampling at pT, p being an integer, one has the freedom to shift the starting time by $qT, q = 0, 1, \ldots, p - 1$, thereby to pick up one of the p components of the attractor. However, there are some subtle points that one must keep in mind when using the SSS method.

First of all, being a method of discrete sampling, it has the same demerit as the discrete Fourier transform (see § 4 below), namely, non-uniqueness in interpretation and incapability to resolve frequency higher than the sampling frequency. Suppose the actual period T^* is related to the sampling period T as

$$T^* = \frac{n}{m} T,$$

where n and m are coprime integers, then sampling at multiples of T will always yield n points (or clusters) in the map for all $m \geq 1$. If n is the product of two integers k and l, then one can change the sampling interval to lT or kT, increasing the resolution thereby. However, if one has misused a k which was not a factor of n, then there would appear a spurious factor k in measuring the period. To be safe, one must go gradually from the fundamental frequency to subharmonics, comparing the results with power spectra analysis whenever available.

Second, by using long enough sampling period it is possible to distinguish a periodic orbit from chaotic islands. However, due to the inhomogeneous structure of attractors there are always sparse regions in a map of chaotic bands that makes it difficult to determine the true "period". A few more trials with longer period will resolve the problem. For instance, SSS at $64T$ gives a cluster of points with a seeming "gap", but SSS at $128T$ repeats the same picture. This will be a good evidence of a 64-band chaotic regime. For the time being, SSS provides the most effective means to recognize chaotic bands of very long period (or "semi-period" as it was called by Lorenz [14]).

Third, when the period under study gets longer and longer, the difference among various points in the SSS map will show up only in less and less significant digits and round-off errors may stretch the image of a point into a strip which may be mistaken as a chaotic island. By going to double (or quadruple if available) precision this case can be resolved easily.

Fourth, another situation that should be distinguished from round-off errors is a converging transient. Especially when the parameter is close to a bifurcation point the transient gets very long due to critical slowing down [4] and one can hardly see the transient dying away in finite computing time. However, the round-off errors usually lead to more or less evenly spaced points, whereas transients show a clear converging pattern. One can even determine the distribution function of these points near a period-doubling bifurcation point [21]). With a little practice these cases can be identified unmistakably.

The last point to be discussed in connection with applying the SSS method is the necessity to distinguish intermittent transitions from round-off errors or transients. Poincaré maps close to an intermittent transition has a quite characteristic looking [19]. It can be checked by plotting the x_{n+k} *vs.* x_n maps near a tangent bifurcation into a period k orbit.

3.2.2   Locating a periodic orbit

Despite of the great efficiency of the SSS method it is still rather time-consuming when determining a periodic orbit from scratch, because one has to throw away at least a few hundreds of "transient" periods to reach the stationary solution. However, the technique of locating periodic orbit simplifies drastically for periodically driven systems, since there is no need to calculate the period. In fact, one has to solve only 6 equations (2 nonlinear plus 4 linear ones) to locate a periodic orbit for Eqs. (2).

Let us describe the method in more general terms. Suppose we have got a point $\mathbf{P}_n$ ($\mathbf{P}_n = (x_n, y_n)$ for Eqs. (2)) in the stroboscopic portrait. Integrating the nonlinear system for a period T (it must be done in, say, $L = 1024$ steps), we get another point $\mathbf{P}_{n+1}$. $\mathbf{P}_n$ and $\mathbf{P}_{n+1}$ are quite close, so we suspect that it is not far from an exact period $\mathbf{P}$. We may write

$$\mathbf{P}_n = \mathbf{P} + \mathbf{W}_n \qquad\qquad \mathbf{P}_{n+1} = \mathbf{P} + \mathbf{W}_{n+1} \qquad\qquad (19)$$

Although $\mathbf{P}$, $\mathbf{W}_n$ and $\mathbf{W}_{n+1}$ are unknown, $\mathbf{W}_{n+1}$ must be evolved from $\mathbf{W}_n$ according to the linearized equation, i.e.,

$$\mathbf{W}_{n+1} = \mathbf{U}(T)\mathbf{W}_n \qquad\qquad (20)$$

Substracting the two equations (19), we get

$$\mathbf{W}_n = \left(\mathbf{U}(T) - \mathbf{I}\right)^{-1} \left(\mathbf{P}_{n+1} - \mathbf{P}_n\right) \qquad\qquad (21)$$

Everything on the right hand side of Eq. (21) can be calculated: $\mathbf{P}_{n+1}$ from $\mathbf{P}_n$ according to the original nonlinear equations, $\mathbf{U}(T)$ from the linearized system. Notice that $\mathbf{U}(T)$ is the product of L factors

$$\mathbf{U}(T) = \mathbf{U}(T, \frac{L-1}{L}T) \cdots \mathbf{U}(\frac{2}{L}T, \frac{1}{L}T)\mathbf{U}(\frac{1}{L}T, 0) \qquad\qquad (22)$$

each calculated from the system, linearized at the end point of the previous step. Being the solution of a linear system, there are two ways to compute $\mathbf{U}(\frac{k}{L}T, \frac{k-1}{L}T)$: either numerically by adding the linear equations for $\mathbf{U}$ to the linear ones with initial values

$$\mathbf{U}(0) = \mathbf{I}$$

at each step, or analytically by using the closed form (for $n = 2$)

$$\mathbf{U}(t,0) = e^{\lambda t} \begin{pmatrix} C(t) + \frac{a_{11}-a_{22}}{2} S(t) & a_{12}S(t) \\ a_{21}S(t) & C(t) - \frac{a_{11}-a_{22}}{2} S(t) \end{pmatrix}$$

where

$$C(t) = \begin{cases} \cosh(st) & D > 0 \\ 1 & D = 0 \\ \cos(st) & D < 0 \end{cases} \qquad\qquad S(t) = \begin{cases} \frac{\sinh(st)}{s} & D > 0 \\ t & D = 0 \\ \frac{\sin(st)}{s} & D < 0 \end{cases}$$

$$\lambda = (a_{11} + a_{22})/2 \qquad\qquad D = \lambda^2 - \det(J) \qquad\qquad s = \sqrt{|D|}$$

Notice that a_{ij}'s are elements of matrix $\mathbf{J}$ introduced in Eq. (7).

Which way to go depends on the complexity of the equations involved and may be determined by trials. We remind the readers once again that it is desirable to monitor the Liapunov exponents (see § 5) when using the orbit locating algorithm to be aware of the stability of the orbit under study.

3.3   Winding Numbers and Symbolic Dynamics

If one casts a glance at a final Poincaré map of a periodic orbit, the immediate information gained is the period, i.e., the number of points in the plot. However, it costs a little more effort to extract such useful information as the winding number or the word describing the period in certain symbolic dynamics, provided, of course, one stays in the right part of the parameter space.

We explain first the winding number on the example of the "bare" circle map

$$\theta_{n+1} = \theta_n + \Omega \quad mod(1)$$

If $\Omega = p/q$ where p and q are integers, then the q points seen in the map must be made in p turns. Therefore, we can measure p in the process of plotting the map and define the winding number as the ratio p/q. It is a useful quantity when dealing with periods related to frequency-locking regimes. For example, if we have followed a period-doubling sequence from period 2 to 4 to 8 with the winding numbers changing from $1/2$ to $2/4$ to $4/8$, we are quite confident about our being in the supercritical part of the corresponding frequency-locked (Arnold) tongue.

Now a few words about symbolic dynamics. In dissipative systems the Poincaré maps often approach one-dimensional objects due to phase volume contraction. When this happens it is sometimes possible to assign certain letters to the numerically observed points to get a word describing the period, just as in the case of one-dimensional mappings [15]. We show this on a numerical example.

For Eqs. (2) a detailed search along the slanting line $A = 0.46 - 0.2\omega$ in the $A \sim \omega$ parameter plane has exposed three different period 5 orbits. Since all points in the SSS portrait spread along the y direction, the stretch along x being much smaller, we can consider only the y_i values as the output from some unknown one-dimensional iterations. (In general, one should introduce a one-parameter parametrization for the curve seen in the SSS portrait or the Poincaré section). Identifying the largest y_i as the rightmost R and attribute y_{i-1} to the central point C, because all

ω		Adjacent periodic points (x_i, y_i)				Word
0.5847	0.433,3.749	0.225,2.284	0.317,3.521	0.336, 3.548	0.285,3.248	
	R	L	R	R	C	RLRR
0.7095	0.294,3.537	0.323,3.982	0.194,1.843	0.230,3.034	0.294,3.858	
	C	R	L	L	R	RLLR
0.8480	0.211,3.364	0.245,4.036	0.285,4.333	0.190,1.478	0.187,2.492	
	L	C	R	L	L	RLLL

Table 1: Assignment of Letters to the Periodic Points

admissible words actually begin with the letters RL where R corresponds to the rightmost point in the interval. Then all other y_i's acquire an unique assignment of letters. In this way we have assigned a word to each of the observed periods. Table 1 shows this procedure for the period 5 orbits.

In fact, all but one orbits with period less or equal to 6 have been seen along the above-mentioned line and they are ordered in the same way as in the logistic map.

Another example of assigning letters to the numerically observed periods is the Lorenz model (1) [3]. Now three alphabets are needed owing to the antisymmetry under $x \to -x, y \to -y$ that makes it closer to the antisymmetric cubic map

$$x_{n+1} = Ax_n^3 + (A-1)x_n$$

rather than the Hénon map. Along the most studied b line in the parameter space 47 primitive periods (i.e., excluding the period-doubled regimes, etc.) out from 53 are described by words made of letters R, M and L and ordered in the same way as in the cubic map.

4   POWER SPECTRA ANALYSIS

Power spectra analysis provides us with the simplest method to distinguish chaotic motion from quasiperiodic, the spectra of the former being noisy broad-bands and the latter – discrete lines without simple frequency interrelationship. However, there are some subtleties in performing the power spectra analysis. We mention a few of them in what follows.

4.1   Preliminaries

Suppose we are given a time series

$$x_1, \ x_2, \ \cdots \ x_n, \ \cdots$$

sampled at equal time interval τ from a certain computer or laboratory experiment. Taking a finite subsequence $x_1, \ x_2, \ \cdots \ , x_N$ from the series and imposing periodic boundary condition $x_{N+j} = x_j \ \forall j$ one calculates the correlation function

$$c_j = \frac{1}{N} \sum_{i=1}^{N} x_i x_{i+j} \tag{23}$$

and then performs a discrete Fourier transform to get the Fourier coeffcients

$$p_k = \sum_{j=1}^{N} c_j \exp\left(\frac{2\pi k j \sqrt{-1}}{N}\right) \tag{24}$$

Roughly speaking, p_k represents the contribution of the kth frequency component to $\{x_i\}$. This was the original definition of a "power spectrum".

With the rediscovery of the fast Fourier transform (FFT) algorithm in 1965, power spectra are calculated more efficiently by directly transforming $\{x_i\}$ into Fourier coefficients

$$a_k = \frac{1}{N} \sum_i x_i \cos(\frac{\pi i k}{N}) \qquad\qquad b_k = \frac{1}{N} \sum_i x_i \sin(\frac{\pi i k}{N}) \tag{25}$$

and then computing

$$\bar{p}_k = a_k^2 + b_k^2 \tag{26}$$

From a number of subsequences of $\{x_i\}$ one gets several sets of $\{\bar{p}_k\}$. By averaging these one obtains the power spectrum $\{p_k\}$. This is necessary in general when one deals with genuine aperiodic processes. In the case of ODE's, when there is a nice "periodic" understructure, a single $\{\bar{p}_k\}$ may yield quite good spectrum and one does not need to average all the way.

We mention again that nowadays it is not necessary to write the FFT program by oneself. There is a large choice of subroutines in the existing libraries. The most important step in doing power spectra analysis is to "design" the spectra before making the FFT.

4.2 Design The Spectrum

There are two aspects in "desgning" a good spectrum: choosing the correct sampling parameters and filtering the data (or smoothing the results). We start from the first one. Let the sampling interval be τ and the total sampling time to make a single spectrum be $L = N\tau$, where N is the number of sampled points. τ and N determine two frequencies: $f_{max} = 0.5/\tau$ is the maximal frequency one can measure using the given sampling interval; $\Delta f = 1/L$ is the frequency difference between two adjacent Fourier coeffcients. In order to eliminate effectively the aliasing phenomenon (see, e.g., [16]) one ought to take $f_{max} = k f_0$, where f_0 is the fundamental frequency of the physical system and k is a multiplier of the order 4 to 8, and to keep only the lower part of the spectrum. We aim at resolving the pth subharmonic of f_0 and wish the subharmonic peak to be formed by s points in the spectrum, i.e., $f_0/p = s\Delta f$. Put together all the above-mentioned relations, we get

$$p = \frac{N}{2ks}$$

Note that this a relation independent on τ and f_0. Take $N = 8192, k = 4, s = 8$, we get $p = 128$. This gives the limit of resolution in power spectra analysis on most medium-size computers. Nevertheless, a broad-band noisy spectrum is still the most practical and convenient criterion to identify chaotic motion in laboratory and computer experiments.

As regards filtering the data, it may be more important when one deals with really aperiodic processes with noisy background. In numerical study of ODE's, it is not so essential to have the data filtered before doing FFT. In most cases simple smoothing the spectrum will do. Anyway, filtering the data and smoothing the spectra are equivalent operations, related by a convolution transformation. There are a few subtelities in performing filtering or smoothing. Whenever the necessity asises, one may consult the literature on time series analysis which exists in plenty.

4.3   Symbolic Dynamics and Fine Structure of the Spectra

When one happens to be in a chaotic regime of the parameter space where the periodic windows are ordered according to symbolic dynamics of two or more letters, the fine structure of the power spectra may tell much about one's exact location.

We skip the details of this elementary usage of symbolic dynamics (see, e.g., [5] for a presentation at length) and concentrate on the case of two letters, say, R and L. All admissible words made of these two letters may be classified into primary and composite ones. A composite word can be decomposed into a product of primary words by using the $*$ - composition rule introduced in [2]. For instance,if $W = P * Q * S$,then P, Q and S describe the motion at finer and finer scales that leads to characteristic fine structure in the power spectra. In practice, this is a question of factorizing non-prime numbers. Take, for example, period 6. $6 = 2 \times 3$ and $6 = 3 \times 2$ correspond to $R * RL = RLR^3$ and $RL * R = RL^2RL$, the former being the period 3 window embedded in the two-band chaotic region and the latter — the period-doubled regime of the period 3 window in the one-band chaotic region.

5   LIAPUNOV EXPONENTS

Since in these lectures emphasis is made on numerical aspects of studying ODE's, we shall only summarize briefly main facts about the Liapunov exponents before turning to practical methods.

A nth order ODE system has n Liapunov exponents. They are real numbers, put in descending order by convention and denoted by

$$LE_1 \geq LE_2 \geq \cdots \geq LE_{n-1} \geq LE_n$$

Among these numbers there may be m_+ positive ones, m_0 zeros, and m_- negative ones, representing the number of stretching, marginal and contracting directions in the phase space:

$$m_+ + m_0 + m_- = n$$

For conservative systems all Liapunov exponents sum up to yield zero:

$$\sum_{i=1}^{n} LE_i = 0$$

(in fact, they cancel in pairs if the system is Hamiltonian), whereas the sum is negative for dissipative systems, reflecting the overall phase space contraction. For an autonomous system there must be at least one zero Liapunov exponent, provided the solution is bounded and does not approach a fixed point. The appearance of the first positive Liapunov exponent signals the transition to chaos. The signature of the Liapunov exponents provides a classification for attractors, e.g.,

$(-, \ -, \ -, \ -)$ *fixed point*

$(\ 0, \ -, \ -, \ -)$ *periodic motion*

$(\ 0, \ 0, \ 0, \ -)$ *quasiperiodic motion on a torus*

$(+, \ 0, \ -, \ -)$ *chaos*

$(+, \ +, \ 0, \ -)$ *superchaos a la Rossler*

etc. Consequently, the fact that a Liapunov exponent passes through zero when varying the control parameters indicates a bifurcation. In this way we can have a phenomenological classification of attractors and transitions between them. Even the knowledge of the largest Liapunov exponent alone would tell us much about the attractor.

5.1 The General Case

We describe a purely geometric way to calculate the Liapunov exponents [17].

To compute the largest Liapunov exponent, i.e., LE_1, choose at random a small vector $\mathbf{W}$, sitting at the orbit and pointing to a near-by location. Integrating the start point one step forward along the exact trajectory, i.e., using the nonlinear equations (5), and integrating the end point of the vector $\mathbf{W}$ according to the linearized equations (7), one gets a new vector $\mathbf{U}(t)\mathbf{W}$. Take the length ratio of these two vectors, then repeat the calculation for the next steps. Averaged along the whole (in practice, sufficient long segment of) trajectory, this ratio would approach the largest exponent LE_1. Written down explicitly, we have

$$LE_1 = \lim_{t \to \infty} \log \frac{\|\mathbf{U}(t,0)\mathbf{W}\|}{\mathbf{W}}$$

From § 3.2.2 we know that $\mathbf{U}(t,0)$ should be calculated as the product of $\mathbf{U}$'s, evaluated at each step.

To calculate the next Liapunov exponent, one should choose two vectors pointing out from the trajectory and integrate the nonlinear as well as linear equations one step forward to get the end points. Then calculate the area ratio of the parallelograms formed by these vectors before and after the step. Averaged along the trajectory, this ratio will give the sum of two largest Liapunov exponents, i.e., $LE_1 + LE_2$, provided the initial vectors were chosen randomly and transversally to each other and to the trajectory. To calculate LE_3, proceed similarly with a parallelopiped, and so on, and so forth. This procedure usually converges quite fast, at least it is so for the largest LE_1 [20]. However, there is one point worthy to mention. Due to the presense of negative exponents, the vectors chosen at the beginning tend to align themselves, causing the area or volume to vanish. To avoid numerical difficulty one should orthogonalize the vectors every few steps. However,if one deals with two-dimensional vectors only, it is not necessary to use the Schmidt's orthogonalization procedure: just use the fact that vector $(-b, a)$ is orthogonal to vector (a, b).

5.2   Liapunov Exponents for Periodic Orbits

We have mentioned in § 3.1.2 and § 3.2.2 that it is desirable to monitor the Liapunov exponents when using the periodic orbit locating procedure, because the method works equally well for both stable and unstable orbits and one can cross the border without noticing it. Fortunately, it is not necessary to average along the trajectory in this case. The calculation of Liapunov exponents can be accomplished by integrating one more period after one has settled down on a periodic orbit. In fact, according to the Floquet-Liapunov theorem the time evolution operator $U(t)$ for a periodic solution of period T can be decomposed:

$$U(T) = K e^{\Lambda T}$$

where

$$K^2 = I$$

Therefore, to get rid of K one must calculate it at $2T$. Then the Liapunov exponents will be given by the eigenvalues of the matrix

$$\Lambda = \frac{1}{2T} \log(U(2T)). \tag{27}$$

In these lectures we have not touched the problem of how to calculate dimensions and entropies as well as the relation between dimension, entropy and Liapunov exponents. One might have included various methods to visualize the trajectory (two- and three-dimensional projections, stereoscopic projections, bifurcation diagrams), to reconstruct the stable and unstable manifolds and to explore the homoclinic and heteroclinic intersections, etc.,but that would lead us too far afield.

References

[1] Curry, J. H., in *Global Theory of Dynamical Systems*, ed. by Z. Nitecki and C. Robinson, p.111, *Lecture Notes in Mathematics*, vol. **819**, Springer-Verlag, 1979.

[2] Derrida, B., Gervois, A., and Pomeau, Y., *Ann. Inst. Henri Poincaré*, **A29**, 305(1978).

[3] Ding Ming-zhou, and Hao Bai-lin, *"Systematics of periodic windows in the Lorenz model and its relation with the antisymmetric cubic map"*, ASITP Preprint 86-015, submitted to *Commun. Theor. Phys.*

[4] Hao Bai-lin, *Phys. Lett.* **86A**, 267(1981).

[5] Hao Bai-lin, *"Elementary Symbolic Dynamics"*, in the Proceedings of the Spring School on *"Order and Chaos in Nonlinear Physical Systems"*, 21 April – 13 June, 1986, to be published by Plenum Press.

[6] Hao Bai-lin, *"Bifurcation and Chaos in the Periodically Forced Brusselator"*, ASITP Preprint 86-011, to appear in a book honouring the retirement of Prof. K. Tomita.

[7] Hao Bai-lin, and Zhang Shu-yu, *Phys. Lett.* **87A**, 267(1982); *Acta Physica Sinica* **32**, 198(1983).

[8] Hao Bai-lin, and Zhang Shu-yu, *Commun. Theor. Phys.* **1**, 111(1982); *J. Stat. Phys.* **28**, 769(1982).

[9] Hao Bai-lin, Wang Guang-rui, and Zhang Shu-yu, *Commun. Theor. Phys.* **2**, 1075(1983); *Acta Physica Sinica* **33**, 1008(1984).

[10] Hénon, M., *Physica* **5D**, 412(1982).

[11] Lambert, J. D., *Computational Methods in Ordinary Differential Equations*, Wiley, 1973.

[12] Li Jia-nan, and Hao Bai-lin, "Bifurcation spectra in a delay-differential system related to optical bistability", ASITP Preprint 85-021, to appear in *Commun. Theor. Phys.*

[13] Lorenz, E. N., *J. Atmos. Sci.* **20**, 130(1963).

[14] Lorenz, E. N., *Ann. N. Y. Acad. Sci.*, **357**, 282(1980).

[15] Metropolis, N., Stein, M. L., and Stein, P. R., *J. Combin. Theory*, **A15**, 25(1973).

[16] Rayner, J. H., *An Introduction to Spectral Analysis*, Pion Limited, London, 1971, p.74.

[17] Shimada, I., and Nagashima, T., *Progr. Theor. Phys.* **61**, 1605(1979).

[18] Sparrow, C., *The Lorenz Equations, Bifurcations, Chaos and Strange Attractors*, Springer-Verlag, 1982.

[19] Wang Guang-rui, Chen Shi-gang, and Hao Bai-lin, *Acta Physica Sinica* **32**, 1139(1983); English translation: *Chinese Phys.* **4**, 284(1984).

[20] Wang Guang-rui, Chen Shi-gang, and Hao Bai-lin, *Computational Physics (China)*, **2**, 47(1985).

[21] Wang You-qin, and Chen Shi-gang, *Acta Physica Sinica* **33**, 341(1984).

An Experimentalist's Introduction to the Observation of Dynamical Systems

Neil Gershenfeld

Applied and Engineering Physics
Cornell University
Ithaca, NY 14853

One of the most powerful and surprising mathematical tools for the experimentalist to come from the modern theory of dynamical systems is a set of theorems that provide a procedure for reconstructing a system's full motion in phase–space from the observation of a single degree of freedom. These theorems provide a remarkable ability to analyze the dynamics of an experimental system without any prior knowledge about the governing equations or relevant degrees of freedom. One of the primary applications of this ability is distinguishing between the deterministic but apparently disordered behaviour of a chaotic system and truly random noise. In this article, the practical application of the theory is described and demonstrated, followed by a more careful treatment of the details of the implementation and a tour through some of the ways to characterize a system's underlying dynamics once the phase–space trajectory has been retrieved (these include the dimensions, Lyapunov exponents, and thermodynamic quantities). Finally, a self-contained introduction is given to the mathematical background necessary to state and describe the underlying proofs.

1. Introduction

Modern dynamical systems research, the study of the evolution in time of the solutions of differential equations, has led to powerful observational tools for the experimentalist. In particular, a remarkable set of theorems (proved by F. Takens) provide a means for the full motion in phase space of an experimental system to be reconstructed ("embedded") from the measurement of a single degree of freedom. This reconstructed picture of the phase space flow will be related to the exact one by a smooth change of coordinates, and so topological information will be preserved although the quantitative coordinates will differ. This reproduction nevertheless allows deep qualitative and quantitative questions to be answered, such as whether or not the behaviour of a disordered system is due to random noise or low–dimensional deterministic but chaotic dynamics. Figure 2.6 illustrates the reconstruction. This theory has been applied to such experimental systems as photoconductors[1], Josephson junctions[2], electron-hole plasmas[3], p-n junctions[4], spin waves[5], lasers[6], Couette-Taylor hydrodynamic flows[7], the Belousov-Zhabotinskii chemical reaction[8], forced surface waves[9], and Rayleigh-Benard convection in mercury[10], liquid helium[11], and water[12].

The theory that describes and justifies the procedure for the reconstruction of a system's dynamics draws heavily on differential topology. The primary results can be described in simpler terms, but to state them precisely and to understand why they are true and how they work one must have some familiarity with this language. This mathematical orientation of the theory has limited its use. Within the community of experimentalists working with chaotic systems the main results are well known and heavily used, however even here the inaccessibilty of the primary references leads to the results frequently being used without much insight into why they work. The theory has found little use beyond this community, even though the applications are potentially quite broad for the ability to measure and analyze the dynamics of an experimental system with access to only one degree of freedom and no *a priori* knowledge about the dynamics.

The purpose of this paper is to provide a self-contained introduction to the theory, aimed at the experimentalist interested in the results but having no relevant theoretical training. The practical implementation of the embedding procedure is described, as well as some of the related tools to analyze the data so produced, and an introduction is given to the supporting mathematics necessary to understand why the procedures work. Extensive use will be made of the example of the analysis of deterministic chaos, because the theory was developed in this context and has found its richest applications here. To help organize the onslaught of new terms and symbols to be defined, the first occurence of each will be underlined.

There are six sections in this article. Although the supporting theory is quite mathematically detailed, the use of the principal results can be simply described; this is done in section 2. It starts by introducing the difference between random noise, deterministic chaos, and quasiperiodicity. The procedure for retrieving a system's dynamics is then described and illustrated, and finally the procedure is applied to the important problem of distinguishing between deterministic chaos and random noise. A simple heuristic proof is given for the central embedding theorem. Section 3 provides further details about the implementation of the embedding procedure.

Once the trajectory in phase space has been successfully measured it must still be interpreted if it is to be of use, and frequently a visual inspection reveals little more than a tangle of spaghetti. Section 4 reviews some of the available procedures for extracting qualitative and quantitative information about the nature of the dynamics from such phase space data. These include the dimensions, Lyapunov exponents, and thermodynamic quantities. The body of the paper closes in section 5 with a discussion of the limitations on the information that is available from this analysis and with some speculations on possible future directions of development.

The underlying theory is much more technical than the preceeding material and is not necessary for understanding it, and so it has been relegated to the appendices. The definitions of any unfamiliar concepts that have crept into the body of the paper may be found in these appendices. Appendix A attempts to provide the necessary mathematical background, starting with set theory and ending with differential topology and dimension theory. Appendix B contains the foundation of this article. Here the background developed in Appendix A is applied to describing the proofs of the theorems on the observation of dynamical systems. The appendix ends with some important theorems related to the evolution of dynamical systems.

2. Survey

2.1 Determinism, Chaos, and Noise

The realization that extremely simple systems can have apparently random behaviour has been a startling step in the evolution of the theory of classical mechanics and differential equations. First consider figure 2.1, which shows the time series and power spectrum for white noise (noise with a flat power spectrum). These data are the measured thermal noise fluctuations in an electrical resistor; both the eye and the power spectrum argue that they are indeed random.

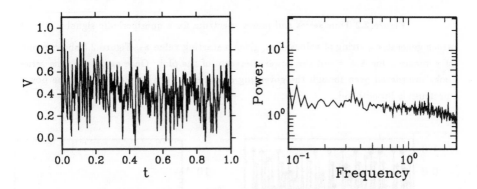

Figure 2.1 Time series and power spectrum for white noise

Now consider the equation

$$y(t) = \sin(t) + \sin(\sqrt{2}\, t) .\qquad(2.1)$$

The time series is shown in figure 2.2; it certainly does not look like the deterministic behaviour of a simple equation. The traditional first step when confronted with an experimental signal that isn't easily interpreted is to examine the Fourier transform, and this is also shown in figure 2.2. Here we see what we already knew, that the signal is composed of just two modes (energy at two frequencies), and the apparent disorder arises because the periods of the modes are incommensurate (their ratio is irrational). This is called quasiperiodicity.

Next, consider the logistic map

$$x_{n+1} = \lambda x_n (1 - x_n) ,\qquad(2.2)$$

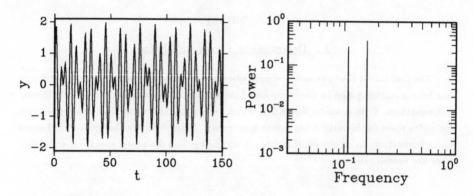

Figure 2.2 Time series and power spectrum for a quasiperiodic signal

which generates a string of values of x_n given a starting value x_0. Figure 2.3 shows a plot of x versus n for $\lambda = 4$ and the power spectrum of the plot. Once again the time series looks disordered even though the governing equation is very simple, yet here the power spectrum is broad-band.

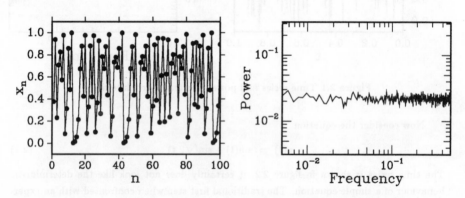

Figure 2.3 Time series and power spectrum for the logistic map

Traditional wisdom, motivated by the study of linear systems, states that modes in the power spectrum correspond to generalized degrees of freedom of the system, and that broad-band power spectra come from infinite-dimensional systems. Such wisdom dramatically fails in the face of the complexity of nonlinear systems. This is an example of a <u>chaotic</u> system, one that is governed by a low-dimensional set of equations but that has a broad-band power spectrum. One of the main goals of this paper will be to show how the experimentalist can determine that an equation such as 2.2 underlies the observed behaviour in figure 2.3. The logistic map has been one of the simplest, most historically important, and most mathematically tractable examples of a chaotic system; see reference [13] for more information.

One of the first sets of differential equations found to show this type of behaviour was the Lorenz system[14]. These equations were obtained by Lorenz as a drastic approximation to the Navier-Stokes equation for a convecting system; although they can be related to physical systems they are best thought of as an intrinsically interesting mathematical model. The system consists of three nonlinear first-order ordinary differential equations:

$$
\begin{aligned}
\dot{x} &= \sigma(y - x) \\
\dot{y} &= \rho x - y - xz \\
\dot{z} &= -\beta z + xy
\end{aligned}
\tag{2.3}
$$

Figure 2.4 shows the numerical solution of these equations for x as a function of time for the values $\sigma = 10$, $\beta = 8/3$, $\rho = 28$. Once again, we see a simple deterministic system producing dynamics that look decidedly non-deterministic.

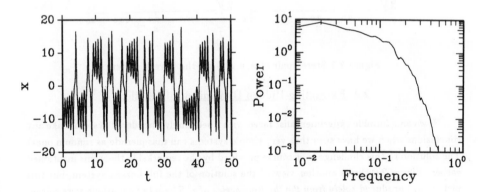

Figure 2.4 $x(t)$ for the Lorenz system and its power spectrum

There are three degrees of freedom for the Lorenz system; plotting all of them as a function of time produces the structure shown in figure 2.5. This view of phase space shows that the system evolution takes place on a well-defined subset of three-dimensional space (an "attractor"), with the random behaviour seen in figure 2.4 coming from the particular path the system follows on the attractor. If one did not already know that the data were generated by the Lorenz system, the plot provides strong grounds to suspect that the observed behaviour is governed by the deterministic dynamics of a system with just a few degrees of freedom. The two images in figure 2.5 are from slightly different angles corresponding to the different views needed for stereo vision; this is an example of a stereo pair. Stereo pairs provide a convenient way to convey three-dimensional information. To view them the focal plane and the point of convergence of the eyes must be separated in order to provide each eye with the appropriate image; this is effected by allowing the two images to fuse into a single one and then focussing this image, a skill that usually requires some practice[15].

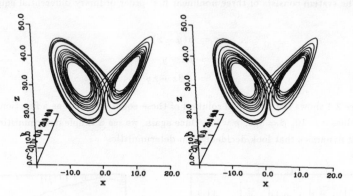

Figure 2.5 Stereo pair of $x, y, z(t)$ for the Lorenz system

2.2 Extracting Hidden Dynamical Information

How can a humble experimentalist faced with a seemingly disordered system make any conclusions when we have seen that very simple systems can masquerade as random ones? The solution to this challenge is of course provided by the remarkable theorems mentioned earlier. Figure 2.6 shows another view of the solution of the full Lorenz system, but this picture was produced *solely from the the time series of x!* The set of equations were solved, the values for y and z were thrown out, and by a procedure to be described a picture very similar to that of the full solution was generated from the values of x. For comparison, a view of the full solution from a similar orientation is shown next to the recreated solution.

The idea that a system's dynamics can be reproduced from a single degree of freedom was independently suggested by two sources[16,17]. In the former paper F. Takens proved the underlying theorems and provided the mathematical framework to be used here. There

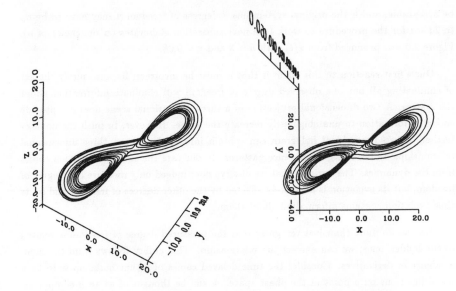

Figure 2.6 $x, y, z(t)$ for the Lorenz system reconstructed from $x(t)$ (left), and $x, y, z(t)$ from the direct calculation (right)

are a number of equivalent implementations of this procedure; we will describe here the use of the most common form (time delays).

Consider an experimental signal $y(t)$ which is a function of the degrees of freedom of the experimental system, for example a measurement of a single chemical species in a reactor containing a number of chemicals. Next, construct a vector Φ out of time-delayed copies of y:

$$\Phi(t) = \{y(t), y(t + \tau), y(t + 2\tau), \ldots, y(t + n\tau)\}, \qquad (2.4)$$

where τ is an arbitrary constant. Φ provides the coordinates of a point in an n-dimensional space, and these coordinates will evolve as t is varied (τ is kept fixed). τ is called the <u>delay time</u> and n is the <u>embedding dimension</u>. The miraculous claim is that the trajectory of Φ in its n-dimensional space is simply related to the trajectory of the initial system in its full phase space. In the chemical example, this means that if the experiment was able to simultaneously measure all the chemicals that are relevant to the reaction, a plot of the data would differ from the one generated by the trajectory of Φ by only a smooth change of coordinates. This change of coordinates will vary throughout the phase-space, but in such a way that the two plots will differ only by local stretching: qualitative and quantitative measures of how the flows "look" will be unchanged. Almost any value chosen for τ will be acceptable, and if the original system has d degrees of freedom n may have to be up to $2d + 1$ for the procedure to work (the next subsection elaborates on the choice of n). Figure 2.6 was produced from $x(t)$ with $n = 3$ and $\tau = 0.08$.

One's first reaction to this claim is that it must be incorrect, because surely the act of eliminating all but one observed degree of freedom will eliminate information about the system. A two dimensional projection of a three dimensional scene does not provide enough information to unambiguously recreate the scene. However, to push the analogy further, a two-dimensional hologram can contain information about three dimensional relationships by recording interference patterns. In our case the extra information comes from the dynamics. The variable that we observe does indeed only measure one degree of freedom, but its evolution is intimately affected by the other degrees of freedom and so its time evolution contains information about them.

We can do more than just recognize that the measured degree of freedom is related to the hidden ones; we can extract the relationship. This ability follows from the independence of derivatives. Consider the time delayed copies of y that make up Φ to be a set of functions on a point in the phase space. Φ can be thought of as an n component function that provides a map from the phase space to a new set of variables. Because the dynamics of the system are not trivial, these n functions are related to each other in a non-trivial way. The local linearization of the function is provided by the Jacobian matrix (the matrix of derivatives of Φ), and because of the complicated relationship between the elements of Φ the Jacobian will almost always have no linearly dependent rows. If a set of vectors locally span a space (all members of the space can be expressed as a sum of the basis vectors), then if they are acted on by such a non–degenerate Jacobian they will span the space they are mapped to. If this condition holds everywhere, then a trajectory will be mapped into a new one that may be stretched a bit but won't fundamentally look different. But this is exactly our situation; this is the key to the ability to recreate the dynamics. As described in Appendix B, the precise name for this recreation is an embedding (assuming some technical conditions are met). Once again: the evolution in time of a single observable is affected by all the degrees of freedom of the system; the fact that the evolution equations are not-trivial insures that the time delayed functions are non-trivially related and hence have different derivatives; and this set of non-trivial functions provides a map from the original coordinates into a new set of basis vectors and so can only stretch but not obliterate trajectories. The requirement that n be at least $2d + 1$ insures that there is enough room to prevent accidental crossing of the trajectories produced by the mapping.

2.3 Distinguishing Between Chaos and Noise

The procedure for distinguishing between random noise and deterministic chaos may now be seen: deterministic chaos occurs in a low-dimensional space, while random noise does not. We will next give a step-by-step recipe for the implementation of a procedure to detect this difference, and illustrate its use by comparing the Lorenz system to the resistor thermal noise. There will be four steps in the process.

Step one: Measure a signal The starting point is the measurement of a signal from the experimental system, $y(t)$. In order to prevent artifacts from the digitization process, the signal must be sampled at a rate equal to twice the highest frequency present, and if necessary the signal must be low-pass filtered to insure that there is a well-defined upper limit to the frequencies.

Step two: Reconstruct the system's dynamics Choose a delay time τ that is the shortest time over which there are clearly measurable variations in y (section 3.1 gives further details on choosing τ). For a given value of n (the choice of n is described in the next section), generate a trajectory in an n-dimensional space by applying equation 2.4 to the data. This process can be pictured as applying a comb to the data, where there are n teeth to the comb and the teeth are separated by a time τ. The trajectory is generated by sliding the data set, possibly sampled at a time finer than τ, past the comb. If τ is too small it will unnecessarily compress the size of the reconstructed trajectory, and if it is too large it will stretch the trajectory so much that the original structure is hard to interpret, and it will also enhance the influence of experimental noise.

Step three: Measure the dimension In later sections of the paper we will consider the measurement of the dimension of an object given a set of points from the object (for example, the determination that a set of points in three dimensions actually lie on a plane). The most useful procedure will be shown to be the evaluation of the correlation sum, defined by

$$C(r) \; = \; \lim_{N \to \infty} \frac{1}{N^2} \times \{\text{number of pairs } (i,j) \text{ whose distance } |x_i - x_j| \text{ is less than } r\} \,, \; (2.5)$$

where N is the number of points in the data set, and the limit is evaluated by increasing N until the answer converges. The correlation sum will usually scale as r to some power:

$$C(r) \; \sim \; Ar^\nu \,. \tag{2.6}$$

ν is the correlation dimension or correlation exponent; we will see that the correlation dimension agrees with our intuitive notion of dimension for simple objects (such as lines, planes, and volumes), and can give a non-integer dimension for a class of more complicated objects (fractals) that can occur in chaotic systems. The scaling law (equation 2.6) will hold over an intermediate region where r is smaller than the size of the object and larger than the smallest spacing between points, and so ν is usually measured by fitting the slope of a log-log plot of $C(r)$ versus r in this region. This definition will work even if the distance between two points is defined, not as the Euclidean distance, but as the maximum value of the separation of each of the components of the points. This provides a gain in speed for the algorithm.

It will frequently happen that the value of ν calculated by choosing a single value for i and then finding all the j's for which $|\vec{r}_i - \vec{r}_j| \leq r$ will be the same as that found by summing over all values of i. This shorter calculation gives the <u>pointwise dimension</u> and provides a great saving over the time needed for the full computation. The pointwise dimension is usually evaluated at a few points, and if the answers lie within the experimental statistics of each other they are accepted. Figure 2.7 shows the measurement of the correlation dimension for the Lorenz system by evaluating ν at one point and fitting the slope of the log-log plot in the intermediate range; the slope of the line shown gives the correct dimension (2.05).

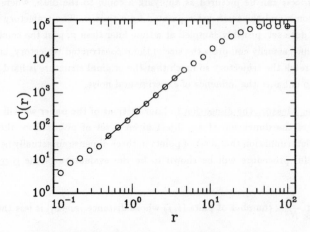

Figure 2.7 Log-log plot of $C(r)$ versus r for the Lorenz system, showing the measurement of the correlation dimension

<u>Step four: Find the embedding dimension</u> The measured correlation dimension ν can never be larger than the embedding dimension n (a sphere can't fit in a plane). This means that if n is chosen to be smaller than is needed to represent the system's behaviour, ν will

be estimated to be erroneously low (an example of this is applying a two-tooth comb to the measurement of an observable from a system whose dynamics lie in three dimensions). As n is increased ν will increase until it reaches its correct value. Increases in n beyond this point won't affect ν (a sphere placed in four dimensions is still a sphere), although the calculations are increasingly slow as the embedding dimension increases. Here then is the rule for choosing n and for distinguishing between deterministic chaos and random noise: plot the correlation dimension versus the embedding dimension; if the curve cuts off then the asymptotic value of ν is reliable, and if the curve keeps growing then the system is truly random within the limits of the experimental observation. Figure 2.8 demonstrates the test for randomness by plotting ν versus n for the Lorenz system and for random noise, showing how the measured dimension for the Lorenz system cuts off while that of the noise increases without bound.

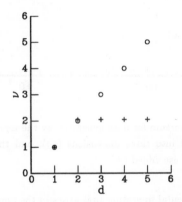

Figure 2.8 Plot of the correlation dimension versus the embedding dimension for the Lorenz system $(+)$ and random noise $(\bigcirc)$, showing the difference between deterministic chaos and random noise

This procedure even provides a clear signature for detecting a deterministic chaotic signal that is contaminated by noise. If there is noise of magnitude ϵ added to deterministic data, a plot of $C(r)$ versus r will have a knee at ϵ; the slope above the knee will give the correct dimension for the deterministic system, but the slope below the knee (due to the random noise) will be equal to the dimension of the embedding space. An example of a knee is shown in figure 2.9, which shows $C(r)$ versus r for the logistic map embedded into three dimension with and without added random noise. As the dimension of the embedding space is increased the slope above $r = \epsilon$ will remain constant, and the slope below ϵ will continue to grow. As long as the noise occurs on a scale smaller than the largest features of the dynamics it may be detected by the presence of such a knee with a slope that continues to grow as the embedding dimension is increased. This argument will also hold if there are different deterministic dynamics on different length scales; in this case there will still be a knee, but the slopes above and below it will saturate.

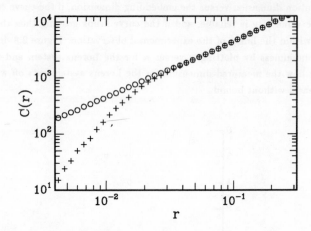

Figure 2.9 Correlation sum for data generated by the logistic map and
embedded into three dimensions ($\bigcirc$), and the same with
random noise added ($+$)

The principal experimental limitation that arises in the application of this procedure is the need to satisfy the limit $N \to \infty$. The number of points needed to satisfy this limit is a strongly increasing function of the dimension; in section 3.2 we show that the number of points needed to measure the dimension of a d-dimensional object can grow as fast as 10^d to 100^d. This will usually limit dimension estimates to values below ~ 10. As previously mentioned, the practical test of the adequacy of the size of the data set is to increase the number of points until the dimension estimate stops changing. If too few points are used in attempting to estimate the dimension of a random process, ν will fall below the line $\nu = d$ as the embedding dimension d is increased.

3. Implementation Details

The basis of this article, as described in the previous chapter and proved in appendix B, is very generally true and simple to state: time-delayed values of an observable y of a dynamical system,

$$\Phi(t) \;=\; \big\{ y(t), y(t+\tau), y(t+2\tau), \ldots, y(t+n\tau) \big\}, \tag{3.1}$$

provide a copy of the phase-space trajectory of the underlying system. However, this result offers no assistance in choosing the values of the embedding dimension (n), the time delay (τ), and the digitization precision. This section describes explicit procedures that assist in these choices. The first two subsections consider the time delay and the embedding dimension. In the last section a reformulation of the embedding procedure into signal processing terms will be presented; this provides an alternate approach for choosing the embedding parameters and includes a convenient routine for finding a linear change of coordinates that provides the "best" representation of the system's dynamics. Although these routines are useful, there is no substitute for interactively manually varying all the relevant parameters and graphically observing the results. This manual analysis is frequently all that is needed, and is a necessary prerequisite to obtaining reliable results from the routines to be described.

3.1 Choosing The Time Delay

To obtain the maximum data rate one will generally record the experimental signal at the fastest physically relevant scale, and then embed this data set by sampling it at some selected delay time. The rate that the data is digitized from the experiment must satisfy the Nyquist criterion[18]: if the maximum frequency in the signal is ω, it must be digitized at a rate 2ω. This requirement guards against the phenomenon of aliasing, where a high frequency signal sampled at a slow rate can be indistinguishable from a low frequency. To satisfy the Nyquist criterion the data must have a well defined upper frequency; this is usually achieved by low-pass filtering the data with the cutoff frequency set above the highest time scale of interest.

Once a data set satisfying the Nyquist criterion has been produced, the time delay τ must chosen. Taken's theorem tells us that the probability for making a bad choice is essentially zero; bad choices are those that, for instance, are commensurate with an aspect of the system's dynamics and hence restrict the region of phase space that is sampled. This is a mathematical statement however, not a practical one. The finite precision available to the experimentalist will set a lower limit on the precision with which phase space can be viewed, and so trajectories that are mathematically distinct may be observationally identical. For τ near zero the trajectories will lie near the diagonal of the embedding space, and as τ is increased they will expand away from the diagonal. This implies that the best choice for τ is one that maximally separates nearby trajectories in phase space. The accumulation of external noise during a measurement and the occurence of the folding in the attractor will mask the relationship between points that should be correlated by the system dynamics; this argues that the delay time should be as small as possible. The best choice for τ will be one that is short as possible relative to the rate at which the dynamical relationship between points is obscured, but as long as possible relative to the rate at which the system output varies.

There is no substitute for gaining some familiarity with the nature of the attractor by simply plotting it for a variety of values of τ, and frequently this procedure will be sufficient to find a satisfactory choice for τ. Routines such as the following one serve to find a value for τ that is in some sense optimal and to automate the selection procedure. A simple procedure to chose τ would be to look for the decay, or the first zero, in the correlation function $\langle y(\tau), y(t+\tau) \rangle$, however this is frequently unacceptable because the experimental system may have periodicities such that the correlation function oscillates for quite a few periods. We will present an algorithm that attempts to satisfy the requirements on τ by the application of information theory[19], however this is by no means a unique solution to the problem. Other procedures that try to separate trajectories may be more efficient or accurate for a given problem.

Information theory[20], associated with C.E. Shannon, provides a theory of message communication in signalling systems. As such, it can be used to quantify the relation between the time delay and the amount of information available from a measurement. Consider a meter that can indicate n different values $v_1, v_2, \ldots, v_n$, and for a given experiment let the probabilities that each of these values will be observed be $P(v_1), P(v_2), \ldots, P(v_n)$. Then, following an argument similar to that used in statistical mechanics, the <u>information entropy</u> I of the system is defined to be

$$I = -\sum_i P(v_i) \log P(v_i) , \tag{3.2}$$

where the sum is over all possible states and the logarithm is conventionally evaluated to the base 2. The function defined above is a standard, but not unique, choice for a function that satisfies the conditions that it is (1) continuous in the probabilities, (2) if all the probabilities are equal it is a monotonically increasing function of n, and (3) if a choice can be broken down into successive sub-choices, then I should be a weighted sum of the entropies of the sub-choices. I provides a measure of the amount of information that is gained by a measurement of v (how surprising the outcome is). If a measurement has a probability of either one or zero then I equals zero, and the entropy is maximized by the greatest uncertainty ($P = 1/2$, remembering the the logarithm is base 2). The entropy can be naturally defined for continuous quantities:

$$I = - \int P(s) \log P(s) ds . \tag{3.3}$$

Our goal is to choose τ so that a measurement of $y(t)$ predicts as little information as possible about a measurement of $y(t + \tau)$. For notational flexibility, define $s = y(t)$ and $q = y(t + \tau)$, and let the probabilities for measuring s be P_s, for measuring q to be P_q, and for measuring s followed by q to be P_{sq}. The <u>mutual information</u> I_m is the number of bits of q that can be predicted, given a measurement of s. It is easily found, by summing over the conditional probabilities in the entropies, to be

$$I_m = \int P_{sq}(s,q) \log \left(\frac{P_{sq}(s,q)}{P_s(s)P_q(q)} \right) ds \, dq . \tag{3.4}$$

The mutual information may be extracted from measured data. The first step is to pass through the data set and and store the values of $x(t), x(t + \tau)$ that occur in boxes in the s, q plane. Then, P_{sq} is estimated to be

$$P_{sq} = \frac{N_{sq}}{N_{\text{total}} \Delta s \Delta q} , \tag{3.5}$$

where N_{sq} is the number of entries in the s, q bin, N_{total} is the total number of points, and $\Delta s, \Delta q$ are the box sizes. P_s and P_q are given by appropriately summing P_{sq}. The box sizes $\Delta s, \Delta q$ must be chosen to be small enough to resolve variations in N_{sq}, but large enough to provide reasonable statistics for the probability estimates. Reference [19] provides a recursive algorithm that varies the box size over the s, q plane as needed to balance the trade-off between resolution and statistics.

If the mutual information is calculated for a range of values of τ, it will typically decay for increasing but small values of τ, and then reach a base value due to external noise that is no longer dependent on τ. The value of τ at which the mutual information first reaches this asymptotic value is the best choice. Choosing τ below this value will compress the volume of phase space the system explores, and choosing τ above this value will add no new information but will introduce extra external noise into the measurement.

3.2 Choosing The Embedding Dimension and Measuring The Dimension

A manifold is a generalized notion of a surface that allows it to be described without reference to an external coordinate system; a real space is the familiar one with orthogonal axes labelled by real numbers. A simple two-dimensional manifold can obviously be embedded into a two dimensional real space, however a Möebius strip (which is a two-dimensional manifold) can't fit into a two-dimensional space but rather needs three, and a Klein bottle (also two-dimensional) needs four. Whitney's embedding theorem, described in Appendix A, guarantees that $2d + 1$ dimensions will suffice to embed a d dimensional manifold into a real space. The value $2d + 1$ may be understood by recognizing that it provides enough room for 2 complete copies of the manifold, with 1 extra dimension left over. If a poorly-chosen mapping leads to crossings of the manifold with itself, we are guaranteed that in $2d + 1$ dimensions there is enough room to stretch the mapping to remove the crossings. This is a worst-case value, however, and it is certainly inconvenient to process experimental data in more dimensions than are necessary. Without any *a priori* information on d, how can the number of embedding dimensions be selected?

The answer comes from noting that for the smooth mappings that we are interested in here (diffeomorphisms), the dimension of a manifold is preserved if it is mapped into a larger space, and clearly a manifold can be no larger than its embedding space. This suggests a procedure to select the embedding dimension: if the experimental signal is embedded by time delays into spaces of increasing dimension, the dimension of the measured object will increase as the embedding dimension is increased until the target space is large enough to contain it. After this point, the calculated dimension will no longer increase with increasing embedding dimension.

The dimension measurement is almost exclusively performed with the correlation dimension ν introduced in section 2.3 or with closely-related quantities to be discussed shortly. There actually exists an infinite set of different fractal dimensions that all share the property used here of being preserved under embeddings[21]; ν is used because it is easily computed experimentally[22,23]. The calculation of most other dimensions entails covering phase space with boxes and counting the points contained in them, a procedure that is impractical in most situations[24]. The global folding of the flow in phase space attractor may bring unrelated regions of the attractor erroneously within a single box, and as more dimensions are added the collection of data sets large enough to fill in the structure of the attractor becomes prohibitive. In practice, these limits are so strong that box-counting routines are rarely useful for dimensions greater than two.

In practical measurements, the scaling of the correlation sum will be violated in the limits of small r (due to the finite precision of the data and the presence of noise) and large r (due to the finite size of the attractor). Note that if noise is present the slope for small r will not even saturate. The intermediate region between these limits is where the scaling should hold, thus the dimension is measured by fitting this intermediate region on a log-log plot of $C(r)$ versus r. For fractal attractors $C(r)$ may have an intrinsic oscillatory component[25]; in such cases the correlation dimension measures an average scaling rate.

There is a still an important simplification necessary to make the calculation of the correlation exponent practical. The evaluation of equation (2.5) is an N^2 algorithm (the number of steps is quadratic in the number of points); N^2 algorithms are infamous in computer science as becoming rapidly useless as N increases. Calculating the correlation sum for a single point, by keeping one point fixed and scanning over all the others, is only an order N algorithm. This gives is the pointwise dimension. The time to do the full double sum will increase over the time to do the single sum for one point by a factor equal to the total number of points, and as the data sets must be large for good statistics this increase in time will be enormous. It is often the case that the dimension is constant over almost all of an attractor. In practice, the pointwise dimension is usually evaluated at a few points on the attractor, and if the values obtained lie within experimental statistics of each other, they are used as an estimate. In the rarer cases where there is a large variation in ν over an attractor, performing the N^2 calculation is still not advisable because then a single value for ν does not provide a useful characterization of the structure; what is of interest here is the distribution of ν over the attractor.

The correlation sum is defined in the limit $N \to \infty$, and we have already indicated that this limit is checked by increasing the size of the data set until the dimension estimate stops changing. We now provide a rough estimate of the number of points that are needed for random noise. Assume that the points are uniformly distributed in a box of side R in d dimensions. The maximum value of r for which the scaling of $C(r)$ will hold is the size of box ($r_{max} \simeq R$), and the minimum value of r is the nearest-neighbor distance between the points ($r_{min} \simeq R/N^{1/d}$). If one wants to fit α decades of r to obtain the correlation dimension, then

$$\begin{aligned}
\alpha &= \log \frac{r_{max}}{r_{rmin}} \\
&= \frac{1}{d} \log N \\
\Rightarrow N &= 10^{\alpha d} .
\end{aligned} \qquad (3.6)$$

As α is typically 1 or 2, this gives the previously quoted result that the number of points needed will be in the range 10^d to 100^d. Note that this exponential growth is quite a severe limitation, rapidly exceeding available computer capacity, but also note that this calculation is rough and is for the case of uniformly distributed random noise. It may be possible to obtain valid dimension estimates with fewer points than suggested here.

To maximize the available computer resources some care should be given to the design of the algorithms and data structures used to implement the dimension calculation. The calculation of the correlation dimension can be vectorized very efficiently for use in an array processor/supercomputer environment; for example the calculations in this article were performed in vector mode on an IBM 3090-600E. On this system the computation of $C(r)$ for a 10^6 point data set takes on the order of 15 seconds.

The preceeding routine determines the dimension necessary for an embedding; the danger associated with too small an embedding dimension is the accidental crossing of trajectories. It is important to recognize that, depending on the questions being asked, such crossings may in fact be acceptable. If the probes of the structure being used are either smart enough or stupid enough to ignore these crossings, savings in computational time and data storage can be realized.

3.3 Singular Systems Analysis

Broomhead and King[26,27] have formulated the embedding procedure in the language of signal processing theory. They apply a singular systems analysis to the reconstruction of an attractor that provides an alternate approach for selecting the embedding parameters to those already covered. A very useful result from their analysis is a simple algorithm for finding a linear change of coordinates that provides the "best" plot of the attractor.

If the observable measured in the ith time step is v_i, then the associated point in an n dimensional embedding space is $x_i = \{v_i, v_{i+1}, \ldots, v_{i+(n-1)}\}$. Let N be the total number of embedded points measured. Broomhead and King first introduce a matrix with rows consisting of the embedded points :

$$ X \;=\; \frac{1}{\sqrt{N}} \begin{bmatrix} x_1^T \\ x_2^T \\ \vdots \\ x_N^T \end{bmatrix}, \qquad (3.7) $$

where x^T is the transpose of the column vector x. Multiplying X on the left by a vector $s^T \in \mathbb{R}^N$ will give some linear combination of all the points on the trajectory. In particular, one can introduce a set $\{c_i\}$ of orthonormal vectors (called singular vectors) in $\mathbb{R}^n$ with corresponding weights $\{\sigma_i\}$ that span the space of all linear combinations of points of the attractor. Let $\{s_i\}$ be the corresponding weighting functions that give these orthonormal vectors:

$$s_i^T X = \sigma_i c_i^T .\tag{3.8}$$

Acting on the right of equation 3.8 with its transpose, and using the orthogonality of the c_i's, it follows that

$$XX^T s_i = \sigma_i^2 s_i .\tag{3.9}$$

XX^T is a real symmetric matrix, hence its eigenvectors s_i must be orthonormal. Taking the transpose of equation (3.8), multiplying on the left by X, and using equation 3.9, it follows that

$$X c_i = \sigma_i s_i .\tag{3.10}$$

From this, just as with equation (3.8), it follows that

$$\Xi c_i = \sigma_i^2 c_i ,\tag{3.11}$$

where $\Xi \equiv X^T X$. Multiplying equation 3.11 on the left by c_i^T and using the orthonormality gives

$$(X c_i)^T (X c_i) = \sigma_i^2 .\tag{3.12}$$

Recognizing that $X c_i$ is the set of projections of the points of the attractor onto c_i, we see that the σ_i may be interpreted as the mean square values of the projections of the attractor onto the orthonormal set $\{c_i\}$. This means that the set $\{c_i \sigma_i\}$ may be viewed as defining the principal axes of an n-dimensional ellipse that, on the average, describes the bounds of the attractor. The c_i's provide the promised coordinates for plotting the attractor. The coordinate transformation $x^T c_1, x^T c_2, \ldots, x^T c_n$ will produce a picture of the attractor with the directions of greatest span of the attractor lined up with the axes of the coordinate system. This is a linear transformation; it would be useful to extend this analysis to find a nonlinear transformation that is better able to follow the structure of the attractor.

Writing Ξ out in detail shows that it may be viewed as the covariance of the components of the attractor:

$$\Xi = \frac{1}{N}\begin{bmatrix} \sum_{i=1}^{N} v_i v_i & \cdots & \sum_{i=1}^{N} v_i v_{i+(n-1)} \\ \vdots & & \vdots \\ \sum_{i=1}^{N} v_{i+(n-1)} v_i & \cdots & \sum_{i=1}^{N} v_{i+(n-1)} v_{i+(n-1)} \end{bmatrix} .\tag{3.13}$$

As Ξ is an $n \times n$ matrix, and the embedding dimension n is usually fairly small, equation (3.11) may be solved to find the c_i's.

Broomhead and King suggest a method to use this theory to find the delay time and the embedding dimension. Given a time τ_p that is the period of the highest frequency present in the signal, they choose the delay time τ and the embedding dimension such that $n\tau = \tau_p$. n is chosen by realizing that as an attractor is placed in higher and higher dimensional spaces it will still need the same number of basis vectors for the n-ellipsoid that describes it. As n is initially increased, new eigenvalues will appear as the attractor reveals its intrinsic structure. Once n is large enough, however, the set of eigenvalues describing the attractor will not change, and the new eigenvalues that appear will be very small ones due to either experimental noise or numerical noise from the routines that find the eigenvalues.

4. Analyzing the Dynamics

The ability to experimentally recover a low-dimensional attractor if it underlies a system's dynamics, however remarkable, is still of no use if the experimentalist cannot understand its structure. For all but the most straightforward systems, a simple visual inspection will not be sensitive enough to characterize the dynamics. In this section we will give an introduction to some of the qualitative and quantitative measures of how an attractor "looks". This pursuit will necessarily draw on a number of mathematical fields. This is an actively evolving area of current research; a recent review is given in reference [28], a useful summary of the variety of ways to measure a dimension and the divergence of trajectories is reference [29] and the current literature is often published in the journal Physica D.

We start by considering how a high–dimensional system can behave like a low– dimenisonal one, a requirement for this form of analysis. We then introduce and relate the concepts of strange attractors, chaos, diverging trajectories, information production, the variety of definitions of the dimension, and the thermodynamic formalism.

4.1 Flows, Maps, and Chaos

A hydrodynamic system has infinitely many degrees of freedom, yet just beyond the onset of convection it can act as a very low dimensional system. The Center Manifold theorem, stated in Appendix B, helps to explain how this reduction can occur. It guarantees that, locally, the linearization of a dynamical system accurately reflects the full nonlinear dynamics. In the local linearization there will be stable, unstable, and neutrally stable degrees of freedom. It will often happen that the unstable degrees of freedom will diverge until they reach a bound, the stable degrees will exponentially vanish, and so the dynamics will collapse down onto the neutral degrees of freedom ("center manifold"). The center manifold, if it exists, will frequently have far fewer dimensions than the full system, and so the system will behave as if it had only this reduced number of degrees of freedom. A more complicated possibility is for the stable and unstable manifolds to cross; in this case the dynamics still occur on a reduced manifold of the full phase space, and we will soon see that this case is intimately related to chaos.

The Lorenz attractor comes from a three-dimensional system, yet its correlation dimension of 2.05 is distinctly less than three. This difference suggests that a full three-dimensional plot of the evolution of the Lorenz system contains contains redundant information about the flow. The Poincaré section is a standard technique, generally applied to near-periodic systems, that is used to dissect the flow and produce a more lucid representation. The section is formed by taking a surface transverse to the flow, and then plotting the intersections of the flow with the surface. This is shown in figure 4.1.

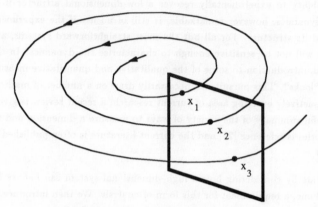

Figure 4.1 The Poincaré section

The Poincaré section replaces the continuous time dynamical system with a map (between crossings of the cut-surface). This map may have a much simpler structure than was obvious from the original system, and serve as a useful step and guide for further analysis. Figure 4.2 shows the intersection of the Lorenz attractor with the plane $z = 27$; it looks one dimensional. From this section one could, for example, analyze the scaling structure of the cross section, or find an effective one dimensional map underlying the dynamics by parameterizing the location of the crossings along the cross section. Note that, because phase space trajectories for differential equations cannot cross, the effective one dimensional map must actually be infinitely interleaved in order to contain the dynamics of all trajectories for all time. Nevertheless, this fractal structure occurs on a very fine scale, and for many questions the dynamics is adequately represented by the effective map.

In various places we have touched on the ideas of deterministic chaos and fractal attractors; we will now consider how such behaviour can arise. Let us start by returning to the logistic map with $\lambda = 4$. The map is shown at the left of figure 4.3. The set of drawings next to the map show the evolution of the interval $[0, 1]$ under the action of the map. In the first stage, $0 \mapsto 0, 1/2 \mapsto 1$, and $1 \mapsto 0$. This retracing back to zero is shown separated vertically for clarity. The interval has been stretched to twice its length and folded over by the map. Applying the map again, the folded interval is streched and folded once more.

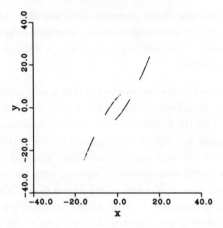

Figure 4.2 The intersection of the Lorenz attractor with a plane at $z=27$

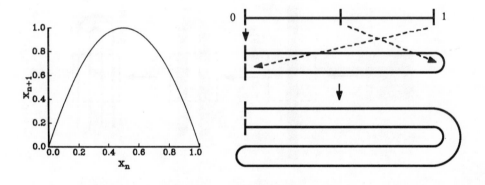

Figure 4.3 The stretching and folding associated with the logistic map

Now consider the fate of closely spaced initial conditions. A point will hop from sheet to sheet as the map is applied, and the sheet the point is on after n iterations depends on whether the n digit of its fractional binary representation is one or zero (the two sheets at the first iteration are the images of points between 0 and 1/2, and between 1/2 and 1; the four sheets at the next iteration correspond to points between 0 and 1/4, 1/4 and 1/2, 1/3 and 3/4, 3/4 and 1, and so forth for further iterations). No matter how close two points are, at some point their binary representations will differ, and then the shredding action of the map will irrevocably separate them. Seen in this light, it is not surprising that the solution is chaotic. This sensitive dependence on initial conditions is often used as a working definition of chaos.

The mechanism present in the previous example of a map underlies chaotic dynamics in differential equations. Smale introduced the topological notion of stretching and folding to dynamical systems with the Smale horseshoe. In two dimensions this takes the form of a box that is elongated and folded; the repeated folding action on the box leads to a partitioning of the box into increasingly fine strips. Running the map forwards and backwards produces a fractal arrangement of squares, with the evolution of a point at each stage of the folding given by the part of the fractal that it is on at that level. Once again, there is a sensitive dependence on the initial condition (two trajectories starting arbitrarily close will eventually get arbitrarily far apart, up to the limit of the size of the attractor). The construction of a Smale horseshoe is shown in figure 4.4.

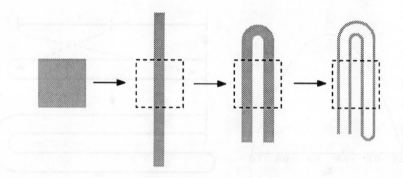

Figure 4.4 The Smale Horseshoe

Such a horseshoe can occur when stable and unstable manifolds cross; the Smale-Birkhoff Homoclinic Theorem, described in Appendix B, elaborates on when this will occur. Such a crossing produces a "strange attractor", with ensuing chaos. The intimate relation between chaos, sensitive dependence on initial conditions, unpredictability, horseshoes, and strange attractors may now be seen.

4.2 Information, Entropy and Correlation

In section 3.1 the information entropy was used to characterize the information content of a signal. This concept is more generally useful in characterizing chaotic systems, because the stretching process associated with a strange attractor reveals ever-finer information about the initial conditions which can be described in terms of information gain. We will reintroduce the definition of the information entropy for an attractor, find that this leads to another scheme for calculating the fractal dimension, quantify the information production rate, and consider the experimental measurement of these quantities. See reference [30] for further information.

Consider a dynamical system, and observe in particular a region C of phase space. Let $\mu(x, C)$ be the fraction of time that a system spends in a region C after starting at x. If μ is the same for almost all x, then μ is said to be the <u>natural measure</u>. μ is a measure in the sense of measure theory, and the following discussion will hold for measures other than the natural one; however, we will only be interested in the natural measure. μ is an <u>invariant measure</u> because the action of the flow will leave it unchanged (this is just a statement of the conservation of probability). Now partition phase space with boxes of size r and index the boxes with b; the probability for the system to be in a box is then $P_b = \mu(b)$. As before, the information entropy of the partition is defined by

$$I(\{b\}) = -\sum_b P_b \log(P_b) . \tag{4.1}$$

I can be interpreted as the average information gain from making a measurement of the system to an accuracy of r, or as the average amount of information required to specify the system's state to an accuracy of r.

The information will increase as r is decreased, and this dependence can be explicitly calculated for the simple case in which the probabilities for the system to be in the occupied boxes are all equal. The number of occupied boxes (the boxes that make up the cover of the attractor), N_c, is given by the capacity as $N_c = Ar^{-d_c}$, where A is the proportionality constant. The normalized probability for a point to lie in a given box for this case of uniform probability is just N_c^{-1}. The information can be calculated from this probability:

$$I = -\sum_{b=1}^{N_c} P_b \log(P_b) = -\log \frac{1}{N_c} = -\log Ar^{d_c} = -\log A - d_c \log r . \tag{4.2}$$

For all attractors studied so far the dependence of I on r is indeed logarithmic, however if the probabilities for the boxes are not equal then the scaling exponent is not equal to the capacity. The general form $I = -\sigma_0 - \sigma \log r$ motivates the definition of a new dimension, the <u>information dimension</u> σ, by

$$\sigma = \lim_{r \to 0} \left[-\frac{\sigma_0}{\log r} - \frac{I}{\log r} \right] . \tag{4.3}$$

Once again, the constant σ_0 is not needed in the infinitesimal limit but affects the formula away from this limit. For manifolds with a uniform probability density the information and capacity dimensions will agree, but if the distribution is nonuniform then the information dimension will be less than the capacity (on the average a measurement will provide less information about the system's state).

The information may also be related to the correlation sum introduced in section 2.3. Remember that the sum is defined for a set of N_p points by

$$C(r) = \lim_{N_p \to \infty} \frac{1}{N_p^2} \{\text{the number of pairs of points } (x_i, x_j) \text{ with } |x_i - x_j| < r\} . \tag{4.4}$$

In this formula the distance between two points may be defined to be the standard Euclidean distance, but it may also be (more simply) defined to be the maximum value of the differences between each of the coordinates of the points (this is called the <u>max norm</u>).

It will be helpful later to break the correlation sum into pieces. Define $C_i(r)$ to be a local version of $C(r)$:

$$C_i(r) = \frac{1}{N_p} \{ \text{ number of points } x_j \text{ with } |x_i - x_j| < r\} . \tag{4.5}$$

Inspecting this formula leads to the conclusion that $C(r)$ is that average over the points of $C_i(r)$:

$$C(r) = \frac{1}{N_p} \sum_{i=1}^{N_p} C_i(r) . \tag{4.6}$$

There is a very important approximation that relates the correlation sums and the probabilities P_b. Let $P_{b(i)}$ be the probability that a random point is in the box containing the point x_i, let the size of the boxes be r, and let the box x_i is in contain $N_{b(i)}$ points. $C_i(r)$ is the number of points within r of x_i divided by the total number of points N_p; the approximation is that

$$C_i(r) \approx \frac{N_{b(i)}}{N_p} = P_{b(i)} . \tag{4.7}$$

The idea behind this approximation is that most points in $b(i)$ will be within r of x_i, and although some will be counted that are further away others will be ignored that are close enough but are in neighbouring boxes. The error made in making this approximation will only be a factor of order unity, and will certainly be no larger than the number of nearest neighbor boxes. Not only is this error small, but for a reasonable attractor the correction term should not depend strongly on r and hence the scaling law should be essentially unchanged. The errors incurred in this approximation can be checked directly for a given attractor by calculating $C_i(r)$ and $P_{b(i)}$ directly; the difference is almost always smaller than the experimental and statistical sources of noise.

This approximation also gives a simple relation for $C(r)$. In the following derivation the notation $i \in b$ in a sum will mean that the sum is over all points x_i that are contained in box b. The derivation proceeds by substituting $P_{b(i)}$ for $C_i(r)$ and replacing the sum over points with a sum over boxes and a sum over points within a given box:

$$
\begin{aligned}
C(r) &= \frac{1}{N_p} \sum_{i=1}^{N_p} C_i(r) \approx \frac{1}{N_p} \sum_{i=1}^{N_p} P_{b(i)} = \frac{1}{N_p} \sum_b \sum_{i \in b} P_b \\
&= \frac{1}{N_p} \sum_b N_b P_b = \sum_b P_b^2 .
\end{aligned}
\tag{4.8}
$$

The correlation sum is just the sum over the covering boxes of the probability that a point is in each box. This might have been inferred directly by an argument similar to that used for $C_i(r)$ by realizing that N_b^2 is the number of pairs of points in the box b.

The information can be evaluated for an experimental system by covering the attractor with boxes and calculating the probabilities, however the preceeding relations allow it to be calculated directly from the $C_i(r)$'s. This may be seen by once again replacing a sum over points with a sum over boxes and a sum over points within each box:

$$
\begin{aligned}
I &= -\sum_b P_b \log P_b = -\sum_b \frac{N_b}{N_p} \log P_b = -\frac{1}{N_p} \sum_b \sum_{i \in b} \log P_b \\
&= -\frac{1}{N_p} \sum_i \log P_{b(i)} = \frac{-1}{N_p} \sum_i \log C_i(r) .
\end{aligned}
\tag{4.9}
$$

The information is the average of the logs of the $C_i(r)$. As with the calculation of the correlation dimension, it will frequently be possible to replace the sum over all the $C_i(r)$ with N_p times a single one because the scaling behaviour is reasonably uniform over the attractor. This simplification reduces an unwieldy N^2 steps calculation to a useful N one; the approximation may be checked by comparing the values of $C_i(r)$ for a few x_i's.

The preceeding discussion has referred to the information contained in a single measurement of the system's state. A series of state measurements performed on the evolving system will generate a series of values that can also be analyzed in information theoretic terms. The stretching associated with a chaotic attractor will continually provide new information by revealing finer details of the initial conditions, while a non-strange attractor will not continue to produce new information, hence the rate of growth of the information contained in a series of measurements performed on a system is a useful characterization of chaos. Once again, cover the attractor with boxes of side length r, and let $b(i)$ be the index of the box the system is in at the ith measurement. n measurements of the system's state, spaced a time τ apart, can be expressed by the sequence of the b's. Letting φ_τ be the map that advances the flow by a time τ, the symbol sequence (string of box indices) generated by the map will be $S_n(i) = \{b(i), \varphi_\tau(b(i)), \ldots, \varphi^{(n-1)\tau}(b(n))\}$. S_n without an index will refer to an arbitrary symbol sequence.

The probability for a sequence to occur, P_{S_n}, leads to a definition for the information of the sequence of n symbols:

$$I_n = -\sum_{S_n} P_{S_n} \log P_{S_n} , \qquad (4.10)$$

where the sum is over all possible symbol sequences. We are interested in the information production rate

$$\frac{\delta I}{\delta n} = \frac{I_{n+1} - I_n}{\delta n} , \qquad (4.11)$$

however the information production rate is usually measured as the average rate in the limit of time going to infinity in order to average out variations:

$$K = -\lim_{\epsilon \to 0} \lim_{\tau \to 0} \lim_{n \to \infty} \frac{1}{n\tau} \sum_{S_n} P_{S_n} \log P_{S_n} . \qquad (4.12)$$

K, the information production rate, is called the <u>Kolmogorov entropy</u>, the <u>metric entropy</u>, or simply the entropy. The limits of $\epsilon, \delta t \to 0$ are in fact not needed to define the information rate for most fractals and will be left off in the rest of this section.

Returning to the simple case of equal probabilities for all symbol sequences leads to the definition of the <u>topological entropy</u> h. If each of the N_c occupied boxes are equally probable to occur, then the probability of a given sequence of n symbols is $N_c^{-n} = Ar^{d_c n}$, and the topological entropy is

$$h = -\lim_{n \to \infty} \frac{1}{n\tau} \sum_{b=1}^{N_c} \frac{1}{N_c^n} \log \frac{1}{N_c^n} = \frac{1}{n\,\tau} \log \frac{1}{N_c^n} = \frac{d_c}{\tau} \log r . \qquad (4.13)$$

The topological entropy is the logarithm of the number of distinguishable symbol sequences, and is a measure of the information production rate available to an observer who doesn't know the probabilities for the messages to occur.

Just as the information is simply related to the correlation sums, so is the information production rate. The correlation sums are extended to count the number of points whose trajectories stay within r of each other:

$$C^n(r) = \lim_{N_p \to \infty} \frac{1}{N_p^2} \{ \text{number of pairs of points } (x_i, x_j) \text{ with}$$

$$|x_i - x_j|, |\varphi_\tau(x_i) - \varphi_\tau(x_j)|, \ldots, |\varphi^{(n-1)\tau}(x_i) - \varphi^{(n-1)\tau}(x_j)| < r \}$$

$$C_i^n(r) = \lim_{N_p \to \infty} \frac{1}{N_p^2} \{ \text{number of points } x_j \text{ with} \tag{4.14}$$

$$|x_i - x_j|, |\varphi_\tau(x_i) - \varphi_\tau(x_j)|, \ldots, |\varphi^{(n-1)\tau}(x_i) - \varphi^{(n-1)\tau}(x_j)| < r \}$$

$$C^n(r) = \frac{1}{N_p} \sum_{i=1}^{N_p} C_i^n(r) \, .$$

Following arguments identical to those used previously, the correlation sums may be approximately related to the probabilities for sequences to occur. Remembering that $P_{S_n(i)}$ is the probability for a measurement of n values to yield the sequence generated by x_i, we have

$$C_i^n = \frac{N_{S_n(i)}}{N_p} = P_{S_n(i)}$$

$$C^n = \frac{1}{N_p} \sum_{i=1}^{N_p} P_{S_n(i)} = \sum_{S_n} P_{S_n}^2 \, . \tag{4.15}$$

These generalized correlation sums provides a means to experimentally measure the Kolmogorov entropy. Consider the following quantity:

$$\Phi^n \equiv \frac{1}{N_p} \sum_i \log C_i^n(r) \, . \tag{4.16}$$

Rewriting this sum shows that Φ is simply related to the entropy:

$$\Phi^n = \frac{1}{N_p} \sum_i \log P_{S_n(i)} = \frac{1}{N_p} \sum_{S_n} \sum_{i \in S_n} \log P_{S_n(i)}$$

$$= \frac{1}{N_p} \sum_{S_n} N_{S_n} \log P_{S_n} = \sum_{S_n} P_{S_n} \log P_{S_n} \, . \tag{4.17}$$

Therefore

$$K = -\lim_{n \to \infty} \frac{1}{n\tau} \Phi^n \, , \tag{4.18}$$

or alternately

$$K = \lim_{n \to \infty} \frac{1}{\tau} (\Phi^{n+1} - \Phi^n) \, . \tag{4.19}$$

As always, these formulas are usually evaluated, subject to suitable checks, by replacing the sum over all points with N_p times just one of the C_i^n's.

We now introduce a useful generalization of the entropy that will relate the topological entropy to the Kolmogorov entropy, and introduce a new family of entropies that will include one that is particularly simply related to C^n. The order-q Renyi entropy is defined by

$$K_q = -\lim_{n\to\infty} \frac{1}{n\tau} \frac{1}{q-1} \log \sum_{S_n} P_{S_n}^q . \tag{4.20}$$

The value of K_q as $q \to 0$ gives the topological entropy because $P_{S_n}^0$ will be zero if any box in the sequence is empty, and will be one if all the boxes in the sequence are populated (remember that the topological entropy is the log of the number of possible sequences). The limit of K_q as $q \to 1$ is just K; this may be seen by the following expansion:

$$\frac{1}{q-1} \log \sum_{S_n} P_{S_n}^q = \frac{1}{q-1} \log \sum_{S_n} P_{S_n} e^{(q-1)\log P_{S_n}}$$

$$\approx \frac{1}{q-1} \log \sum_{S_n} P_{S_n} \left(1 + (q-1)\log P_{S_n}\right)$$

$$= \frac{1}{q-1} \log \sum_{S_n} P_{S_n} + (q-1)P_{S_n} \log P_{S_n} \tag{4.21}$$

$$= \frac{1}{q-1} \log \left(1 + \sum_{S_n}(q-1)P_{S_n} \log P_{S_n}\right)$$

$$\approx \sum_{S_n} P_{S_n} \log P_{S_n} .$$

K_2 is the entropy that can be easily calculated. Referring back to equation 6.17 and the definition of K_2, it immediately follows that

$$K_2 = -\lim_{n\to\infty} \frac{1}{n\tau} C^n$$

or

$$K_2 = -\lim_{n\to\infty} \frac{1}{\tau} \log \frac{C^n}{C^{n+1}} . \tag{4.22}$$

From the definition of the q-th order entropies it may be seen that $K_i \geq K_j$ if $i \leq j$, and so if K_2 is measured to be nonzero then K must also be. This means that a measurement of K_2 will be sufficient to determine whether or not the system continually generates new information.

4.3 Lyapunov Exponents

The divergence of nearby trajectories underlies the sensitive dependence on initial conditions in a strange attractor. The Lyapunov exponents quantify the relationship between nearby trajectories, and they will provide another indicator of the presence of chaos that will be closely related to those already examined.

Consider the growth of an arbitrary perturbation δx about a point x. The growth rate of this perturbation will be exponential, with the rates locally given by the eigenvalues of the Jacobian matrix:

$$\frac{dx}{dt} = f(x)$$
$$\frac{d}{dt}(x + \delta x) = f(x + \delta x)$$
$$\frac{df}{dt} + \frac{d}{dt}\delta x \approx f(x) + (D_x f)\delta x$$

(4.23)

Each of the eigenvectors of the Jacobian matrix $D_x f$ will locally grow at a rate $e^{\lambda_i t}$, where the λ_i's are the eigenvalues of the matrix. Taking the flow to be φ_t, the growth of δx is given by the Jacobian, T_x^t, of the flow:

$$\delta x(t) = T_x^t \delta x(0) = (D_x \varphi_t)\delta x(0) .$$

(4.24)

The <u>Lyapunov exponents</u> are given by the asymptotic growth rate of the eigenvalues of T_x^t:

$$\{\lambda_i\} = \lim_{t \to \infty} \frac{1}{t}\|T_x^t\| .$$

(4.25)

The existence of this limit is assured by the multiplicative ergodic theorem of Oseledec, and Grassberger and Procaccia[31] provide a detailed model for the divergence of nearby trajectories in terms of stochastic fluctuations about a mean exponential rate.

This definition is easily extended to discrete time systems, where the flow φ_t is replaced by the nth iterate of the map φ^n. T_x^n is defined to be the Jacobian of φ^n, and the chain rule alows this to be written as the product of the derivatives of the map at the points along the trajectory:

$$T_x^n = D_{x_0}\varphi^n$$
$$= D_{x_0}(\varphi \circ \ldots \circ \varphi)$$
$$= (D_{x_{n-1}}\varphi)\ldots(D_{x_1}\varphi)(D_{x_0}\varphi) .$$

(4.26)

A dynamical system may have some positive exponents, corresponding to directions associated with chaotic stretching, some negative ones, corresponding to directions of contraction, and some zero ones, corresponding to directions in which trajectories at most converge or diverge at a rate slower than exponential; the distribution of the signs of the exponents is itself a useful description of the nature of the dynamics. The positive exponents are traditionally ordered as $\lambda_1 \geq \lambda_2 \geq, \ldots, \geq \lambda_n$.

The Lyapunov exponents may be equivalently defined as the growth rate of an infinitesimal volume element. Consider an infinitesimal n-ball surrounding a point; as the point evolves under the flow the n-ball will move with it, and as it evolves it will be stretched into an n-ellipsoid. Because the ball starts out infinitesimally small it will remain infinitesimal, and so the growth will once again be given by the linearized system. Calling the principal axes of the ellipsoid p_i, the Lyapunov exponents are given by

$$\lambda_i = \lim_{t \to \infty} \frac{1}{t} \log \frac{p_i(t)}{p_i(o)} \ . \tag{4.27}$$

There is a third equivalent definition. Consider an arbitrary line. The line grows at a rate given by the largest exponent, because the growth rate of the component of the line along the eigendirection associated with the largest exponent will rapidly overwhelm the other directions ($e^{\lambda_2 t}/e^{\lambda_1 t} \to 0$ as $t \to \infty$). Similarly, an arbitrary area will grow as $e^{(\lambda_1 + \lambda_2)t}$, and a d dimensional object grows as the sum of the first d eigenvalues. This hierarchy of growth rates serves to define the exponents once again.

The sum of all the positive exponents measures the growth rate of the total expanding volume. This expansion of volume is just what the information production associated with a positive Kolmogorov entropy refers to, hence it should not be surprising that the entropy is just the sum of the positive exponents:

$$K = \sum_i \lambda_i^+ \ . \tag{4.28}$$

This is called the <u>Pesin identity</u>, and is true (using terms introduced in the appendices) if the probability measure is invariant with respect to the flow (which is a diffeomorphism), and if the measure is smooth along the unstable manifolds (with respect to Lebesgue measure). For the unphysical cases that do not satisfy these requirements, the entropy will be less than the sum of the positive eigenvalues.

The relation between the entropy and the information dimension σ suggests that the Lyapunov exponents should be related to the information dimension, and this is indeed the case. There is a conjecture, due to Kaplan and Yorke, that

$$\sigma = j + \frac{\sum_{i=1}^j \lambda_i}{|\lambda_{j+1}|} \ , \tag{4.29}$$

where j is defined by the condition that

$$\sum_{i=1}^{j} \lambda_i > 0 \ \text{ and } \ \sum_{i=1}^{j+1} \lambda_i < 0 \,, \tag{4.30}$$

and the exponents are indexed from largest to smallest. This conjecture is known to be violated in some cases, however it has been found to hold in all physically plausible systems tested. The bounds of its validity are an open question.

A Poincaré section, when valid, captures the dynamics of a system by converting the flow into a discrete map. Letting the mean time between crossings of the surface of section be $\langle \tau \rangle$, the time used in evaluating the limit of T_x^n will equal the time t rescaled by $\langle \tau \rangle$, and so the Lyapunov exponents may be calculated directly from the section:

$$\lambda_i(\text{section}) \ = \ \langle \tau \rangle \lambda_i(\text{flow}) \,. \tag{4.31}$$

If a flow is known analytically, the Lyapunov exponents can be calculated directly. If the governing equations are known, but are not analytically soluble, the Lyapunov exponents can still be calculated because the linearized system is known and can be applied to small perturbations about a given numerical solution. Information about the exponents can in fact also be deduced from experimental measurements, however the lack of knowledge of the governing equations makes this a more difficult problem. We will describe two approaches to extracting the exponents: following the evolution of observed trajectories that are close to each other, and fitting the data to find T_x^t. Because of the exponentially growing dominance of the largest eigenvalue and the exponential decay associated with negative exponents, these procedures will in general only provide one or a few of the largest exponents. For many purposes this limitation is not severe, because the detection of a positive exponent is sufficient evidence for the presence of a strange attractor. The description of both routines assumes that an attractor has already been measured according to the procedures of section 3.

The third definition of the exponents (the growth rates of lines, planes, ...) provides the motivation for the method of following trajectories[32]. If the governing equations were known this definition could be used by following the evolution of a set of small basis vectors about a central point. The growth rate of a single vector gives λ_1, the growth rate of the area spanned by two vectors gives $\lambda_1 + \lambda_2$, and so forth. This calculation may be done easily because the linearized system applies to the neighborhood of the central point; the only subtlety is that the exponential dominance of the largest eigenvalues requires that the basis vectors be re-orthogonalized periodically in order to remain resolvable by the finite precision of the computer. With experimental data the trajectories in the neighborhood of a point cannot be followed arbitrarily, because the initial conditions of the experiment cannot usually be chosen to such a precision. If the data set is large enough, however, there should be enough nearby trajectories for the definition to still be applicable.

The first step in measuring λ_1 is to pick a point to serve as the central one to be followed. The data set is then scanned to find a second point on a neighbouring trajectory. The distance between these points is followed as the central point is stepped through the data set, and λ_1 is given by the exponential growth rate of this vector. The growth of the spacing between the two points must be monitored to guard against two pitfalls. If the distance grows anomolously quickly, it probably means that the two points do not lie on the same region of the attractor, and their initial nearness is just an artifact of the way the attractor is folded. In this case the routine must back up and look for a new second point. If the points are acceptable, the spacing will nevertheless eventually grow to such a large size that the folding of the attractor will invalidate the estimate. In this case a new second point must be chosen, and it must not only be close to the central point, but it also should lie close to the direction of the old second point. The latter constraint is applied in order that the calculation approximates the limit $t \to \infty$; if the second point is replaced by one not lying along the same direction then the exponent is being calculated by averaging a finite time evolution. This may possibly still give the same experimental answer, but such agreement is not guaranteed and should be verified if possible.

$\lambda_1 + \lambda_2$ is measured in a similar fashion. Once a central point is chosen, two neighbours are picked to define an area. The evolution of these points are followed, subject to the same checks as before that they lie on the same region of the attractor. The points will need to be replaced once the spacing grows too large, or when the dominance of the largest eigenvalue drives the two vectors too close together. The replacement points are now chosen so that the area they define has a normal aligned as close as possible to that of the preceeding area. This procedure can be extended to as many n-tuples of points as the resolution of the data set allows.

Eckmann and Ruelle provide an alternate method that finds the Lyapunov exponents by finding T_x^t from a nonlinear least-squares fit to the data[28]. Their procedure starts once again with the selection of a point x_i to follow, and also a time τ must be chosen. The evolution of the data will be replaced with a discrete-time map with a time step of τ; a good choice for τ is the time it takes for two points to separate sufficiently far that they must be replaced in order to remain in the infinitesimal limit. Let n be the number of time steps of the data sampling rate corresponding to a time step of τ. A final parameter to be selected is r, which defines the length scale beyond which the folding of the attractor affects the scaling. r can, for example, be taken to be the cut-off of the scaling region on a $C(r)$ versus r plot.

The next step, once these parameters are chosen, is to find all the points x_j such that $|x_i - x_j| \leq r$ (the points start in the same small neighborhood) and $|x_{i+n} - x_{j+n}| \leq r$ (the points remain within the neighborhood after a time τ). Since the separation between x_i and x_j is small enough for the linearized equations to hold, they can be related by

$$x_{i+n} - x_{j+n} = T^\tau_{x_i}(x_i - x_j) . \tag{4.32}$$

$T^\tau_{x_j}$ is found from a nonlinear least-squares fit of equation 4.32 to all the x_j satisfying the above constraints. Once $T^\tau_{x_i}$ is found the calculation is repeated for x_{i+n}, and this process is continued until the data set is traversed. The set of matrices $\{T^\tau\}$ can then be used to find the Lyapunov exponents according to equation (4.25) For this algorithm the limitation on the number of exponents that can be found stems from the numerical uncertainty in routines to find eigenvalues. Because the sum of the eigenvalues is usually not very much greater than zero these matrices will usually be nearly singular, and this means that standard procedures for finding eigenvalues may fail. There are a number of routines available for nearly singular matrices; Eckmann and Ruelle propose a decomposition of T into upper-triangular matrices that results in the eigenvalues appearing on the diagonal sorted by size, and reference [33] has a good introduction to handling nearly singular matrices.

This method of finding the Lyapunov exponents offers the advantages over the method of following trajectories that the algorithm does not need to be nearly so sophisticated in its selection of acceptable points to use, that it uses all the possible points in the neighborhood of the central point on the first pass, and that it doesn't require the approximation involved in replacing a second point with a nearer one that lies as closely as possible in the same direction. There has not yet been enough experimentation to determine if these advantages are realized in practice.

4.4 Generalized Dimension

So far we have used the capacity d_c, the information dimension σ, and the correlation dimension ν. To help organize this abundance of definitions, one would like a periodic table of the definitions. Such an organization was provided by Hentschel and Procaccia[21]. Following arguments similar to those used with the Renyi entropies, they introduce a generalized dimension D_q that is defined for all q and gives the dimensions that we have studied for $q = 0, 1, 2$. This framework will be quite helpful in unifying the various dimensions, however the correlation dimension will still be the only one that can be efficiently calculated in most cases.

Start by defining a new correlation sum C_n that measures the higher order correlations (pairs, triples, ...):

$$C_n(r) = \lim_{N_p \to \infty} \frac{1}{N_p} \{\text{the number of } n-\text{tuples of points } (x_i, x_2, \ldots, x_n)$$

$$\text{whose distances } |x_i - x_j| \text{ are less than } r \text{ for all } i, j\} . \qquad (4.33)$$

The relation between $C_n(r)$ and the probability that a box is occupied, P_b, is found by noting that N_b^n is the number of n-tuples in the box:

$$C_n(r) = \sum_b P_b^n , \qquad (4.34)$$

and the scaling of $C_n(r)$ introduces a generalized correlation dimension:

$$C_n(r) \sim r^{\nu_n} . \qquad (4.35)$$

$C_n(r)$ provides a means to calculate the dimension ν_n for any integer n. To introduce a dimension with non-integer n's, a form reminiscent to the Renyi entropy is used:

$$D_q = \lim_{r \to 0} \frac{1}{q-1} \frac{\log \sum_b P_b^q}{\log r} . \qquad (4.36)$$

Following the same arguments used in relating the Renyi entropies to h, K and K_2, it may be seen that D_q for $q = 0, 1, 2$ gives the capacity, the information dimension, and the correlation dimension:

$$\lim_{q \to 0} D_q = d_c$$

$$\lim_{q \to 1} D_q = \sigma \qquad (4.37)$$

$$\lim_{q \to 2} D_q = \nu ,$$

and relating these to the ν_n's gives

$$D_q = \frac{\nu_q}{q-1} . \qquad (4.38)$$

Equation (4.38) provides the desired definition of a generalized dimension; it measures the spatial scaling for various levels of local correlation.

Hentschel and Procaccia make use of a particular, very general, family of rescaling transformations to define the fractals that they study. They assume that at each stage of the iterative construction of the fractal a box of size l^d (d is the dimension of the space the fractal is embedded in) is replaced by M_1 boxes of size $(l/s_1)^d$, M_2 boxes of size $(l/s_2)^d$, ..., M_R boxes of size $(l/s_r)^d$. The probability for the ith box to be occupied is P_i. They show that, when the scaling parameters are known, D_q may be calculated directly by solving the implicit equation

$$1 = \sum_{I=1}^{R} M_i P_i^q s_i^{(q-1)D_q} . \tag{4.39}$$

Although their results assume the presence of this scaling structure, they have found the theory to be remarkably accurate for fractals which not scale in this way. This theory of generalized dimensions is closely related to a recent analysis of the scaling of fractals in terms of the singularities of measures[34].

The analytical calculation of D_q may be demonstrated for the Cantor set (described in Appendix A.3). For the Cantor set generated by removing the middle thirds of each of the intervals D_q is constant for all q and is equal to the capacity calculated previously. The result is more interesting if an asymmetrical Cantor set is used, such as that shown in figure 4.5.

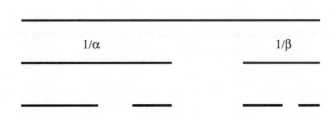

Figure 4.5 The construction of an asymmetrical Cantor set

Here $R = 2$, $M_1 = M_2 = 1$, $s_1 = \alpha$, and $s_2 = \beta$. Assuming that the probability density is uniform, the probabilities to fall in either piece is given by the length of the piece over the sum of the lengths:

$$P_1 = \frac{\frac{1}{\alpha}}{\frac{1}{\alpha} + \frac{1}{\beta}} \quad P_2 = \frac{\frac{1}{\beta}}{\frac{1}{\alpha} + \frac{1}{\beta}} . \tag{4.40}$$

These values may be used in equation (4.39) to find D_q:

$$1 = \left(\frac{\beta}{\alpha + \beta}\right)^q \alpha^{(q-1)D_q} + \left(\frac{\alpha}{\alpha + \beta}\right)^q \beta^{(q-1)D_q} . \tag{4.41}$$

Plugging in $\alpha = \beta = 3$ verifies that $D_q = \log 2 / \log 3$ for all q. The solution of equation 4.41 for $\alpha = 2.5$ and $\beta = 6.25$, found numerically, is plotted in figure 4.6. These values turn out to provide a good approximation to Feigenbaum's attractor associated with the logistic map.

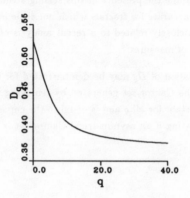

Figure 4.6 The generalized dimension for the asymmetrical Cantor set
with $\alpha = 2.5$ and $\beta = 6.25$

4.5 Thermodynamics

The preceeding material can be integrated by casting it into the language of statistical mechanics and thermodynamics[34,35]. As before, we start by assuming that a measurement has led to a set of points in phase space. Cover the space with boxes of length l, let p_i be the probability for a point to be in the ith box ($= N_i/N$), and define a scaling exponent α_i by

$$p_i \sim l^{\alpha_i} . \tag{4.42}$$

For the duration of this section we will be interested in the limit of $l \to 0$, however to reduce typographic clutter the limits will be not be indicated.

The connection with statistical mechanics is made by noticing that a sum over the p_i's raised to a power q, such as we have frequently used, is just a partition function:

$$\sum_i p_i^q \sim \sum_i l^{q\alpha_i} \equiv \sum_i e^{-\beta \epsilon_i} = \mathcal{Z}, \tag{4.43}$$

where q has been identified with the temperature, and α with the energy. These suggestive identifications are made by mathematical analogy, but they are not directly related to the true underlying thermodynamic quantities. The thermodynamic entropy is the logarithm of the number of states $n(\alpha)$; f is the symbol conventionally used in the dynamical systems literature for this quantity:

$$S(\beta) \equiv f(\alpha) = -\ln n \Rightarrow n(\alpha) \sim l^{-f(\alpha)}. \tag{4.44}$$

This has a very natural interpretation: $f(\alpha_i)$ is the dimension of the set of boxes which have a scaling exponent α_i. The last thermodynamic quantity that is frequently used is the free energy A:

$$\mathcal{Z} = e^{-\beta A} \equiv l^\tau = \sum_i p_i^q$$

$$\Rightarrow \tau = \frac{\ln \sum_i p_i^q}{\ln l} = (q-1)D_q \tag{4.45}$$

Once again, there is a natural interpretation – the free energy maps into the generalized dimension D_q.

The mapping into statistical mechanics is thus

$$f \Leftrightarrow S \text{ (entropy)} \qquad \tau \Leftrightarrow -\beta A \text{ (free energy)}$$
$$\alpha \Leftrightarrow \epsilon \text{ (energy)} \qquad q \Leftrightarrow -\beta \text{ (temperature)}$$

These can be related through the appropriate Maxwell relations and Legendre transformations:

$$\alpha = \frac{\partial \tau}{\partial q}$$
$$f = q\alpha - \tau \tag{4.46}$$

References [36] and [37] describe the measurement of these quantities from experiments, and references [38] and [35] discuss the extraction of an effective transfer matrix from data.

The thermodynamic viewpoint described here gracefully recognizes an underlying mathematical parallel to bring the well-developed machinery of statistical mechanics to bear on the question of characterizing measured attractors. An active and intriguing question is what can be deduced about the underlying Hamiltonian from the measurement of the analogical thermodynamic quantities.

5. Speculations and Limitations

This completes the introduction to the theory of observing dynamical systems. A very abstract description of a dynamical system in the language of differential topology has lead to a remarkable ability to observe and analyze the dynamics of an experimental system. This theory has a great potential both for future generalization and application, and for erroneous or disappointing misuse. We close with a list of some of the limitations of the theory and speculations on possible future directions of development.

1. The mathematical proofs used in the appendices are generally on a very strong footing, with the primary uncertainty being the relation between experiment and statements that are only true generically. The mathematical footing for the analytical tools is much weaker, although most of what has been described has been amply verified in practice. The lack of satisfying proofs, and the presence of counterexamples (such as the violation of the Kaplan-Yorke conjecture) will probably be shown to be irrelevant under the limitations imposed by physical experiments, however greater rigor would be helpful.

2. A number of the practical considerations, such as choosing time delays for the functions used in the embedding, or selecting the delay time from the decay rate of the mutual information, provide good answers but have not been shown to provide optimal ones. It would be useful to have a better understanding of how to make these choices optimally.

3. A number of the routines described can be applied in a cookbook fashion, and will generate numbers even when important requirements are not met. A good example is the calculation of the dimension, which can easily be implemented but must be done carefully in order to obtain a reliable result. Reference [39] discusses further the risks in dimension estimation. There is often no substitute for interactively graphically examining data. Similarly, there often are short-cuts that may be difficult to rigorously justify but that are practically verifiable and offer significant savings. Examples include calculating the pointwise dimension rather than the correlation sum, and using an embedding dimension that is smaller than what is needed to prevent all accidental crossings.

4. The dimension calculation is one of the central components of the theory, and the exponential growth in the number of points needed to measure a dimension as the dimension is increased is a severe constraint on this calculation. Straight-forward approaches to combatting this growth rate include efficient program design and the use

of larger computers, however the exponential rate is so fast that it will quickly outpace these improvements. To push the calculation of the dimension to higher dimensions, possibilities include the use of specialized hardware (because the algorithm is so simple), and research into finding approximate routines that use additional information about the data to reduce the number of points needed.

5. For the embedding procedure to work, all the relevant variables must be in communication. The procedure does not provide a truly free lunch; it is not possible to extract information about a degree of freedom that is not intimately coupled to the experimental probe.

6. The embedding procedure offers no assistance in understanding a system which consists of many copies of a simple underlying unit. If a laser beam is passed through a line of identical non– or weakly–interacting convecting cells, the simple deterministic dynamics of each of the cells will almost surely be hidden in the apparent high dimension of the arrangement. A challenging goal is to learn how to distinguish between an N–dimensional system and one which consists of N 1–dimensional systems.

7. It must be remembered that the embedding procedure is a powerful technique to extract otherwise inaccessible dynamical information, but is not a substitute for careful experimental design. Some questions, such as pattern formation, clearly require spatially resolved information. Even where embedding is plausible, there is still a tradeoff between adding extra physical probes and gaining more quantitative state information.

8. The discussion of deterministic disorder throughout this paper has been restricted to what is called weak turbulence, which is the deterministic chaos associated with a low-dimensional system. The theory described here currently offers little insight into strong turbulence, which is the disorder of, for example, the fully-developed turbulence of a high Reynold's number fluid flow. The interactions of eddies and vortices in such flows are certainly governed by deterministic rules, but to the embedding procedures strong turbulence generally behaves like random noise. The ultimate goal of this area of research is to extend the deep insights into the randomness of such simple systems as the logistic map to the more physically meaningful randomness we are surrounded by outside of the laboratory.

9. Most of the tools described for analyzing measured dynamics have been associated with the detection of chaos. There are a number of other important aspects about how an attractor "looks", such as topological indices, that are potentially applicable

to data generated by embedding and have not been generally used. Also, there are many experimentally interesting questions other than detecting the presence of chaos that can benefit from the ability to produce a qualitative picture of dynamics without any *a priori* information and with access to only one degree of freedom.

10. The collection of analytical tools leading up to the thermodynamic formalism are, to an unbiased author, undeniably elegant. The historical jury is still out on whether the sophistication of the formulation will be matched by a comparable practical usefulness.

10. A frequent hindrance to the application of this theory is a misperception of what "qualitative" means. The embedding provides a similar, but not an identical, copy of the original system. We have repeatedly seen how certain quantitative values, such as the dimension, remain invariant under such a transformation and hence are available. Quantitative information is also available about the relation between experimental parameters and the topological shape of the attractor; a view of, for example, the evolution of limit cycles and fixed points as a forcing parameter is varied can guide the quantitative modelling of the system's behaviour. Finally, qualitative information (qualities), rather than numbers, are the true end goal of the observer of nature.

Acknowledgements

The author would like to gratefully acknowledge helpful discussions with John Guckenheimer, Ben Wittner, Gemenu Gunaratne, Hao Bai-lin, Watt Webb, Jim Sethna, Shelby F. Nelson, Winfried Denk, and Paul Tjossem. This work was supported by the NSF under grant number DMR-8414796 and the Cornell National Supercomputer Facility, and by an IBM Graduate Research Fellowship.

Appendix A: Mathematical Background

Differential topology generalizes the familiar notions of vector calculus to more abstract spaces in order to separate the essence of the structures investigated from the coordinates used to describe them; we will need this in order to describe the evolution of dynamical systems in its most natural setting. This appendix provides a self-contained introduction to the supporting mathematical theory and the language used to describe it. All unfamiliar terms and concepts needed later are described here; because of the breadth of material quickly covered it is probably best to skim this section and then use it as a dictionary when needed rather than read it as a tutorial. The reader is cautioned that this Appendix is presented to quickly introduce a large number of potentially unfamiliar terms and concepts, however for more precise and pedagogical treaments the excellent available references should be consulted. [40] provides a very readable introduction to differential topology, particularly transversality arguments, for manifolds embedded in $\mathbb{R}^n$ (terms to be defined shortly), reference [41] cleanly presents more details of the full theory, and reference [42] is a superb encyclopedic source. Most of the following background material is taken directly from these references.

A.1 Sets, Maps & Manifolds

The first step is to consider sets and mappings. An element x of a set X is denoted by $x \in X$, and a set contained in a larger set by $X \subset Y$. A set is frequently defined by symbols such as $\{x \in S \mid \forall y \ \exists z \text{ with } x + y < z\}$, which is tranlated "the set of all elements x contained in the set S such that, for all y, there exists a z with $x + y < z$." The null set $\emptyset$ is the set with no members. Given an ordering that allows the relative magnitude of two elements in a set to be determined, the maximum element of the set is the supremum (denoted sup), and the minimum element is the infimum (denoted inf). If the elements of a set can be indexed with integers, the set is countable. The intersection of two sets $X \cap Y$ is $\{x \mid x \in X \text{ and } x \in Y\}$ and the union $X \cup Y$ is $\{x \mid x \in X \text{ or } x \in Y\}$. Two sets are disjoint if their intersection is the null set. A set S with a subset N removed is denoted $S \backslash N$. A partition of a set is a collection of non-empty disjoint subsets of a set whose union is the set; and the intersection of two partitions, $A \vee B$, is the set of all intersections of all elements of the partition. The complement of $X \subset Y$ is the set off all points in Y not contained in X (ie, $Y \backslash X$).The direct product of two sets X and Y is denoted $X \times Y$ (X cross Y) and is defined by $X \times Y = \{(x, y) \mid x \in X, y \in Y\}$.

A vector space (linear space) is a "module for which the ring of operators is a field," which means (roughly) that it is a set for which the operations of addition of elements and multiplication of an element by a scalar are defined and have the familiar properties such as commutativity and invertibility. The space (field) of <u>real numbers</u> is denoted by $\mathbb{R}$, the familiar n-dimensional real <u>Euclidean space</u> by

$$\mathbb{R}^n = \underbrace{\mathbb{R} \times \mathbb{R} \times, \cdots, \times \mathbb{R}}_{n \text{ times}}, \qquad (A.1)$$

and the space of <u>natural numbers</u> (non-negative integers) by $\mathbb{N}$. The set $\{x \in \mathbb{R} \mid a \leq x < b\}$ is written $[a, b)$. The <u>diagonal</u> of the direct product of a set with itself is the set of all elements for which each coordinate has the same value (for example, the diagonal of $\mathbb{R} \times \mathbb{R}$ is $\{(x, x) | x \in \mathbb{R}\}$).

Given two sets X, Y, a <u>mapping</u> between them is written by $f : X \to Y$ and associates every element $x \in X$ with a uniquely determined element $y \in Y$. The action on elements is denoted by $y = f(x), f : x \mapsto y$. X is the <u>domain</u> and the subset of Y that f maps X into is the <u>range</u>; Y is also said to be the <u>target</u> space. y is the <u>image</u> of x, and x is the <u>preimage</u> of y. Given a point $y \in Y$, the set of all points $x \in X$ such that $f(x) = y$ is the <u>level set</u> of f for y. Two points in a level set are said to have the same <u>level</u>. Given two maps f, g, their <u>composition</u> is denoted by $f(g(x))$ or $f \circ g$, and the composition of a map f with itself n times is written f^n.

A map $f : X \to Y$ is <u>linear</u> if $f(\alpha x + \beta y) = \alpha f(x) + \beta f(y) \forall x, y \in X$, where α and β are constants. The <u>kernel</u> of a linear map is $\{x \in X \mid f(x) = 0\}$. The set of all linear maps between two vector spaces X and Y is denoted $L(x, y)$, and is itself a vector space (because the addition and multiplication of functions are easily defined and shown to have the required properties). The space of linear maps between a vector space and the reals, $L(V, \mathbb{R})$, is the <u>dual</u> space to V and is denoted V^*. An element of a dual space is called a <u>linear form</u>. For example, if V is $\mathbb{R}^n$, then a linear function on an element $x \in \mathbb{R}^n$ is given by

$$x \mapsto \sum_i a_i x_i. \qquad (A.2)$$

The space of all linear forms on $\mathbb{R}^n$ is conveniently labelled by the a_i's, and hence is equivalent to $\mathbb{R}^n$.

If for every $y \in f(x)$ there is only one x such that $y = f(x)$ then the map is said to be <u>one-to-one</u> or <u>injective</u>. If f is injective then it has an <u>inverse mapping</u> f^{-1} defined in the obvious fashion: if $f(x) = y$ then $f^{-1}(y) = x$. If the range of f is the entire space Y, then the map is said to be <u>onto</u> or <u>surjective</u> and is written $f(X) = Y$. If f is both one-to-one and onto then it is <u>bijective</u>. For example, $y = 2x$ is a bijective map of the reals onto the reals, while $y = \sin(x)$ is neither injective ($\sin(0) = \sin(2\pi) = 0$) nor surjective ($-1 \leq \sin(x) \leq 1$).

The next set of ideas relate to the notion of distance. A <u>metric space</u> is a set for which there is a measure of the distance between points and the neighborhood of a point can be defined. The following discussion can be phrased entirely in terms of Hausdorff topological spaces, which do have neighbourhoods but do not have metrics, however this further abstraction makes the following discussion less transparent and dynamical systems usually do have a natural metric. The distance function (<u>metric</u>) d need not be related to a Euclidean distance – it need only satisfy

$$
\begin{aligned}
&d(x,y) \geq 0 \\
&d(x,y) = 0 \text{ iff } x = y \\
&d(x,y) = d(y,x) \\
&d(x,z) \leq d(x,y) + d(y,z) \ .
\end{aligned}
\qquad (A.3)
$$

Two metrics d and d' on a set S are <u>equivalent</u> if there exists a constant C such that, for all $x, y \in S$, $Cd(x,y) \geq d'(x,y) \geq C^{-1}d(x,y)$. A <u>neighborhood</u> of a point x is an open set (often taken to be arbitrarily small) containing x. An <u>open ball</u> of radius r centered at a point x_0 is $\{x \mid d(x,x_0) < r\}$.

The metric allows the interior, exterior, and boundary of a set to be defined. Take S to be a subset of a metric space X. S is <u>bounded</u> if there exists a number M such that $d(x,y) \leq M \forall x, y \in S$. A point $x \in S$ is a <u>limit point</u> if every neighborhood of x contains a point $y \neq x$ such that $y \in S$. S is <u>closed</u> if every limit point of S is contained in S. The <u>closure</u> of a set S is denoted $\overline{S}$ and is the union of S with all its limit points; it is the smallest closed set containing S. The <u>support</u> of a function f defined on a set is the smallest closed set outside which f vanishes everywhere; it is the closure of the set $\{x | f(x) \neq 0\}$. A point $x \in S$ is an <u>interior point</u> if there is a neighborhood N of x such that $N \subset S$. S is <u>open</u> if every point contained in S is an interior point. The interior of a grape without its skin is an open set; the grape with the skin is closed.

An <u>open cover</u> of S is a collection of open subsets of S such that S is contained in their union, a <u>subcover</u> is a subset of a cover that is itself a cover, a <u>finite cover</u> is a cover with a finite number of elements, and a <u>refinement</u> is a subcover for which each of the subsets making up the subcover is contained in some corresponding element of the cover. S is <u>compact</u> if every open cover of S has a finite subcover, and S is <u>paracompact</u> if every covering has a locally finite refinement. Bounded, closed subsets of $\mathbb{R}^n$ are compact.

Having introduced distance, continuity and smoothness can be described. A mapping $f : X \rightarrow Y$ between metric spaces is <u>continuous</u> at $x \in X$ if for every neighborhood N

of $f(x)$ contained in Y ($N \subset Y$) there exists a neighborhood $M \subset X$ of x such that $f(M) \subset N$; if f is continuous at all $x \in X$ then it is said to be continuous on X. An isomorphism is a bijective map. A homeomorphism is a bijective map f which is bicontinuous (f and f^{-1} are continuous): each element of X is uniquely associated with an element of Y and there are no jumps in the association.

An m-dimensional surface in $\mathbb{R}^n$ is defined by relations between the coordinates of $\mathbb{R}^n$; this notion generalizes to a manifold which does need to be placed in $\mathbb{R}^n$. A manifold is a metric space for which every point has a neighborhood homeomorphic to $\mathbb{R}^n$; this homeomorphism provides coordinates for the manifold by identifying elements of the manifold with the corresponding elements in $\mathbb{R}^n$. The topological dimension of a manifold M is that of the corresponding real space, and is denoted $\dim(M)$. A submanifold is a manifold contained in another one. The codimension (codim) of a submanifold M in a manifold N is given by $\operatorname{codim}(M) = \dim(N) - \dim(M)$.

An open set U in X along with a homeomorphism φ to an open set in $\mathbb{R}^n$ is called a chart (U, φ), and a collection of charts that covers X is an atlas. φ is called a coordinate system; φ^{-1} is a parametrization. The charts, whose existence is guaranteed by the definition of a manifold, provide the connection from the abstract manifold to real coordinates that permit calculations needing coordinates to be performed. Frequently these coordinates in $\mathbb{R}^n$ are used to index points in X without reference to a chart; this use will be clear from the context. The guaranteed presence of a chart insures that this notational flexibility is safe.

The definition of a manifold must be extended to include boundaries. A halfspace of $\mathbb{R}^n$ is a subset defined to be $H = \{x \in \mathbb{R}^n \mid \lambda(x) \geq 0\}$ where $\lambda : \mathbb{R}^n \to \mathbb{R}$ is a linear map. For example, a half space is obtained by restricting $\mathbb{R}^n$ to all the reals greater than or equal to 10. The boundary of H, denoted ∂H, is defined to be the kernel of λ (the set of all values of H that map into 0). A manifold with boundary is a manifold whose charts take open subsets of the manifold into open subsets of a halfspace of $\mathbb{R}^n$. The boundary of the manifold M, denoted ∂M, is the inverse image of the boundary of the half space under the chart ($\partial M = \varphi^{-1}(\partial H)$).

A.2 Differential Topology

A.2.1 *Derivatives and Tangent Spaces*

Having introduced coordinates, derivatives can now be taken. A function is said to be C^k if its first k partial derivatives exist and are continuous. Two charts on X (U, φ) and

(V, ψ) with overlapping neighborhoods of definition $(U \cap V \neq \emptyset)$ provide a map $\varphi \circ \psi^{-1}$: $\mathbb{R}^n \to \mathbb{R}^n$; if these maps are all C^k then the manifold is said to be C^k. Frequently k is not mentioned explicitly, but is assumed to be large enough (the manifold is smooth enough) for the given context, and the manifold is simply said to be a <u>differentiable manifold</u>. We will only be concerned with differential manifolds (dynamical systems are by definition differentiable).

The charts provide a natural way to calculate the derivative of a map between manifolds. If $f : X \to Y$ with $\dim(X) = n$, $\dim(Y) = m$, $x \in X$, and the charts on X and Y are φ and ψ, then $h = \psi \circ f \circ \varphi^{-1}$ is a map between reals and its derivative is a linear map given by the Jacobian matrix. The derivative of h provides a <u>local representation</u> for the derivative of f:

$$
D_x f = \begin{pmatrix} \frac{\partial h_1}{\partial x_1}(x), & \cdots, & \frac{\partial h_1}{\partial x_n}(x) \\ \vdots & & \vdots \\ \frac{\partial h_m}{\partial x_1}(x), & \cdots, & \frac{\partial h_m}{\partial x_n}(x) \end{pmatrix} . \tag{A.4}
$$

Some authors use d instead of D. If h is C^k then f is C^k also; a <u>diffeomorphism</u> is a C^k bijection. The <u>rank</u> of the map is the rank of the Jacobian matrix, which is equal to the number of independent rows or columns. If the rank of the Jacobian matrix equals the dimension of the domain, the Jacobian matrix or the map is said to have <u>full rank</u>. If the Jacobian matrix has full rank then a set of basis vectors which span the domain of the map will be mapped into a set which span the range.

In the same way that the derivative of a function provides its best linear approximation, the best linear approximation to a manifold M at a point x is given by its <u>tangent space</u>, denoted by $T_x(M)$. This is a central idea in differential topology. Let us start by considering a 2-d manifold M placed in $\mathbb{R}^3$; in this case it is easy to visualize the tangent space, which is just a plane. Let C be a curve in $M(C : \mathbb{R} \to \mathbb{R}^3, C : t \mapsto x)$. The tangent vector to C at x is a vector in $\mathbb{R}^3$ given by

$$
v_x^C = \frac{dx}{dt} . \tag{A.5}
$$

The set of tangent vectors to all curves passing through a point is the tangent space to the manifold at that point. Figure A.1 shows the tangent space in this case. The tangent space to $\mathbb{R}^n$ itself at a point is the set of all vectors passing through that point; this is just $\mathbb{R}^n$ again.

This description of a tangent space must now be extended to describe a manifold without reference to an ambient space; this extension is provided by the chart. With

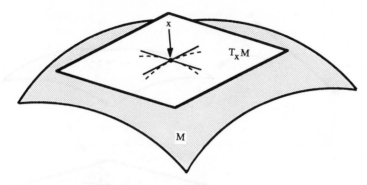

Figure A.1 The tangent space in $\mathbb{R}^3$

a chart φ on a manifold M a curve in M may be described in terms of its coordinates $(\varphi \circ C : \mathbb{R} \to \mathbb{R}^n,\ t \mapsto p)$. The tangent to the curve is given by

$$v_X^C = \frac{d}{dt}(\varphi \circ C) . \tag{A.6}$$

This is a tangent vector in $\mathbb{R}^n$ which depends on the chart; there are straight-forward (although possibly less intuitive) definitions of tangent vectors that are independent of the chart. Once again, the set of all tangent vectors passing through x is the tangent space $T_x(M)$. This relation is indicated in figure A.2. As the tangent vectors provide local approximations to curves that span $\mathbb{R}^n$, the tangent space has the same dimension as $\mathbb{R}^n$ and hence the same dimension as the manifold. The tangent space is isomorphic to $\mathbb{R}^n$. A simple calculation shows that the tangent vectors caluated with a new chart are a linear combination of those calculated with an old one, hence the tangent space spanned is the same. This means that the tangent space is intrinsic to the manifold, rather than dependent on the particular chart.

Tangent spaces are intimately related to the derivative of a map between manifolds: the derivative is a map between the tangent spaces ($f : M \to N$, $f : x \mapsto y$, $D_x f : T_x M \to T_y M$). To see this, let C be a curve in M ($C : \mathbb{R} \to M, C : t \mapsto x$), and φ and ψ be the charts on M and N. f maps the curve on M onto one on N; let v and w be the respective tangent vectors. w can be calculated as above:

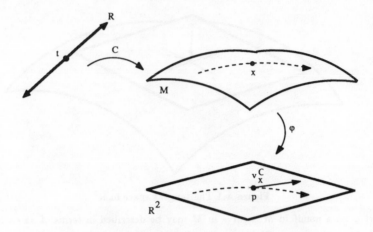

Figure A.2 A tangent vector on an arbitrary manifold

$$
\begin{aligned}
w_y^C &= \frac{d}{dt}(\psi \circ f \circ C) \\
&= \frac{d}{dt}[(\psi \circ f \circ \varphi^{-1}) \circ (\varphi \circ C)] \\
&= D_x f \circ \frac{d}{dt}(\varphi \circ C) \quad \text{(chainrule)} \\
&= D_x f \circ v_x^C
\end{aligned}
\qquad (A.7)
$$

We see that a tangent vector to M is a linear function of a tangent vector to N, with the linear map between the tangent spaces given by the derivative of the map between the manifolds. If $D_x f$ is surjective then x is said to be a <u>regular value</u>, and a point that isn't a regular value (the range is not spanned) is a <u>critical value</u>. At a critical point the rank of the Jacobian matrix of the map is less than the dimension of the range of the map.

Duals to tangent spaces arise naturally in considering maps into real spaces. Let f be a map from a manifold M to $\mathbb{R}^n$, $x \in M$, $\dim(M) = m$ and the components of f be $(f_1(x), f_2(x), \ldots, f_n(x))$. $D_x f : T_x M \to T_{f(x)}\mathbb{R}^n$, and since we showed above that the tangent space to $\mathbb{R}^n$ is $\mathbb{R}^n$, we see that $D_x f : T_x M \to \mathbb{R}^n$. Remembering that a elements of a dual space are linear functions from the space to $\mathbb{R}$, we see $(D_x f)_i \in T_x^*(M)$. The dual to a tangent space is called a <u>cotangent space</u> and is populated by <u>covectors</u>.

The tangent spaces at various points overlap; they are pulled apart in the <u>tangent bundle</u> TM of the manifold $M \subset \mathbb{R}^n$, defined by

$$
TM = \{(p, v) \mid p \in M, \ v \in T_p(X)\} . \qquad (A.8)
$$

Each element of the tangent bundle consists of an element of the manifold and a tangent vector; the elements of the bundle are in the set $M \times \mathbb{R}^n$. For a fixed p, (p, v) for all v is the tangent space of p, and $(p, 0)$ for all p is a copy of the manifold. A <u>cross section</u> X of a tangent bundle is a smooth map that associates a tangent vector to each point of the manifold $(X : p \in M \mapsto v \in T_p M)$. For any smooth map $f : X \to Y$, a <u>global derivative map</u> between the tangent bundles is defined by

$$df : TX \to TY, \ df : (p, v) \mapsto \left(f(p), df_p(v)\right) . \tag{A.9}$$

A.2.2 *Dense Sets, Measure Zero, and Transversality*

In this section we introduce some central ideas that help to distinguish between reasonable and abnormal mathematical behaviour.

A subset S of a set X is <u>dense</u> if every point of X is a limit point of S, a point in S, or both (i.e., $\overline{S} = X$). A <u>residual subset</u> is one that contains the intersection of countably many dense open sets. A <u>separable</u> metric space is one that has a countable dense subset. A subset is said to have <u>measure zero</u> if it has a countable cover of solids of arbitrarily small volume. A subset of measure zero consists of distinct points; these points can be covered by a countable number of open balls with arbitrarily small radii. This is to be contrasted with a subset of non-zero or positive measure; in this case the radii of the open balls cannot be shrunk arbitrarily. Measure zero is an elementary concept taken from measure theory. A property of a set that is true at all points except for a subset of measure zero is said to be true <u>almost everywhere</u>, and a property that is true on an open and dense subset is said to be <u>generic</u>. These two ideas provide a means to measure the size of a subset and so to determine if it is important observationally or just a mathematical idiosyncracy; the former characterizes the size in terms of measure theory, and the latter in terms of topology.

As an example, let X be all the points on a dart board and consider subsets of the dart-board. A measure zero subset of the dart-board can be covered by an arbitrarily small area. If we throw infinitesimal darts at this dart-board, a simple calculation shows that the probability of hitting the measure zero set is $1/\infty = 0$ and the probability of landing in the positive measure set (the rest of the dart-board) is $\infty/\infty = 1$. If the dart-thrower were infinitely lucky and did land on the measure zero set, thermal and quantum fluctuations would still insure that the dart did not stay on the set. For this reason, measure zero conditions are usually taken to be unobservable and irrelevant to physical experiments; this is important because some of the theorems we will use provide results that are true everywhere but on a set of measure zero. An important caveat is that a property that is mathematically true on a set of measure zero may appear to have a positive measure due to experimental restrictions on the digitization precision and the number of samples.

If S is a dense subset of the dart-board, then an infinitesimal dart that hits the dart-board is guaranteed to be infinitesimally close to S, but not necessarily on S. Hence a dense set is less attractive to an experimentalist (dart-thrower) then a positive measure set, however proving that a set is dense is often much easier than proving that it has a non-zero measure. While there do exist mathematical objects which are dense but have measure zero, the constraints imposed by real physical systems are such that dense and positive measure sets are often indistinguishable.

A particularly useful theorem that will be referred to later provides a good example of the nature of measure zero results:

> **Sard's Theorem** The set of critical values of a smooth map of manifolds has measure zero.

Remember that critical values are ones for which the rank of the Jacobian of the map is less than full rank; this theorem assures us that these troublesome places will be isolated. The proof is relatively straightforward but lengthy; see reference [40] for details.

Transversality classifies stable and unstable intersections between manifolds. Two submanifolds X and Y in Z are said to intersect <u>transversely</u>, denoted $X \pitchfork Y$, if

$$T_p(X) \oplus T_p(Y) = T_p(Z) \ \forall \ p \in X \cap Y \ . \tag{A.10}$$

$\oplus$ is the <u>direct sum</u>; it means that each vector in $T_p(Z)$ can be written as the sum of a vector in $T_p(X)$ and a vector in $T_p(Y)$ (that is, the tangent spaces to X and Y at p together span the space Z). Figure A.3 illustrates transversal and nontransversal intersection of one-dimensional manifolds in two dimensions. Transversal intersections are <u>stable</u>: small perturbations of the manifolds will not remove the transversality. A nontransveral intersection is a much more precarious arrangement because almost any movement of the manifolds will make the manifolds transversal.

Given that $X \pitchfork Y$ in Z, adding an extra dimension to Z (and therefore its tangent space) will break the equality in the transversality condition. This leads to the

> **Intersection Theorem** If two compact submanifold manifolds X and Y in Z intersect, and $\dim(X) + \dim(Y) < \dim(Z)$, then the intersections may be removed by arbitrarily small perturbations.

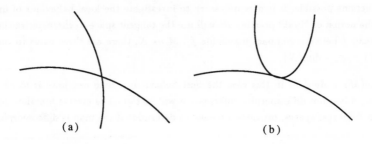

Figure A.3 Transversal (a) and nontransversal (b) intersections

Figure A.4(a) shows the transversal intersection of two one-dimensional manifolds in $\mathbb{R}^2$, and figure A.4(b) shows an arbitrarily small perturbation removing the intersection when a third dimension (out of the page) is added.

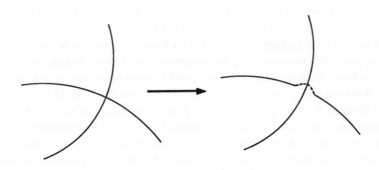

Figure A.4 The removal of transversality by the addition of an extra dimension

A.2.3 *Embedding*

The evolution of dynamical systems occurs on compact manifolds. As we are interested in reproducing these manifolds by experimental observations and certainly want the "best" reproductions possible, it is now necessary to investigate the local behaviour of maps to make the notion of "best" precise. We will use the tangent space to characterize the map. For a map f between compact manifolds $f : M \to N$, there are three cases to consider: $\dim(M) =, <, > \dim(N)$.

$\dim(M) = \dim(N)$ In this case the best behaviour a map can have is to be diffeomorphic; if it is not diffeomorphic information will be lost. The inverse function theorem, familiar from real spaces, provides a means to determine if the map is diffeomorphic.

> **The Inverse Function Theorem**   If $D_x f : T_x M \to T_y N$ is an isomorphism, then f is a local diffeomorphism.

The proof of this theorem is the same as that for the standard Euclidean case, suitably generalized through the use of the charts. The derivative map is linear (the Jacobian matrix), and fails to be an isomorphism when the determinant is zero (it is not full rank), hence to determine if f is locally diffeomorphic one must simply evaluate the Jacobian.

$\dim(M) < \dim(N)$ Here the map can no longer be a diffeomorphism, because N has more degrees of freedom than M. To preserve the structure of M in the larger space, the derivative of the map must be invertible and hence injective. In this case the map is said to be a local <u>immersion</u>, and if it is an immersion everywhere then it is simply an immersion. For the map to be an immersion, the Jacobian matrix must be of full rank. For maps into reals, this can be rephrased in terms of the dual spaces. We showed earlier that for a map $f : M \to \mathbb{R}^n$, with $\dim(M) = m$ and the action of f on an element $x \in M$ given by $f(x) = \big(f_1(x), f_2(x), \ldots, f_n(x)\big)$, that $(D_x f)_i$ is a covector in $T_x^*(M)$. The derivative with respect to the coordinates of M of the f_i's are vectors which make up the rows of the Jacobian matrix, and if the Jacobian matrix is to be of full rank then m of these vectors must be linearly independent at x, and this in turn means that they must span $\mathbb{R}^m$. But we know that $T_x M$ is isomorphic to $\mathbb{R}^m$ and hence so is $T_x^* M$, and so we arrive at the condition that, for the map from a manifold to a real space to be an immersion, the covectors must span the cotangent space.

To insure that the image of an immersion is a manifold we must add further conditions. Figure A.5(a) shows a map of the circle S^1 into $\mathbb{R}^2$ that is an immersion but that doesn't have a manifold for an image. The center of the figure-eight does not look like $\mathbb{R}^1$ and hence it is not a manifold; the problem arises beacuse the center of the figure-eight has two pre-images (the map is not injective). Requiring an immersion to be injective is not sufficient, however; figure A.5(b) demonstrates a map that is an injective immersion but the image is still not a manifold. The problem here is that the points $\pm\infty$ are mapped into an infinitesimally small neighborhood of the origin, once again preventing a diffeomorphism into $\mathbb{R}^1$ (any neighborhood of the origin, no matter how small, will still contain the images of $\pm\infty$). This latter problem is handled by requiring that the map be <u>proper</u>. A map is proper if the preimage of every compact set is compact. An injective immersion that is proper is an <u>embedding</u>; the following theorem states that this is the best behaviour f can have (see the references for a proof).

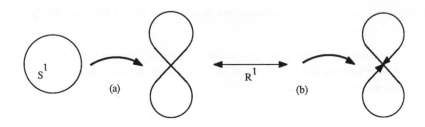

Figure A.5 (a) An immersion that is not injective (b) An injective immersion that isn't proper

Embedding Theorem An embedding $f : X \to Y$ maps X diffeomorphically onto a submanifold of Y

Note that if X is compact then every map on it is proper, hence embeddings of compact manifolds are just injective immersions.

$\mathbf{dim}(M) > \mathbf{dim}(N)$ In this case we must necessarily lose information about M; the strongest condition possible is that the derivative of the map is surjective. If $D_x f$ is locally surjective onto $T_x N$ then f is a local <u>submersion</u>.

A.2.4 *Embedding in real spaces*

Embedding manifolds into $\mathbb{R}^n$ is a special case that deserves further attention because of its relevance to experimental observation. Given a manifold of dimension d, how big must the dimension n of the target real space be? The following example illustrates the

Figure A.6 The embedding of a two-dimensional manifold in $\mathbb{R}^2$

subtlety of this question. A two-dimensional simple manifold can be embedded in two dimensions, as shown in figure A.6.

Now complicate the manifold by associating two sides as indicated in figure A.7. This is a called Möebius strip and in order to be embedded without intersecting itself it needs a third dimension to fit the twist.

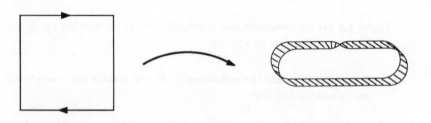

Figure A.7 The embedding of a Möebius strip in $\mathbb{R}^3$

Even more complicated is the Klein bottle, formed by joining the remaining faces of a Moebius strip without a twist. The Klein bottle requires four dimensions for an embedding. Figure A.8 shows a three dimensional representation; where the bottle passes

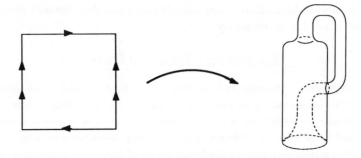

Figure A.8 The embedding of a Klein bottle in $\mathbb{R}^4$

through itself it is understood to detour into a fourth dimension.

We have seen two-dimenisonal manifolds that require two, three, and four dimension. Can they be any worse? The answer is provided by

> **Whitney's Embedding Theorem**   Every d-dimensional compact manifold embeds in $\mathbb{R}^{2d+1}$

The intersection theorem provides a good intuitive explanation of why this should be true. The problem in the previous set of examples was that extra dimensions were necessary to prevent intersection of the manifold with itself. Consider two identical copies of the same manifold X. If we place these copies in a space with dimension equal to that of X they are sure to intersect, but if we place them in a space with dimension greater than twice that of X the intersection theorem tells us that any remaining intersections will not be transverse (two copies of X cannot span a space whose dimension is greater than twice the dimension of X) and any intersections can be removed with arbitrarily small perturbations. This guarantees that if we place a single copy of the manifold in a space with greater than twice its dimension, any accidental intersections can be removed via small perturbations. This is not yet a rigorous proof however. In this, as well in many other cases in differential topology, proving a local statement (a manifold can be locally embedded in $\mathbb{R}^{2d+1}$) is much easier than proving the global one (the whole manifold can be embedded). One form of the detailed proof makes use of Sard's theorem. After initially proving the theorem Whitney was later able to reduce the upper bound from $2d + 1$ to $2d$ by restricting the theorem to paracompact manifolds, however that is a much more difficult proof to understand. It is very unlikely that a manifold that isn't paracompact will be encountered in the laboratory.

A.3 Dimension Theory and Fractals

Around the turn of the century a bestiary of pathological mathematical objects were discovered that necessitated a refinement of the concept of the topological dimension. We will need these ideas, both because some of these pathologies can be present in dynamical systems, and because they provide a useful way to help characterize an arbitrary metric space. In this section we introduce the formal theory of dimension; in section 6 we return to the more practical question of how to numerically measure the dimension of a given manifold.

One of the first, and simplest, objects to require a subtler notion of dimension than the topological one is the Cantor set. Figure A.9 shows the first few stages in the construction of a particular Cantor set, the middle-thirds one. It is defined recursively as the limit of the process of removing the middle third from the interval [0,1], the middle thirds from the remaining intervals, and so forth.

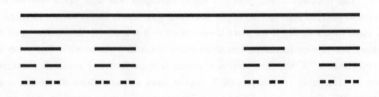

Figure A.9 The Construction of the Middle-Thirds Cantor Set

What is the dimension of this object? Intuitively its dimension is greater than that of a set of isolated points, but less than that of the full line segment; we are led to contemplate a dimension that is not an integer. We shall introduce a definition for a dimension that agrees with the topological one for "nice" cases, and can give a non-integer value otherwise. The theory was initially developed for separable metric spaces (all we will ever need for dynamical systems), and after much effort extended to more general metric spaces. The theory for separable spaces was admirably summarized in 1941 in reference [43], and a personalized and aesthetically satisfying view of the theory and the broad range of applications of non-integer dimensions is reference [44].

Let S be a compact metric space, and let $N(r, S)$ be the minimum number of open balls of radius r needed to cover S. The associated picture is one of stringing beads over the space, and if we then let the bead size shrink it is reasonable to expect that the number of beads will scale as some power of the radius of the beads for small values of the radius:

$$N(r, S) \simeq Ar^{-d_c} , \tag{A.11}$$

(where A is a constant) and so the capacity (sometimes called the limit capacity) is defined to be this scaling exponent:

$$d_c(S) = \limsup_{r \to 0} \frac{\log N(r, S)}{\log(1/r)} . \tag{A.12}$$

The sup is required if the limit does not smoothly approach an asymptotic value, and is frequently unnecessary. This definition is valid only in the infinitesimal limit. Away from this limit the constant prefactor A will add a correction, and the standard form of the definition given above may be modified to include this:

$$\log N(r, S) \underset{r \to 0}{=} d_c(S) \log(1/r) + \log A . \tag{A.13}$$

A moment's reflection will show that the limit capacity for lines, planes, and volumes is just the expected values 1, 2 and 3. Now consider the middle-thirds Cantor set once again, and index the stage in the construction by i. For $i = 0$ the set is just the segment $[0,1]$; for $i = 1$ the set consists of 2 segments of length $1/3$; for $i = 2$ it is 4 segments of length $1/9$; and for the nth level it consists of 2^n segments of length $1/3^n$. Choosing $1/3^n$ for the radius of the open balls used to cover the nth level we immediately see that the limit capacity for the middle-thirds Cantor set is $log(2)/log(3) \approx 0.631$. This is our desired goal: the limit capacity agrees with the topological dimension for cases where we hope it will, and differs for other cases. A set which has a non-integer dimension is called a fractal.

There is an alternate, equivalent, definition of the capacity that lends itself to an easy generalization. The size of the cover raised to the capacity, r^{d_c}, can be interpreted as the volume of the cover actually available for use used in covering the d_c-dimensional manifold (if a plane is covered with spheres, only a two-dimensional slice of the spheres is used).

The largest d-dimensional object that can be covered by a given cover can thus be defined to be

$$V(r) \; = \; \sum_i r_i^d \, , \qquad\qquad (A.14)$$

where the sum is over all the members of the cover. Now consider this equation, with d not initially known. If the manifold is covered as in the definition of the capacity, then

$$V(r) \; = \; \sum_i r_i^d \; = \; N(r)r^d \; \simeq \; Ar^{-d_c}r^d \, . \qquad\qquad (A.15)$$

In the limit that r goes to zero, $V(r)$ will diverge if d in equation (A.15) is less than d_c, will go to zero if $d > d_c$, and will reach a limiting value of A if $d = d_c$. This can be used as an alternate definition of the capacity: d_c is the unique value of d in equation (A.14) below which, in the limit $r \to 0$, $V(r)$ diverges and above which $V(r)$ vanishes.

A cover for which the size can vary between elements will be a closer match to the manifold than one with r fixed. The Hausdorff dimension is defined just as the capacity is in the preceeding definition, with the single difference that r is taken to be an upper limit on the cover size, and $V(r)$ in equation (A.14) is the infimum over all possible covers. $V(r)$ is the smallest value obtained from any possible collection of covering elements, with all elements smaller than r.

To be useful in characterizing the structure of manifolds the fractal dimension should be preserved under diffeomorphisms and embeddings, and this is indeed the case. This may be seen by recognizing that the dimension is defined as the scaling exponent in the limit of small r, but in this infinitesimal limit diffeomorphisms (and embeddings) can be replaced by their linear approximations, which will leave the scaling unchanged. In fact, dimensions are preserved under a less restrictive condition: if a metric is replaced by one that is metrically equivalent, the dimension is unchanged[16]. This also follows because the bound imposed on equivalent metrics prevents any change in the scaling properties. The dimension is manifestly not preserved under submersions (maps into smaller spaces) however, because a manifold cannot have a dimension greater than that of a space that it is embedded in.

Appendix B: Dynamical Systems

B.1 The Topological Description of Dynamical Systems

Dynamical systems are conveniently described in terms of the geometrical language that we have developed; reference [45] is an excellent introduction to this and other aspects of the theory. Although they are often specified explicitly in terms of a real space, they can be generalized to exist on arbitrary manifolds. In the real case a dynamical system with continuous-time evolution (such as the Lorenz set) is specified by a system of ordinary differential equations

$$\dot{x} = f(x) \tag{1}$$

where $x \in \mathbb{R}^n$ is a point in phase space and $f : \mathbb{R}^n \to \mathbb{R}^n$. This can be rephrased more geometrically by saying the f is a <u>vector field</u> on $\mathbb{R}^n$; f associates a tangent vector with each point in phase space. A particular solution to the set of differential equations translates into a particular curve through the vector field which has a tangent vector at each point equal to that specified by the vector field. Such a solution curve or <u>integral curve</u> is denoted by the <u>flow</u> $\varphi_t(x_0)$ which provides the value of x at time t given an initial condition x_0. In the discrete-time case (such as the logistic map) the dynamical system is specified by a map $\varphi : \mathbb{R}^n \to \mathbb{R}^n$ and the flow is given simply by φ^n, where n indexes the discrete time.

The preceding discussion is easily rephrased in terms of manifolds. Let a compact manifold M be the phase space of a system. A dynamical system on M is a map $\varphi : M \to M$ (discrete time) or a vector field (continuous time). The vector field associates each point in the manifold with an element of the tangent space at that point; it is a cross-section of the tangent bundle. The time evolution of the dynamical system is given by the flow $\varphi_t(x_0)$, and an <u>observable</u> on the dynamical system is a smooth function $y : M \to \mathbb{R}$.

We will need to describe the asymptotic behaviour of a system. An <u>invariant set</u> S of a flow φ on a manifold M is a subset of M defined by

$$S = \{x \in M \mid \varphi_t(x) \in S \; \forall x \in S \text{ and } \forall t\} . \tag{2}$$

The <u>positive limit set</u> (or <u>limit cycle</u>) $L^+(x)$ of a point x is the set of <u>limit points</u> that the flow approaches with an initial condition of x:

$$L^+(x) = \{y \in M \mid \exists t_i \to \infty \text{ with } \varphi_{t_i}(x) \to y\} . \tag{3}$$

Here the t_i are a sequence of values of t, and the notation means that as this sequence tends towards ∞, the flow comes arbitrarily close to y. Negative limit sets are defined for flows going backwards in time, and frequently positive limit sets are simply referred to as

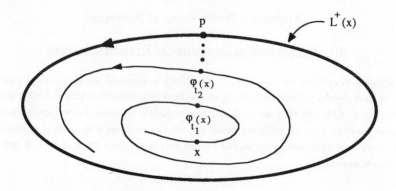

Figure B.1 The limit set for a damped driven oscillator

limit sets. Figure B.1 shows the limit set and a sequence of points for a damped, driven oscillator.

An <u>attracting set</u> or <u>attractor</u> adds the notion of a domain of attraction to a limit set. A closed invariant set S is an attracting set if it has a neighbourhood U such that $\varphi_t(x) \in U \; \forall t$ and $\varphi_t(x) \to S$ (the flow comes arbitrarily close to all the members of S) as $t \to \infty$. The <u>domain</u> or <u>basin</u> of attraction is the set of all initial conditions satisfying the above definition. We will usually consider attractors or limit sets in the following discussion, because the alternatives of settling to equilibrium or diverging are much less interesting.

B.2 Embedding Dynamical Systems (The Central Theorems)

The mathematical background is now complete and we are ready to state the theorems of Takens and indicate the nature of the proofs. The system to be considered is a manifold M in which the system state is defined, a map $\varphi : M \to M$ (discrete time) or a vector field X on M (continuous time) which defines the dynamics, the flow $(\varphi)^i$ (discrete time) or $\varphi_t(x_0)$ (continuous time), and an observable $y : M \to \mathbb{R}$. We will show that M can be embedded in $\mathbb{R}^n$ through the observable y, that the limit set of the flow is reproduced, and that the fractal dimension is unchanged by the embedding.

The first three theorems cover the embedding of M by time delays. The first one is for the discrete time case:

Theorem 1. (Takens) Let M be a compact manifold of dimension m. For pairs (φ, y), $\varphi : M \to M$ a smooth diffeomorphism and $y : M \to \mathbb{R}$ a smooth function, it is a generic property that the map $\Phi_{(\varphi,y)} : M \to \mathbb{R}^{2m+1}$, defined by

$$\Phi_{(\varphi,y)}(x) \;=\; \left\{ y(x), y\big(\varphi(x)\big), \ldots, y\big(\varphi^{2m}(x)\big) \right\}$$

is an embedding; by "smooth" we mean at least C^2.

For each point $x \in M$, $\Phi(x)$ gives a point in $\mathbb{R}^{2m+1}$, and this mapping is an embedding. The $2m + 1$ in the proof should be familiar from Whitney's theorem. We are going to embed our manifold into a real space, hence $2m + 1$ is the largest that this space need be in order to fit the manifold. This is a worst-case bound (for example, the Klein bottle); it is often the case that the dimension need only be m.

To show that Φ is an embedding it is first necessary to show that it is an immersion, and we have seen that this is the case if the covectors span the cotangent space. Here Takens applies transversality to argue that for a generic M, y, and φ this will indeed be the case. This is the key concept of the theory. For only a pathological or trivial choice of M, y, and φ will the derivatives of the iterates of the map be linearly dependent, and if they are then an arbitrary perturbation in the space of possible manifolds, functions, or observables will remove the linear dependence The real world of thermal and quantum noise guarantees the experimentalist this perturbation.

After showing that the map is an immersion (a local property) it is necessary to show that it is injective (a global property). As mentioned earlier this is often a difficult leap for proofs in differential topology. Takens makes this leap with a partition of unity, a technical tool that is frequently used for this purpose. The technical details are necessary to make the proof rigorous, but remember that the heart of the argument is the observation that the derivatives of the iterates of the map will generically be linearly independent. The theorem says nothing about just how independent the vectors are; we will later return to this important distinction between what is true mathematically (Φ is an embedding) and what is true experimentally (experimental resolution may be such that the embedding cannot be resolved).

The transversality arguments used are quite flexible; a wide variety of functions will serve just as well as time delays. All that is necessary is to find $2m + 1$ functions that are generically independent. In the next theorem, Takens shows that derivatives will work. Here φ_t again denotes the flow of X; this time smooth means at least C^{2m}.

Theorem 2. (Takens) Let M be a compact manifold of dimension m. For pairs (X, y), X a smooth vector field and y a smooth function on M, it is a generic property that the map $\Phi_{X,y} : M \to \mathbb{R}^{2m+1}$, defined by

$$\Phi_{X,y}(x) = \left\{ y(x), \frac{d}{dt}y(\varphi_t(x)), \ldots, \frac{d^{2m}}{dt^{2m}}y(\varphi_t(x)) \right\} ,$$

is an embedding.

Time delays are the most commonly used functions due to their simplicity, however an interesting open question is whether other choices for the set of functions might in some sense be optimal.

The above theorems provide diffeomorphisms that map points in M into those in $\mathbb{R}^{2m+1}$. This is not yet useful for examining laboratory data, because in the lab all that is typically available is a long sequence of values of the observable sampled at a fixed time interval with a fixed initial condition. The next two theorems fit such realistic data sets into the preceeding framework.

The process of sampling at discrete times introduces a discrete-time mapping, φ_τ, from the continuous-time flow, φ_t, by repeatedly advancing the flow with $t = \tau$. Our hope is that the observed dynamics produced by the embedding of this map will be the same as that generated by the continuous flow, i.e. that the limit sets are identical. The first step in realizing this hope is to show that the limit sets for the continuous-time flow, and the discrete-time map derived from the flow, are identical for most choices of the time increment. "Most" here means a residual (open and dense) subset of time increments, which means that any increment chosen will be acceptable or will be infinitesimally close to one that is acceptable. The only bad choice for the sampling time is one that is commensurate with a feature of the system such that the discrete map will always hop over a region of the limit set, and in this case a small perturbation of the sampling time will remove the problem. The proof of the following theorem is relatively short and straight-forward. It should not be a surprising result: even if one moves in discrete steps along the system's solution trajectory, one is still moving along the trajectory.

Theorem 3. (Takens) Let M be a compact manifold, X a vector field on M with flow φ_t and p a point in M. Then there is a residual subset $C_{X,p}$ of positive real numbers such that for $\tau \in C_{X,p}$, the positive limit sets of p for the flow φ_t of X and for the diffeomorphism φ_τ are the same. In other words, for $\tau \in C_{X,p}$ we have that each point $q \in M$ which is the limit of a sequence $\varphi_{t_i}(p)$, $t_i \in \mathbb{R}$, $t_i \to +\infty$, is the limit of a sequence $\varphi_{n_i \tau}(p)$, $n_i \in \mathbb{N}$, $n_i \to \infty$.

Note that this proof provides only a residual subset, not a more desirable full measure one. The set is almost certainly of full measure for any physical system, however proving this is more difficult and has not yet been done.

The preceding simple theorem, along with the earlier ones, implies a corollary that finally gives the central result on observing dynamical systems. We have seen that discrete-time maps on the manifold lead to an embedding in $\mathbb{R}^{2m+1}$ through the observable and time delays, and we have just seen that the conversion of a continuous-time flow into a discrete-time one by choosing a fixed sampling interval gives the same limit set. These results together imply that the discretely-sampled data set embedded into $\mathbb{R}^{2m+1}$ has the same limit set as real system.

Corollary 4. (Takens) Let M be a compact manifold of dimension m. We consider quadruples, consisting of a vector field X, a function y, a point p, and a positive real number τ. For generic such (X, y, p, τ) (more precisely: for generic (X, y) and τ satisfying generic conditions depending on X and p), the positive limit set $L^+(p)$ is diffeomorphic with the set of limit points of the following sequence in $\mathbb{R}^{2m+1}$:

$$S_{X,y,p,\tau} = \left\{ \left(y\big(\varphi_{k\tau}(p)\big), y\big(\varphi_{(k+1)\tau}(p)\big), \ldots, y\big(\varphi_{(k+2m)\tau}(p)\big) \right) \right\}_{k=0}^{\infty} .$$

Here diffeomorphic means that there is a smooth embedding of M into $\mathbb{R}^{2m+1}$ mapping $L^+(p)$ bijectively to the set of limit points of this sequence.

The final theorem states that the limit capacity of the limit set of the flow of the real system is the same as the limit capacity of the embedded flow; this follows immediately from the previous observation that dimensions are preserved by embeddings. Using the same definitions as the previous theorems,

Theorem 5. (Takens) $d_c\big(L^+(p)\big) \;=\; d_c\,[L^+(S_{X,y,p,\tau})]$.

The preceeding discussion has assumed that the observable is a scalar quantity, but immediately generalizes to the case that the measuring process yields a vector of non-trivially related quantities. In this case the embedding procedure is done on the vector of observables, and the number of time delayed copies of the vector needed for the embedding is reduced by a factor equal to the number of elements in the vector.

B.3 Center Manifolds and Horseshoes

This section covers the theorems referred to in section 4.1. The first topic is the relation between a nonlinear system and its linearization. We are considering a dynamical system $\dot{x} \;=\; f(x)$; the linear approximation to this system at a point is $\dot{x} \;\approx\; D_x f(x)$, where $D_x f$ is the Jacobian matrix at x. The theory of linear differential equations tells us that the solution to a linear system $\dot{x} \;=\; Ax$ is given by

$$x(x_0, t) \;=\; e^{tA} x_0 \;, \tag{4}$$

where the exponential of an operator is defined by the power series expansion of the exponential. Regardless of the complexity of the nonlinear system, we are sure to be able to solve the linear system, and hence are interested in the relation between the linear and nonlinear systems.

The operator A will have eigenvectors v_i with corresponding eigenvalues λ_i, with the evolution of an eigenvector being

$$x(v_i, t) \;=\; v_i e^{\lambda_i t} \;. \tag{5}$$

Eigenvectors with negative real part eigenvalues will be stable, those with positive real part eigenvalue will be unstable, and those with zero real part eigenvalues will be neutrally stable. This argument carries over to discrete time dynamical systems, with the stability bounds that the real parts of the eigenvalues be $>, =, < 0$ replaced by the requirements that the magnitudes of the eigenvalues be $>, =, < 1$. This division of eigenvalues suggests the introduction of the stable eigenspace, the unstable eigenspace, and the center eigenspace, as the spaces spanned by the corresponding sets of eigenvectors. These are invariant spaces (an element in the space will remain in it under the action of the flow). In describing a dynamical system a reference to the eigenvalues at a point means the eigenvalues of the linear system (Jacobian matrix) at that point.

The first step in relating the linear and nonlinear systems is given by

Theorem (Hatrman-Grobman) If $D_x f$ has no zero or purely imaginary eigenvalues then there is a homeomorphism defined on some neighborhood of x locally taking orbits of the nonlinear flow φ_t into those of the linear flow $e^{D_x f t}$. The homeomorphism preserves the sense of orbits and can also be chosen to preserve parametrization by time.

This means that, as long as there is no center eigenspace, the nonlinear and linear systems will look the same locally. A <u>fixed point</u> is one where $f(x) = 0$; the Hartman-Grobman theorem is relatively obvious away from a fixed point, but perhaps more surprising at one. A <u>hyperbolic</u> fixed point is one with no eigenvalues with zero real part.

In the full nonlinear system there will still be invariant manifolds $W^{s,u,c}$ associated with stable, unstable, and center directions, however they will differ from the linear eigenspaces. The center manifold theorem states that the linear eigenspaces will be tangent to the nonlinear invariant manifolds, and so they provide a good local representation of the flow at a point.

Theorem (Center Manifold) Let f be a C^r vector field on $\mathbb{R}^n$, with stable, unstable, and center eigenspaces E^s, E^u, E^c at a fixed point x. Then there exist C^r stable and unstable invariant manifolds W^u and W^s tangent to E^u and E^s at x, and an invariant C^{r-1} center manifold tangent to E^c at 0. The stable and unstable manifolds are unique, but the center manifold need not be.

Figure B.2 illustrates the center manifold theorem.

This provides the needed guarantee we can locally consider the linearization. We next consider how the horseshoe introduced in section 4.1 can arise from the flow on a vector field. First some defintions: a <u>hyperbolic</u> fixed point for a map is one which has no eigenvalues of unit modulus (that is, no eigenvalues with neutral stability), and a <u>homoclinic orbit</u> is an closed orbit which connects a fixed point to itself. Figure B.3 shows a homoclinic orbit and the dynamics that arise from it. On the left a limit cycle is shown along with sections of its stable and unstable manifolds W^s and W^u. p is a point on the limit cycle. On the right in figure B.3 is shown the Poincaré section of the flow at p; this replaces the flow with a map f which has a fixed point at p. The homoclinic orbit connecting p with itself, arising from the intersection between W^u and W^s at q, may be seen here. The dynamics of figure B.3 is described by the Smale-Birkhoff homoclinic theorem[45]. This theorem is phrased in terms of the language of symbolic dynamics, which we have not developed, however it can be roughly stated in familiar terms.

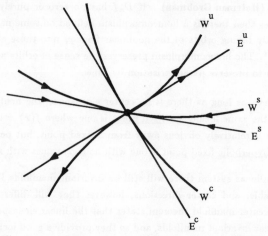

Figure B.2 Illustration of the center manifold theorem in two dimensions

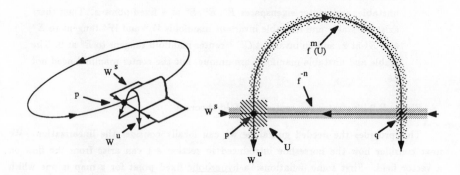

Figure B.3 A Homoclinic Connection

Theorem (Smale-Birkhoff Homoclinic Theorem)   Let $f : \mathbb{R}^n \to \mathbb{R}^n$ be a
diffeomorphism such that p is a hyperbolic fixed point and there exists a
point $q \neq p$ of transversal intersection between $W^s(p)$ and $W^u(p)$. Then f
"has a horseshoe."

The technically correct substitute for "has a horseshoe" is "has a hyperbolic invariant
set λ on which f is topologically equivalent to a subshift of finite type;" for our purposes
these mean the same thing. To see the source of the horseshoe, consider the evolution of

a neighborhood U of p. As the map as applied to U it will stretch out along the unstable manifold, and as the inverse of the map is applied U will stretch out along the stable manifold. There will be values m and n such that $f^m(U)$ and $f^{-n}U$ will extend through and beyond q; the images of U for these values are also shown in the figure. To see the horseshoe, follow $f^{-n}(U)$ as f is applied. After n iterations it will be mapped into U, and after m more iterations it will intersect itself in two strips (one at p and one at q). Applying the map $m + n$ times more will compress these strips into U and then stretch them out as before, generating two more strips in the process, and the repetition of this process will generate a fractal set of strips in U. This is the same repeated subdividing that we saw in the preceeding discussion of the Smale horseshoe; any two arbitrarily close points that are in one of the strips will eventually be irrevocably separated by the map. A shredding mechanism similar to this is present in the Lorenz set. A <u>strange attractor</u> is one that has a transversal homoclinic orbit. The inter-relationship between a homoclinic orbit, a horseshoe, the sensitive dependence on initial conditions, a fractal attractor, and chaos may now be seen.

References

[1] S.W. Teitsworth and R.M. Westervelt, Phys. Rev. Lett. **56** 516 (1986).

[2] M. Iansiti, Hu. Qing, R.M. Westervelt, and M. Tinkham, Phys. Rev. Lett. **55** 746 (1985).

[3] G.A. Held and C. Jeffries, Phys. Rev. Lett. **55** 887 (1985).

[4] R. Van-Buskirk and C. Jeffries, Phys. Rev. A **31** 3332 (1985).

[5] G. Gibson and C. Jeffries, Phys. Rev. A **29** 811 (1984).

[6] A.M. Albano *et. al.*, J. Opt. Soc. Am. B **2** 47 (1985).

[7] A. Branstater *et. al.*, Phys. Rev. Lett. **51** 1442 (1983).

[8] R.H. Simoyi, A. Wolf, and H.L. Swinney, Phys. Rev. Lett. **49** 245 (1982).

[9] S. Ciliberto and J.P. Gollub, J. Fluid Mech. **158** 381 (1985).

[10] J. Stavans, F. Heslot, and A. Libchaber, Phys. Rev. Lett **55** 596 (1985).

[11] A. Libchaber, Physica B&C **109-110** 1583 (1982).

[12] M. Giglio, S. Musazzi, and U. Perini, Phys. Rev. Lett **53** 2402 (1984).

[13] P. Collet and J.-P. Eckmann, *Iterated Maps on the Interval as Dynamical Systems* (Birkhauser, Cambridge, 1980).

[14] E.N. Lorenz, J. Atmos. Sci. **20**, 130 (1963).

[15] W.W. Wood, J.H. Wilson, R.M. Bendow, and L.E. Hood, *Biochemistry, a Problems Approach* (Benjamin/Cummings, Menlo Park, 1981) p. 60.

[16] F. Takens, in *Dynamical Systems and Turbulence, Warwick, 1980,* Vol. 898 of *Lecture Notes in Mathematics,* edited by D.A. Rand and L.S. Young (Springer, Berlin, 1981), p.366.

[17] N.H. Packard, J.P. Crutchfield, J.D. Farmer, and R.S. Shaw, Phys. Rev. Lett. **45**, 712 (1980).

[18] A. Oppenheim, R. Schafer, *Digital Signal Processing* (Prentice-Hall, Englewood Cliffs, 1975).

[19] A.M. Fraser, H.L.Swinney, Phys. Rev. Lett. **33**, 1134 (1986)

[20] C.E. Shannon and W.Weaver, *The Mathematical Theory of Communication* (University of Illinois Press, Urbana, 1949).

[21] H.G.E Hentschel and I. Procaccia, Physica **8D** 435 (1983).

[22] P. Grassberger and I. Procaccia, Phys. Rev. Lett **50** 346 (1983).

[23] P. Grassberger and I. Procaccia, Physica **9D** 189 (1983).

[24] H. Greenside, A. Wolf, J. Swift, and T. Pignataro, Phys. Rev. A **25** 3453 (1982).

[25] L.A. Smith, J.-D. Fournier, and E.A. Spiegel, Phys. Lett. **114A** 465 (1986).

[26] D.S. Broomhead and G.P. King, Physica **20D** 217-236 (1986).

[27] D.S. Broomhead and G.P. King, in *Nonlinear Phenomena and Chaos*, S. Sankar ed. (Adam Hilger, Bristol, 1986).

[28] J.-P. Eckmann and D. Ruelle, Rev. Mod. Phys. **57** p. 617 (1985).

[29] J.D. Farmer, E. Ott and J.A. Yorke, Physica **7D** 153 (1983).

[30] J.D. Farmer, Z. Naturforsch **37a** 1304 (1982).

[31] P. Grassberger and I. Procaccia, Physica **13D** 34 (1984).

[32] A. Wolf, J.B. Swift, H.L. Swinney, and J.A. Vastano, Physica **16D** 285 (1985).

[33] W.H. Press, B.P. Flannery, S.A. Teukolsky, and W.T. Vetterling, *Numerical Recipes* (Cambridge University Press, Cambridge, 1986)

[34] T.H. Halsey *et. al.*, Phys. Rev. A **33** 1141 (1986).

[35] M.H. Jensen, L.P. Kadanoff, I. Procaccia, Phys. Rev. A **36** 1409 (1987).

[36] A. Cumming and P.S. Linsay, Phys. Rev. Lett. **59** 1633 (1987).

[37] M.H. Jensen *et. al.*, Phys. Rev. Lett **55** 2798 (1985).

[38] M.J. Feigenbaum, M.H. Jensen and I. Procaccia, Phys. Rev. Lett. **57** 1503 (1986).

[39] J. Guckenheimer, Cont. Math. **28** 357 (1984).

[40] V. Guillemin, A. Pollack, *Differential Topology* (Prentice-Hall, Englewood, 1974).

[41] M.W. Hirsch, *Differential Topology* (Springer-Verlag, New York, 1976).

[42] Y. Choquet-Bruhat, C. DeWitt-Morette, M. Dillard-Bleick, *Analysis, Manifolds and Physics* (North-Holland, New York, 1982).

[43] W. Hurewicz, H. Wallman, *Dimension Theory* (Princeton University Press, Princeton, 1941).

[44] B. Mandlebrot, *The Fractal Geometry of Nature* (W.H. Freeman and Company, San Francisco, 1983).

[45] J. Guckenheimer, P. Holmes, *Nonlinear Oscillations, Dynamical Systems, and Bifurcations of Vector Fields* (Springer-Verlag, New York, 1983).

Index to Terms